Springer Proceedings in Materials

Volume 17

Springer Proceedings in Materials publishes the latest research in Materials Science and Engineering presented at high standard academic conferences and scientific meetings. It provides a platform for researchers, professionals and students to present their scientific findings and stay up-to-date with the development in Materials Science and Engineering. The scope is multidisciplinary and ranges from fundamental to applied research, including, but not limited to:

- Structural Materials
- Metallic Materials
- Magnetic, Optical and Electronic Materials
- Ceramics, Glass, Composites, Natural Materials
- Biomaterials
- Nanotechnology
- Characterization and Evaluation of Materials
- Energy Materials
- Materials Processing

To submit a proposal or request further information, please contact one of our Springer Publishing Editors according to your affiliation:

European countries: **Mayra Castro** (mayra.castro@springer.com)
India, South Asia and Middle East: **Priya Vyas** (priya.vyas@springer.com)
South Korea: **Smith Chae** (smith.chae@springer.com)
Southeast Asia, Australia and New Zealand: **Ramesh Nath Premnath** (ramesh.premnath@springer.com)
The Americas: **Michael Luby** (michael.luby@springer.com)
China and all the other countries or regions: **Mengchu Huang** (mengchu.huang@springer.com)

This book series is indexed in **SCOPUS** database.

More information about this series at https://link.springer.com/bookseries/16157

Slah Msahli · Faten Debbabi
Editors

Advances in Applied Research on Textile and Materials - IX

Proceedings of the 9th International Conference of Applied Research on Textile and Materials (CIRATM)

Springer

Editors
Slah Msahli
Laboratory of Textile Engineering
ISET of Ksar Hellal
Ksar Hellal, Tunisia

Faten Debbabi
National Engineering School of Monastir
Monastir University
Monastir, Tunisia

ISSN 2662-3161 ISSN 2662-317X (electronic)
Springer Proceedings in Materials
ISBN 978-3-031-08844-5 ISBN 978-3-031-08842-1 (eBook)
https://doi.org/10.1007/978-3-031-08842-1

This Springer imprint is published by the registered company Springer Nature Switzerland AG
The registered company address is: Gewerbestrasse 11, 6330 Cham, Switzerland

Organization

Organizing Committee of CIRATM-9

General Chair

Slah Msahli, Tunisia

Co-chairmen

Laadhari Neji, Tunisia
Blaga Mirela, Romania
Ferreira Fernando, Portugal
Vasiliadis Savvas, Greece
Jeanmichel Laurence, France
Haji Aminoddin, Iran
Visileanu Emilia, Romania
Khorchani Karima, Tunisia

President of the Organizing Committee

Debbabi Faten, Tunisia

Coordination and Planning

Chriaa Rim, Tunisia

Responsible for Registration

Khlif Helmi, Tunisia

Responsible for the Reviewing and the Editing of Papers

Ghazzah Mouna, Tunisia
Asma El Oudiani, Tunisia

Responsible for Promotion

Ben Mlik Yosr, Tunisia

Scientific Committee of CIRATM-9

Cherkaoui Omar, Morocco
Chemani Bachir, Algeria
El-Sayed Hosam, Egypt
Belgacem Mohamed Naceur, France
Vassiliadis Savvas, Greece
Yordan Kyosev, Germany
Hu Jinlian, Hong Kong
Colovic Gordana, Serbia
Ladhari Néji, Tunisia
Ben Abdessalem Saber, Tunisia
Douib Sofien, Tunisia
Meksi Nizar, Tunisia
Babay Amel, Tunisia
Göktepe Fatma, Turkey
Kalaoglu Fatma, Turkey
Kiekens Paul, Belgium
Cherif Chokri, Germany
Grancaric Ana Marija, Croatia
Haj Taib Amine, Tunisia
Boukerrou Amar, Algeria
Ben Hassen Mohamed, Tunisia
Jaouachi Boubaker, Tunisia
Hes Lubos, Czech Republic
Marsal Felix, Spain
Msahli Slah, Tunisia
Sakli Faouzi, Tunisia
Fayala Feten, Tunisia
Zlatina Kazlacheva, Bulgaria
Kazani Ilda, Albania
Marmarali Arzu, Turkey
Lisichkov Kiril, Macedonia
Andrej Demsar, Slovenia
Visileanu Emilia, Romania
Fangueiro Raúl, Portugal
Ferreira Fernando Nunes Batista, Portugal
Haji Aminoddin, Iran
Harizi Taoufik, Tunisia
Baffoun Ayda, Tunisia
Ghith Adel, Tunisia
Boussu François, France
Guiseppe Rosace, Italy
Elsayed Elnashar Ahmed, Egypt
Koncar Vladan, France
Sonnier Rodolphe, France
Horchani Karima, Tunisia
Tomljenović Antoneta, Croatia
Dallal Mohamed, Maroc
Krifa Mourad, UK
Mokhtar Ferid, Tunisia
Chedly Boudokhane, Tunisia

Contents

Advance in Technical Textile

Production and Characterization of Keratin/Tragacanth Gum Nano Hydrogels for Drug Delivery in Medical Textiles

Nazanin Mansouri Shirazi, Niloofar Eslahi(✉), and Adeleh Gholipour-Kanani

Department of Textile Engineering, Science and Research Branch, Islamic Azad University, Tehran, Iran

niloofar.eslahi@srbiau.ac.ir

Abstract. The main goal of this research is to improve the drug delivery of textiles using smart nanohydrogels. Keratin has been considered as the main source of protein due to its biocompatibility, biodegradability, mechanical resistance and bioavailability. Tragacanth gum has wound healing and antimicrobial properties, which can be used in medical applications. In this study, keratin was extracted from protein-based chicken feather by using reduction hydrolysis and nanogel of keratin and tragacanth was produced by using chemical cross-linking method. Then, cinnamon as an antibacterial herbal extract was added to the nanogels and its release was investigated. According to the results, semi-spherical nanogels were obtained, more cinnamon was released from the nanogels by increasing time and the cinnamon extract had antibacterial properties.

Keywords: Keratin · Tragacanth gum · Hydrogel · Cinnamon

1 Introduction

Hydrogels are hydrophilic three-dimensional networks which swell in contact with water but do not dissolve. They come in many forms, including sheets, microparticles, nanoparticles, coating structures and films. For this reason, hydrogels can be used in various research areas such as sensors, tissue engineering and biomolecular separation (Amin et al. 2009). Nanogels are one of the techniques in nanotechnology, which has been most prevalent in drug delivery with high drug loading, biocompatibility and biodegradability. Nanoscaled drug delivery systems serve as drug delivery vehicles able to protect these encapsulated drugs from the physiological environment and release it within specific tissues on account of the enhanced permeability and retention effect (Sun et al. 2017).

Keratin is a natural protein with a cysteine-rich structure. It is found in different sources such as human hair, wool, feathers, horns, and nails and produced in various forms such as film, sponge, powder and hydrogel. Bird feathers, as one of the important sources of keratin having disulfide bonds, hydrogen bonds and hydrophobic interactions, can be used in biomedical applications (Eslahi et al. 2013). Additionally, the abundant functional groups of keratin, such as carboxyl, amid or sulfhydryl, facilitates

S. Msahli and F. Debbabi (Eds.): CIRATM 2020, SPM 17, pp. 3–6, 2022.
https://doi.org/10.1007/978-3-031-08842-1_1

easy modification with biomolecules that could improve the stability, solubility, as well as biocompatibility. These excellent properties make keratin a promising material for drug carrier substances (Li et al. 2012).

In recent years, tragacanth gum has been used as a superabsorbent hydrogel, antimicrobial nanocapsules, wound dressings, skin scaffolds and drug release systems (Meghdadi Kasani et al. 2016). Tragacanth gum is a natural polysaccharide that obtains from different species of Astragalus plant. This polysaccharide composes of an insoluble but water-swellable fraction called bassorin and a water-soluble fraction called tragacanthin. Tragacanth gum is almost insoluble in water, but swells to form a stiff gel due to its structure (Ghayempour et al. 2016).

Essential oils are well known inhibitors of microorganisms. Cinnamon oil is a natural preservative and flavouring substance, which can be used as an interesting alternative to other chemical preservatives. There have been a number of reports of substances in cinnamon that can inhibit the growth of molds, yeasts and bacteria (Matan et al. 2006).

The main goal of this research is to improve the drug delivery of textiles with the help of smart nano hydrogels. For this purpose, keratin was extracted from protein-based chicken feather by using reduction hydrolysis and nanogel of keratin and tragacanth was produced by using chemical cross-linking method. Then, cinnamon as an antibacterial herbal extract was added to the nanogels and the loaded nanogels were padded onto the cotton fabric. Finally, the fabricated hydrogels were tested by different analysis.

2 Experimental

At first, keratin protein was extracted from cleaned chopped feather by using 10 g/l sodium sulfide for 3 h at 60 °C. After being dialyzed and freeze-dried, the purified keratin powder was mixed with Tragacanth gum under N_2 atmosphere at different ratios of 1:0, 1:1, 1:2, 2:1 and 0:1 (gum:keratin). Hydrogen peroxide (2 ml) was also added to the mixtures drop by drop to induce chemical cross-linking and the mixtures were stirred for 24 h at 37 °C. Then, the samples were dialyzed to remove excess H_2O_2 and finally freeze-dried.

To load the drug, cinnamon as an antibacterial herbal extract was added to the nanogels and its release was investigated at different times by UV-visible spectrophotometer. Besides, the morphology and size of the nanogels were evaluated by using FESEM and DLS. Finally, the cotton fabrics were immersed in the prepared nanogels for 24 h and then dried in open air. Antimicrobial test against *E. coli* and *S. aureus* bacteria was performed using well-diffusion method and the zone of inhibition (ZOI) was measured.

3 Results and Discussion

Figure 1 illustrates the FESEM images of the nanogels. As can be seen, semi-spherical nanoparticles were successfully produced. Table 1 shows the average size and PDI of the nanogels obtained from DLS. According to the obtained results, the sample GK2:1 (ratio of gum to keratin = 2:1) shows the minimum size and PDI and therefore this sample is chosen as the optimum one.

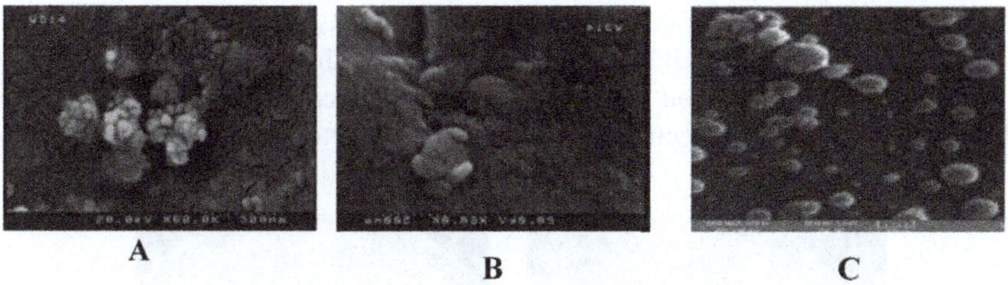

Fig. 1. FESEM images of the samples: A: Gum, B: Keratin, C: Gum/Keratin.

Table 1. Average size and PDI of different samples.

Sample codes	Size (nm)	PDI
GK11	88	0.521
GK12	171	0.509
GK21	80	0.100

In the drug release test, lambda max (λ_{max}) of the cinnamon extract was first measured, showing the peak at 674 nm. Then, at this wavelength, the absorbance of the samples were measured at different times. The release results showed a typical two-phase profile. The initial burst release may be ascribed to the fast release of surface located drug, whereas the second phase is mainly associated with drug diffusion. It is worth mentioning that there is no significant difference between two samples (Fig. 2).

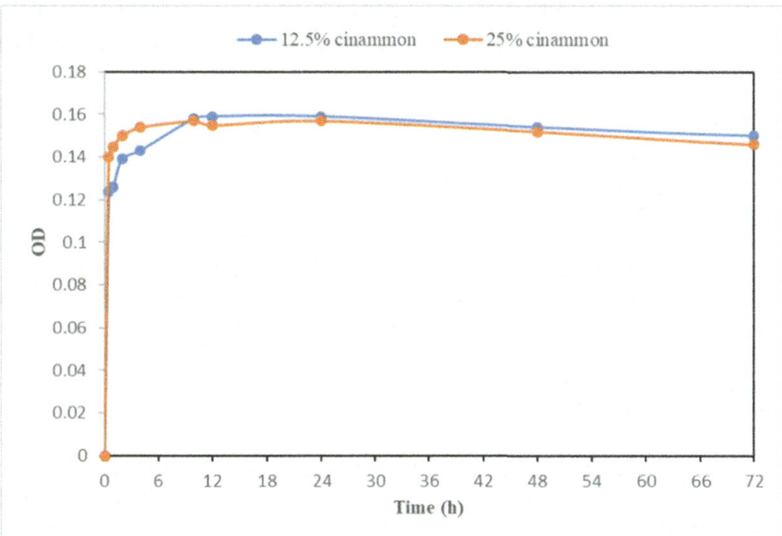

Fig. 2. Drug release profile of (12.5% and 25%) cinnamon-loaded nanohydrogels.

The antibacterial activity of the prepared sample was tested against *S. aureus* as Gram-positive and *E. coli* as Gram-negative bacteria and the results are shown in Fig. 3. The plates containing cinnamon-loaded nanohydrogel indicated ZOIs around 8 and 9 mm for *E. coli* and *S. aureus*, respectively, owing to the antibacterial properties of cinnamon.

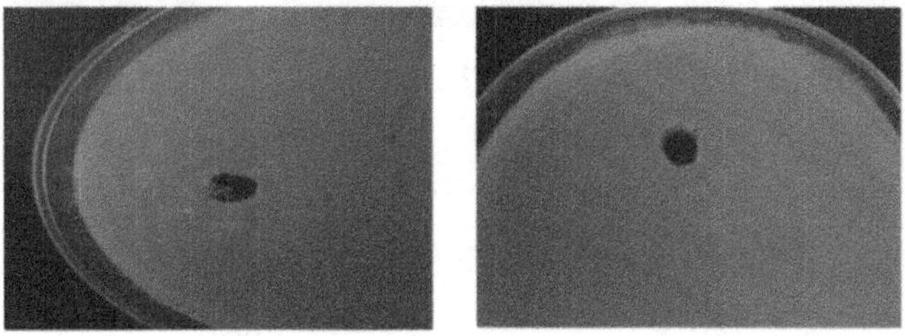

(a) **(b)**

Fig. 3. Antimicrobial activity of the sample against: (a) *E. coli* and (b) *S. aureus.*

4 Conclusion

In this study, we investigated the feasible fabrication of keratin and tragacanth gum nanohydrogels by using chemical cross-linking method. Then, cinnamon as an antibacterial herbal extract was added to the nanogels and its release was investigated. Cotton fabric was also treated with the drug-loaded nanogels and its antibacterial properties were evaluated.

According to the obtained results, semi-spherical nanogels were obtained and more cinnamon was released from the nanogels by increasing time. The antibacterial test against both gram positive and gram-negative bacteria also revealed that the treated fabric had antibacterial activity and therefore it could be used in wound dressing applications.

References

Amin, S., et al.: Hydrogels as potential drug delivery systems. Sci. Res. Essays **3**(11), 1175–1183 (2009). ISSN: 1992-2248

Eslahi, N. et al.: An investigation on keratin extraction from wool and feather waste by enzymatic hydrolysis. Prep. Biochem. Biotechnol. **43**(7), 624–648 (2013). ISSN: 1082-6068

Ghayempour, S. et al.: Encapsulation of aloe vera extract into natural tragacanth gum as a novel green wound healing product. Int. J. Biol. Macromol. **93**, Part A, 344–349 (2016). ISSN: 0141-8130

Li, Q. et al.: Biological stimuli responsive drug carriers based on keratin for triggerable drug delivery. J. Mater. Chem. **22**(37), 19964–19973 (2012). ISSN: 0959-9428

Matan, N., et al.: Antimicrobial activity of cinnamon and clove oils under modified atmosphere conditions. Int. J. Food Microbiol. **107**(2), 180–185 (2006). ISSN: 0168-1605

Meghdadi, R. et al.: Antimicrobial and restorative properties of tragacanth nanofibers. Polym. J. **7**(3), 98–107 (2016). ISSN: 2252-0449

Sun, Z. et al.: Bio-responsive alginate-keratin composite nanogels with enhanced drug loading efficiency for cancer therapy. Carbohydr. Polym. **175**, 159–169 (2017). ISSN: 0144-8617

In-Situ Synthesis of Silver Nanoparticles Using Natural Dyed Wool Fabric

Aminoddin Haji[1] and Majid Nasiriboroumand[2](✉)

[1] Department of Textile Engineering, Yazd University, Yazd, Iran
[2] Department of Handmade Carpet, Shahid Bahonar University of Kerman, Kerman, Iran
Nasiri@uk.ac.ir

Abstract. Among different methods to synthesize silver nanoparticles, the biological method has been extensively investigated. In this study, an in-situ approach was used to synthesis silver nanoparticles (AgNPs) on the surface of wool fabric. The pomegranate peel was employed as a colorant and reducing agent. The change of surface morphology was investigated using scanning electron microscopy. Using this synthesis method, dispersed silver nanoparticles on the surface of wool fabric was seen and antibacterial efficiency against Escherichia coli bacteria was appeared. CIELAB colorimetric data showed an increase in greenness and a decrease in blueness that can be attributed to the presence of AgNPs on the fabrics.

Keywords: Natural dye · Pomegranate · Wool · Silver nanoparticles · In-situ

1 Introduction

Natural dyes are being reintroduced to textile dyeing industry due to growing attention on harmful effects of synthetics dyes, such as water pollution, sustainability of raw materials, and environmental aspects. Compared to the synthetic dyes, natural dyes are preferred because of its environment-friendly nature, lower toxicity, antibacterial properties, biodegradability, and harmonizing natural shades (Barani et al. 2013). Pomegranate peel is one of the most famous natural components that have been used for dyeing of wool and cotton and presents a yellow colour with high wash, light and rubbing fastness. Because of high amount of tannin, it is not needed to use mordant when pomegranate peel is used (Goodarzian 2010).

The use of nanoparticle in textile industry is growing very fast. Silver nanoparticles have been employed as antibacterial agent in various range of material. Silver nanoparticles can be synthesis by chemical reduction of silver ion to silver atom. Currently, there is a growing need to develop environmentally benign nanoparticle synthesis in which toxic chemicals are excluded. As a result, researchers in the field of nanoparticles synthesis and assembly have turned to biological systems for inspiration. Synthesis of silver nanoparticles from plants, microorganism, enzymes and polysaccharides has been reported (Siddiqi et al. 2018).

S. Msahli and F. Debbabi (Eds.): CIRATM 2020, SPM 17, pp. 7–11, 2022.
https://doi.org/10.1007/978-3-031-08842-1_2

Polyphenolic compounds can handle reduction of silver ions to silver nanoparticles. The mechanism for bio-reduction of silver ions to silver atom has been presented in Fig. 1. A polyphenolic compound which has two hydroxyl groups in ortho-position can easily reduce silver to silver atom (Nasiriboroumand et al. 2015). Pomegranate contains a considerable amount of tannin and other polyphenolic compounds with ortho-di hydroxyl group which are able to reduce silver ions to silver atoms and as a result, silver nanoparticles. In this study we report a facile method for in-situ synthesis of silver nanoparticle. We show that the woolen fabric which had been colored by pomegranate peel is able to absorb silver ions from silver nitrate solution and reduce to silver nanoparticles.

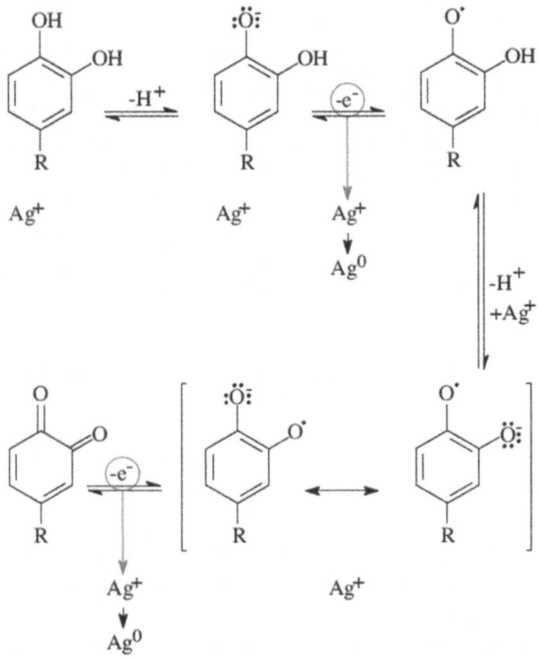

Fig. 1. Tentative mechanism for reduction of Ag+ by polyphenols having two hydroxyl groups in ortho-position.

2 Experimental

2.1 Dyeing Procedure

A plain-woven wool fabric (Iran Merinos, Iran) with an area weight of 240 g/m^2 was used in this study and scoured with 1% non-ionic detergent at 50 °C for 20 min. The L:G (Liquor to Good ratio) of the scouring bath was kept at 40:1. The scoured material was thoroughly washed with tap water and dried at room temperature. The scoured material was soaked in clean water for 30 min prior to dyeing. The dye solution was prepared by adding 10 g of pomegranate peel powder in 1 L distilled water and kept in room temperature for 24 h before dyeing. Dyeing process was started with L:G of 40:1 at 40 °C and temperature was raised to the boil in 20 min and resumed at this temperature 1 h. The dyed material was then rinsed with water thoroughly, squeezed and dried.

2.2 In-Situ Synthesis of AgNPs

First, an aqueous solution of 500 ppm AgNO3 (Merck, Germany) was prepared in ultrapure water. The pH of the solution was adjusted to 8 using sodium hydroxide (Sigma Aldrich, Germany). 1g of dyed wool was immersed in 500 ppm $AgNO^3$ solution with a material to liquor ratio of 1:50 for 2 h at 25 °C. Afterwards, the wool fabric was rinsed with distilled water and dried at room temperature.

2.3 Characterization

The surface morphology of wool fabrics and presence of AgNPs were investigated by SEM (Philips XL 30). Wool samples were coated with gold in a sputter coating unit for 5 min before SEM observation. EDS was used for the qualitative and quantitative elemental analysis of samples. The elemental composition of the material was analyzed with an EDX-720/800HSEDX spectrometer (Shimadzu Europa, Germany). Color coordinates of the dyed fabrics (CIE L*, a*, b*) were determined with Datacolor SF300 spectrophotometer under illuminant D65 using the 10o standard observer.

2.4 Antibacterial Test

A quantitative evaluation of antibacterial efficiency of AgNPs loaded wool fabrics was done against gram-negative bacteria Escherichia coli (ATCC 11229). This test was carried out according to the AATCC test method 100–2004. The square specimens of 4.8 ± .1 cm were prepared. Each swatch was individually placed in a sterile lid of Petri dish and inoculated with 0.5 mL of bacterial suspension (107CFU/mL) for 24 h. After inoculation, each sample was placed in a 40 mL solution containing saline and 2 g/L of a nonionic surfactant (Triton X-100 from Sigma–Aldrich) and was shaken vigorously for 1 min. To measure the number of bacteria at $t = 0$, the samples were placed in saline and surfactant solution as soon as possible after inoculation (zero contact time). The total bacterial count was determined by serial dilution and pour plate method using CASO-agar medium. The antibacterial efficiency of loaded samples as term of bactericidal was calculated using Eq. (1), where E is the percentage of bacteria reduction, A is the number of bacteria recovered from the inoculated treated test immediately after inoculation and B is the number of bacteria recovered from the inoculated treated test after 24 h.

$$E = (A - B)/A \times 100 \tag{1}$$

3 Results and Discussion

The SEM images of in-situ synthesized AgNPs on wool fibers are presented in Fig. 1. It can be clearly seen that the surface morphology of AgNPs-loaded wool fibers was changed compared with the untreated wool and presence of AgNPs on the fibers is clearly distinguished. Using this synthesis method, dispersed particles on the surface of wool fabric were formed. The EDX results show these particles are silver (not shown here). Therefore, it could be concluded that silver nanoparticles are formed on the surface of wool fibers (Fig. 2).

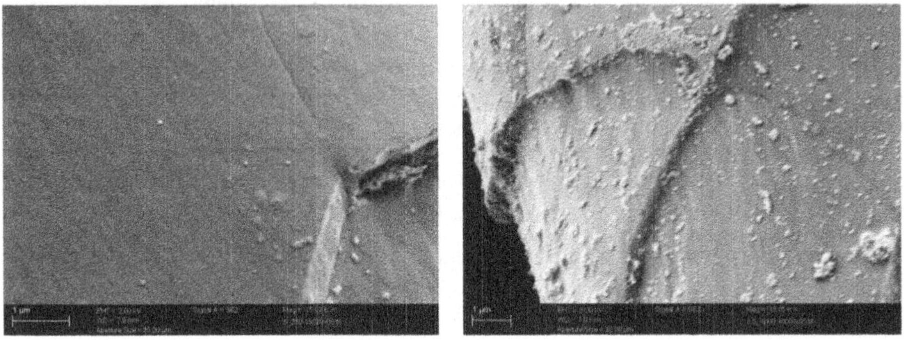

a b

Fig. 2. SEM images of (a) dyed wool fibers and (b) Ag loaded wool fibers

The antibacterial efficiency of samples is presented in Table 1. Pristine wool did not present antibacterial efficiency against bacteria and after 24 h, the number of bacteria colonies was increased. These shows that wool surface is a suitable media for growing bacteria. The dyed fabric shows slightly antibacterial efficiency. This result demonstrated that pomegranate has antibacterial efficiency. But it was not efficiency to completely preserve the wool against bacteria. The fabric after treatment in silver nitrate solution shows excellent antibacterial efficiency. This confirms loading of silver ions and AgNPs on wool fabric.

Table 1. Antibacterial efficiency wool fabrics

	Initial number of bacterial colonies (CFU)	Number of bacterial colonies (CFU)	R%
Pristine wool	2×10^5	2.25×10^7	0
Dyed untreated fabric without Ag	2×10^5	1.18×10^5	44%
Dyed wool-Ag500	2.25×10^5	Less than 10	99%

The influence of loading of silver nitrate solution onto wool fabrics after dyeing on color change of fabrics was evaluated. The color changes were expressed via CIELAB color coordinates. Colorimetric data are given in Table 2. The color of fabrics after Ag loading turned from yellow to a brown, which is in accordance with the changes obtained in literature data. The loading of Ag NPs onto fabrics after dyeing with pomegranate induced a considerable color change: the fabrics became darker, less red and less blue. The increased greenness and decreased blueness are attributed to the presence of Ag NPs on the fabrics. This is suggested to be due to higher dielectric constant of the surrounding medium caused by the inter particle coupling of Ag NPs agglomerated on the fiber surface (Ili et al. 2009).

Table 2. Colorimetric data for untreated wool, dyed wool and fabrics loaded with silver after dyeing

	L^*	a^*	b^*
Pristine wool	78.44	0.53	20.85
Dyed untreated fabric without Ag	62.08	6.24	39.43
Dyed wool-Ag500	57.53	3.12	28.72

4 Conclusions

1- Pomegranate dyed wool is able to absorb silver ion in silver nitrate solution and reduce it to silver nanoparticles.
2- The SEM and EDX results confirmed the presence of silver nanoparticles on the fibers surface.
3- The presence of silver and silver nanoparticle on the wool surface causes an antibacterial efficiency.
4- The change in color of studied fabrics from yellow to brown, the increased greenness and decreased blueness can be attributed to the presence of Ag NPs on the fabrics.

References

Barani, H., et al.: Optimization of dyeing wool fibers procedure with Isatis tinctoria by response surface methodology. J. Nat. Fibers **9**, 73–86 (2012). https://doi.org/10.1080/15440478.2012.681485

Goodarzian, H.: Wool dyeing with extracted dye from pomegranate (punica granatum L.) peel. Word Appl. Sci. J. **8**(11), 1387–1389 (2010)

Ili, V., et al.: The study of coloration and antibacterial efficiency of corona activated dyed polyamide and polyester fabrics loaded with ag nanoparticles. Fibers Polym. **10**(5), 650–656 (2009)

Nasiriboroumand, M., et al.: Novel method for synthesis of silver nanoparticles and their application on wool. Appl. Surf. Sci. **346**, 477–483 (2015)

Siddiqi, K.S., et al.: A review on biosynthesis of silver nanoparticles and their biocidal properties. J. Nanobiotechnol. **16**(1), 14 (2018)

Development of a Hybrid Textile Material for Cardiovascular Application

Maleke Zidi[1(✉)], Foued Khoffi[1], Yosri Khalsi[2], Abdel Tazibt[2], Frédéric Heim[3], and Slah Msahli[1]

[1] Textile Engineering Laboratory, LGTex, University of Monastir, Monastir, Tunisia
Maleke.zidi01@gmail.com
[2] Centre de Recherche d'innovation et de Transfert de Technologie, TJFU, Bar-Le-Duc, France
[3] LPMT, University of Haute-Alsace, Mulhouse, France

Abstract. Cardiovascular disease accounts for a significant percentage of mortality and morbidity in the ageing population and has estimated increase in the coming years. There is an urgent clinical need for improved cardiovascular devices, which mainly include vascular bypass grafts, vascular stents and heart valves, which will promote desirable blood biomaterial interactions with a high patency. Textile polyester material is characterized by outstanding folding and strength proprieties combined with proven compatibility. It could therefore be considered as an important material in various biomedical applications ranging from conventional cell growth.

The purpose of the present work is to develop a hybrid textile material for cardiovascular applications. Moreover, different approaches in modifying a material surface to create optimal interactions with blood is also studied.

Keywords: PET · Nonwoven · Cardiovascular graft · Medical application · Surface modification

1 Introduction

Cardiovascular diseases (CVDs) are the leading cause of disability, limiting the activity and eroding the quality of life of millions of both middle age adults and elderly each year. The Global Burden of Disease (GBD) study estimated that 31% of all deaths worldwide (17 million deaths) were cause by CVDs in 2017 (Allender and Rayner (2017). Among them, more than 7 million are due to ischemic cardiomyopathies, which leads mainly to acute myocardial infarction and chronic heart failure (Reed et al. 2016). Although heart transplantation mains the only radical treatment for end-stage heart failure, its indications are limited by organ shortage and the complications associated with major immunosuppression. Mechanical assist devices are still primarily used as bridges to transplant (or recovery) and despite its ability to provide symptomatic relief, biventricular resynchronization fails in 20–30% of patients (Albouaini et al. 2009).

Finally, none of the large trials implemented over the last decade to investigate new drugs has yielded a positive outcome leading to an increased survival of heart failure patients.

S. Msahli and F. Debbabi (Eds.): CIRATM 2020, SPM 17, pp. 12–17, 2022.
https://doi.org/10.1007/978-3-031-08842-1_4

There is an urgent clinical need for improved cardiovascular devices, which mainly include vascular bypass grafts, vascular stents and heart valves, which will promote desirable blood biomaterial interactions with a high patency.

Textile Poly (ethylene terephthalate), PET, has been widely used for synthetic graft manufacturing due to its excellent bulk properties, such as good mechanical properties and chemical inertness (Wang et al. 2005). However, its surface properties represent an impediment, as its hydrophobic nature fails to provide the much-needed biocompatibility. When in contact with blood, hydrophobic surfaces induce a specific and/or prolonged protein adhesion (Ma et al. 2007) that can lead to thrombus formation and ultimately graft occlusion or blockage. This is one of the main reasons of graft failure and an unwanted consequence of inadequate graft surfaces. Hence, the appropriate tailoring of the surface properties of PET grafts is of prime importance when targeting biomedical applications.

The purpose of the present work is to develop a hybrid textile material for cardiovascular applications. Moreover, different approaches in modifying a material surface to create optimal interactions with blood is also studied.

The goal of modifying the surface of biomaterial is to create a specific chemical and physical environment that offers a favorable cellular response in hard or soft tissue. In cases where tissue integration is desired the physical environment includes macro, micro and even nanoscale features that allow for cells adhere, proliferate and migrate. However, it is important to note that in some instances, textured surfaces are detrimental to the function of the device such as cardiovascular implants. In this work, the surface modification is carried out by the introduction of a nonwoven fabric. It allows in particular modification of the roughness of surfaces and thus the interaction of a given material with its environment.

2 Material and Machines

2.1 Materials

Poly (ethylene terephthalate), PET, or most simply ‘polyester’, is a linear macromolecular homopolymer formed from step reaction polymerization. It is nominally produced by the polymerization of terephthalic and ethylene glycol. The Fig. 1 below illustrates the microscopic view of the diameter of the PET fiber used, which in turn ranges from 44 μm.

PET fiber has many medical applications like sutures and vascular grafts. The properties such as hardness, stiffness, bio-chemical, and dimensional stability, and biocompatibility make the material more promising for biomedical applications.

PET fabrics with different textures using weaves (plain and satin) relatively fine for applications in the medical field. The table below summarizes the characteristics of the two fabrics produced (Table 1).

Dry Laid PET nonwoven of the present invention is made by carding and bonded mechanically by needle punching. The table below summarizes the characteristics of the nonwoven (Table 2).

The mechanical properties of needle-punched nonwovens are dependent upon laying techniques used for the production of web structures or the initial web structure.

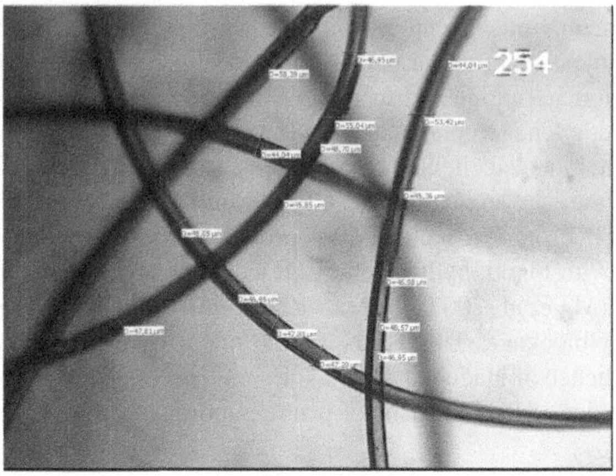

Fig. 1. Microscopic view of the diameter of PET fiber

Table 1. Characteristics of the fabrics

Fabric	F1	F2
Weave	Plain	Satin weave of 5
Warp cound (dtex)	78	78
Weft cound (dtex)	78	78
Warp density (yarn/cm)	88	88
Weft density (yarn/cm)	34	34
Thickness (mm)	0.26	0.33
Air permeability (L/m^2/s)	83.7	283.4

Table 2. Nonwoven characteristics

Thickness (mm)	2.3
Needle punching density (P/cm^2)	70
Air permeability (l/m^2/s)	1188.57

2.2 Machines

Card: MESDAN SPA ITALY (Fig. 2), the main objective of using the card is to separate entangled tufts of fibers from bales and to deliver the individual fibers in the form of a web. The principle of carding is mechanical action, in which the fibers are held by one surface while another combs them out.

Fig. 2. LGtex laboratory carding machine, MESDAN SPA ITALY 337A

Needle punching machine: TEC TEX GERMANY (Fig. 3) is a nonwoven consolidation machine, which is mounted on a beam in which is given an up and down reciprocating motion by means of an eccentric crank mechanism. As a result, the fibers are mechanically interlocked, thereby providing the mechanical strength.

Fig. 3. Needle-punching machine TEC TEX

A needle-punching machine can process a wide range of fibrous webs from different systems, which influences the fiber arrangement within the web. Depending on the feed rate employed, the web density can be varied, and with higher feed rates, thicker and denser nonwovens are formed. The resultant properties of needles punched nonwovens directly relate to the web density; for instance, lower feed rates increases the permeability characteristics of the nonwovens. It is due to the presence of fewer fibers per unit volume. Relatively fewer fibers in the structure results in the formation of larger pores, especially at lower stroke frequencies. However, increases in both feed rate and stroke frequency reduce the permeability of the needle-punched nonwovens, as the pore size decreases with the increase in the number of fibers (Russell and Treloar 2007).

A heat press machine (HASHIMA, Japan) performs the thermal assembly of the fabric and the nonwoven. (Fig. 4).

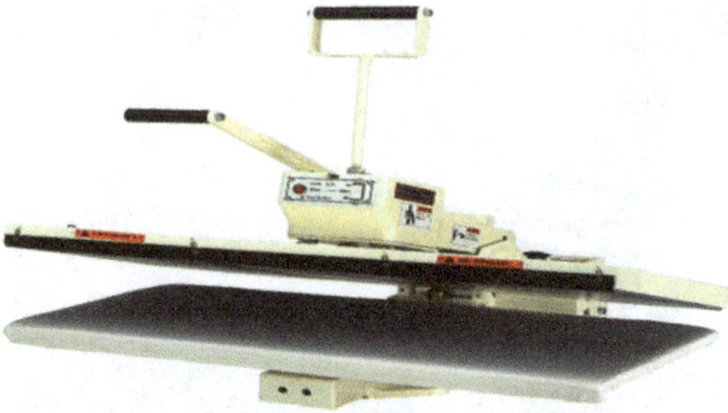

Fig. 4. Heat press machine HASHIMA

3 Results and Discussion

As the primary objective of the assembly of the fabric and the nonwoven is to build a tissue engineered patch and to recreate the natural 3D environment most suitable for an adequate cell and tissue growth, an important aspect of this commitment is to mimic the fibrillar structure of the ECM (Extra Cellular Matrix), which provides essential guidance for cell organization.

This assembly is carried out by two methods: thermal assembly by the heat press machine and mechanical assembly by the needling machine.

3.1 Mechanical Assembly

In this case, the assembly is carried out by mechanical treatment using the needle-punching machine (Table 3).

Table 3. Mechanical assembly characteristics

Composition	Nonwoven + plain fabric	Non-woven + satin weave of 5 fabric
Density (P/m^2)	70	70
Thickness (mm)	2.57	2.75
Air permeability (L/m^2/s)	136	338
Description	The assembly was done but the fabric was damaged because of the high density of the needles	

3.2 Thermal Assembly

In order, to establish between the two textile fabrics, a protein gelatin was used to link them by high pressure (4 bar) (Table 4).

The value of the air permeability of the satin weave is more important considering the satin weave contains more floats. When the number of floats increases, the air permeability of the woven increases also. The higher value of floats number causes increases

Table 4. Thermal assembly characteristics

Composition	Nonwoven + plain fabric	Nonwoven + satin weave of 5 fabric
Temperature (°C)	150	150
Thickness (mm)	0.75	0.77
Air permeability (L/m^2/s)	54.3	131
Description	The assembly was done but it required the intervention of another material which is gelatin	

the air permeability of the fabric. As known, decreasing the number of floats results a tightly woven structure. Therefore, it is thought that the air permeability of the woven is reduced.

All these observations helped to better understand the behavior of the nonwoven on the treated surface and the assembly by supercritical nitrogen jet is envisaged.

4 Conclusion

The control of interaction between implanted material and blood or endothelium is crucial for the successful clinical performance of cardiovascular devices. Much endeavor has been made to resolve some of the issues associated with earlier implants. The diverse surface modification strategies and techniques have been applied in order to regulate blood-implant responses.

In this work, the surface modification is carried out by adding a layer of a PET nonwoven. It allows in particular modification of the roughness of surfaces and thus the interaction of a given material with its environment. This assembly is carried out by two methods: thermal assembly by the heat press machine and mechanical assembly by the needling machine. The physical assembly by supercritical nitrogen jet is also envisaged.

References

Allender, S., Rayner, M.: Coronary heart disease statistics; British Heart Foundation, Heart Statistics (2017)

Reed, G.W., Rossi, J.E., Cannon,C.P.: Acute myocardial infarction. Lancet (2016). https://doi.org/10.1016/S0140-6736(16)30677-8

Albouaini, K., Egred, M., Rao, A., Alahmar, A., Wright, D.J.: Cardiac resynchronisation therapy: evidence based benefits and patient selection. Eur. J. Intern. Med. (2009)

Wang, J., et al.: Surface characterization and blood compatibility of poly (ethylene terephthalate) modified by plasma surface grafting. Surf. Coat. Technol. **196**, 307–311 (2005)

Ma, Z.W., Mao, Z.W., Gao, C.Y.: Surface modification and property analysis of biomedical polymers used for tissue engineering. Colloids Surf. B-Biointerfaces **60**, 137–157 (2007)

Russell, T., Treloar, T.: Process Control in Textile Manufacturing (2007). pp. 279–299 (2013)

The Production of Hybrid Nano-fiber Scaffolds for Tissue Engineering of the Mouse Uterus

Mahdieh Dehghan[1,2], Mohammad Khajeh Mehrizi[1](✉), and Habib Nikukar[2,3]

[1] Textile Department, Faculty of Engineering, Textile College, Yazd University, Yazd, Iran
mkhajeh@yazd.ac.ir

[2] Medical Nanotechnology and Tissue Engineering Research Center, Yazd Reproductive Sciences Institute, Shahid Sadoughi University of Medical Sciences, Yazd, Iran

[3] Department of Advanced Medical Sciences and Technologies, School of Paramedicine, Shahid Sadoughi University of Medical Sciences, Yazd, Iran

Abstract. In this study, a novel polycaprolactone/gelatin/polydimethylsiloxane (PCL/G/PDMS) scaffold with improved mechanical properties and controllable porous structure was prepared through electrospinning method. In this research, the biocompatibility was evaluated, using mouse uterine cells. MTT, Hematoxylin-Eosin (H&E) staining results indicated that the growth of endometrial cells developed in the PCL/G/PDMS scaffold. The results also showed that the mouse embryo good adaptation with the scaffold and the scaffold transplantation to the mouse uterus showed that the uterine tissue was completely perfused after one month. Our data suggests that the PCL/G/PDMS scaffold is a new suitable scaffold material for tissue engineering. The purpose of this project is to eliminate of problems similar to damaged uterus, Rennet uterus (surrogacy), Lack of womb for women's using a scaffold like uterine tissue.

Keywords: Tissue engineering · Nano fiber · Uterus · Embryo · Scaffold

1 Introduction

Scientists in regenerative medicine and tissue engineering are currently applying the principles of cell transplantation, materials science, and bioengineering to build biological alternatives that restore and maintain normal function in diseased and damaged tissues. Some of the therapeutic approaches stemming from tissue engineering efforts have now successfully entered the clinical setting, indicating the promise of a regenerative medicine for the future (Atala 2009). Uterine infertility is a factor of approximately 1:500 women of childbearing age (Bandstein and Hellström 2016). A study aimed at reconstructing engineered uterine tissues (EUTs) containing a smooth muscle layer resembles a normal uterine wall (Lü et al. 2009). Also, endometrial cell sheets help as a new treatment to restore functional endometrium (Takagi et al. 2014). Thus, tissue engineering programs using biomass and stem cells may replace the need for a living donor and may prevent the suppressive treatment of the required immune system (Hellström et al. 2014).

S. Msahli and F. Debbabi (Eds.): CIRATM 2020, SPM 17, pp. 18–22, 2022.
https://doi.org/10.1007/978-3-031-08842-1_5

2 Materials and Methods

2.1 Animal

For uterine cell isolation, syrian mices (25 weeks) (Yazd Reproductive Sciences Institute) were used. All mouses were kept under specific conditions at the Animal Center of Yazd Reproductive Sciences Institute in accordance with facility guidelines. These maintained under controlled temperature (25 ± 38 C), proper humidity (50 ± 5%) and a 12 h light/dark cycle. The Ethical Review Committee of the Institute approved the experimental protocol.

2.2 Donor Uterus Harvest of Mice

Mices were unconscious with isoflurane. The mice were placed on its back and the abdomen was cleaned with 70% ethanol. A celiotomy was performed via cut the middle line from the symphysis pubis to the xiphoid process (Miyazaki and Maruyama 2014). Then both uterine horns were separated and placed in the DMEM.

2.3 Fabrication of Scaffolds

The scaffolds were produced by electrospinning method (Dehghan et al. 2020). To design and fabrication a three-dimensional scaffold of mice uterine tissue, a trilayer scaffold was produced. The inner layer consists of C3G5D2 nanofibers, the middle layer is PNIPAM nanofibers and the outer layer is C3G5D2 nanofibers The PNIPAM in water solution at 37 °C shranks significantly (Apsite et al. 2017) and bends to toward thin C3G5D2 scaffold and a trilayer tubular structure forms spontaneously. The C3G5D2 nanofibers are random and the PNIPAM nanofibers are aligned.

2.4 Isolation of Mouse Uterine Tissue Cells

At first, the uterine horn was separated from the immature syrian mice, and then it was washed with PBS and were minced. Minced uterine horn was treated with DMEM containing 0.1% collagenase type I at 37 °C for 1 h with shaking every 10 min. After an hour, the minced uterine horn was washed with PBS. The minced tissues were explanted inside the dish and the DMEM containing 10% FBS, 1% penstrep and 0.5% fungizon used as culture medium. Initially, 1000 μL of DMEM was added to the dish to attach the tissues to the dish and were incubated in a humidified incubator at 37 °C with 5% CO_2 concentration. After 24 h, 1000 μL of DMEM was added to the dish. The non-adherent tissues were removed after 48 h of culture and DMEM was added and the non-adhere cells were removed every 72 h. After 7 days, the cells were gradually extracted from the tissue and attached to the surface of the dish. On the 15 days, the cells were washed with PBS, incubated with 0.05%/0.02% (v/v) trypsin/EDTA solution and pelleted by centrifugation at 200 g for 5min. The cells were cultured in flasks and were incubated in a humidified incubator at 37°C with 5% CO_2 concentration. Cell culture was continued until the third passage. Cells were harvested at days 3, 7, and 10 for growth and evaluated using H&E staining and MTT test. By triplicate assay the activity and survival of the cells on nanofibers were calculated as a percentage of samples to controls.

2.5 In Vitro Fertilization and Embryo Assessment

The sperm was collected from mices 8–12 weeks for in vitro fertilization (IVF). The sperms were released into the medium and dispersed for 15 min at 37 C. After dispersal, the sperm concentration was determined to achieve the final concentration of $1 * 10^6$ sperm/ml. Then, the insemination dishes were incubated for 1–2 h before addition of oocytes. The melted eggs were transferred from each group separately to 100 μl drops of G-IVF medium. After 5 h of incubation with spermatozoa, the oocytes were washed and cultured in G1 medium. The embryonic development progression was monitored every 24 h for 3 days until blastocyst stage (Mohammadzadeh et al. 2018). Then blastocyst was placed on the C3G5D2 scaffold.

3 Results and Discussions

The uterine tissue culture of the mouse and its cell extraction in the DMEM medium to the third passage are shown in Fig. 1. Attachment, proliferation and morphology of cells on scaffolds with H&E staining on 1, 3 and 7 days were shown in Fig. 2. After 7 days, a pervasive network of cells on the scaffolds proliferated and attached.

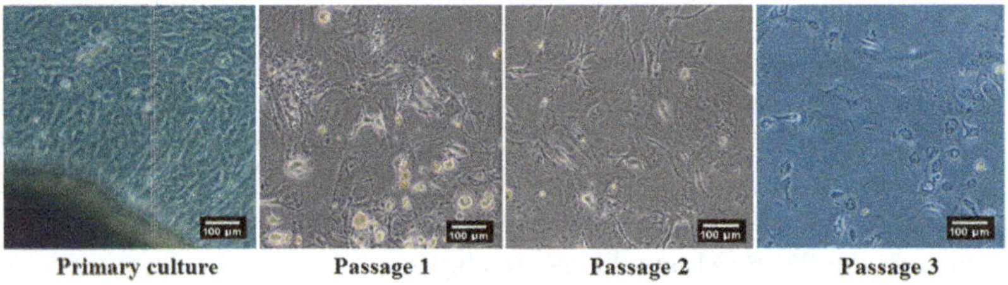

Fig. 1. The uterine tissue culture of the mouse and its cell extraction in the DMEM medium.

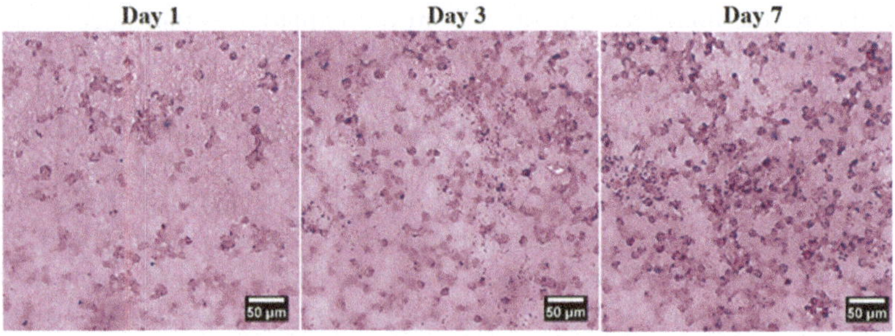

Fig. 2. H&E staining micrographs of uterine cell cultured on PCL/G/PDMS scaffold.

Figure 3 shown mouse embryos which were attached on scaffold and good compatibility with PCL/G/PDMS scaffold. 3D tubular scaffold for mouse uterine tissue produced and transplanted to mouse. After one month, the transplanted uterine tissue is stained by H&E and progesterone receptor (PRs) and is shown that the mouse uterine tissue is formed (Fig. 4).

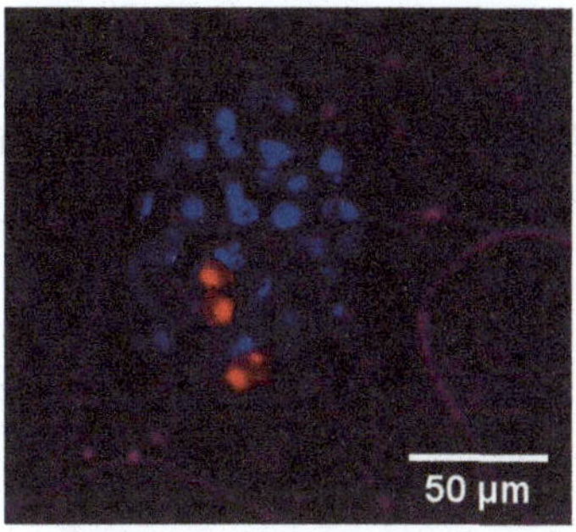

Fig. 3. Mouse embryo attachment on PCL/G/PDMS scaffold

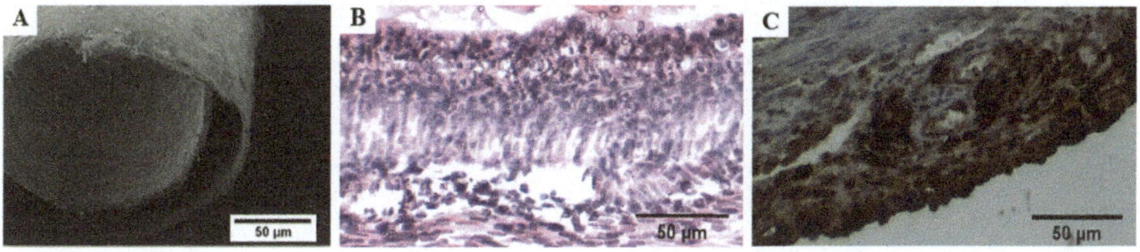

Fig. 4. 3D tubular scaffold for mouse uterine tissue

4 Conclusion

The self-assembly of endometrial cells and stromal cells could make the reconstructed tissues closer to natural tissues in structure. Transplantation of engineered tissue in vivo and birth of offspring could be the next confirmatory step. These findings could be the first phase for the uterine tissue engineering, ready for animal study and then a clinical trial.

References

Atala, A.: Engineering organs. Curr. Opin. Biotechnol. **20**(5), 575–592 (2009). ISSN: 0958-1669

Bandstein, S., Hellström M.: Tissue engineered uterine tissue supports pregnancy in a rat model. In: Programme in Medicine, Gothenburg, Sweden (2016)

Lü, S.-H., et al.: Reconstruction of engineered uterine tissues containing smooth muscle layer in collagen/matrigel scaffold in vitro. Tissue Eng. Part A **15**(7), 1611–1618 (2009). ISSN: 1937-3341

Hellström, M. et al.: Towards the development of a bioengineered uterus: comparison of different protocols for rat uterus decellularization. Acta Biomater. **10**(12), 5034–5042 (2014), ISSN: 1742-7061

Miyazaki, K., Maruyama, T.: Partial regeneration and reconstruction of the rat uterus through recellularization of a decellularized uterine matrix. Biomaterials **35**(31), 8791–8800 (2014), ISSN: 0142-9612

Dehghan, M. et al.: Modeling and optimizing a polycaprolactone/gelatin/polydimethylsiloxane nanofiber scaffold for tissue engineering: using response surface methodology, J. Text. Inst. **112**(3), 482–493 (2021). ISSN: 0040–5000

Psite, I., et al.: Porous stimuli-responsive self-folding electrospun mats for 4D biofabrication. Biomacromolecules **18**(10), 3178–3184 (2017). ISSN: 1525-7797

Mohammadzadeh, M. et al.: Influential effect of age on oocyte morphometry, fertilization rate and embryo development following IVF in mice. Middle East Fertil. Soc. J. **23**(2), 117–120 (2018). ISSN: 1110-5690

Wound Healing Property of the Hydroalcoholic Extract of *Teucrium Polium L.*

Salma Fadhel[1](✉), Fadhel Jaafar[1], Mohamed Ali Lassoued[2], Adberrahmen Merghni[3], Jalel Dahas[4], and Neji Ladhari[1]

[1] Textile Engineering Laboratory of ISET Ksar Hellal, University of Monastir, Monastir, Tunisia
Salmajamel92@gmail.com

[2] Laboratory of Galenic Pharmacy, Laboratory of Pharmacology, Research Unit URSAM, Faculty of Pharmacy of Monastir, Monastir, Tunisia

[3] Laboratoire des Maladies Transmissibles et Substances biologiquement actives (LR99ES27), Faculté de Pharmacie de Monastir, Université de Monastir, Monastir, Tunisia

[4] Medical Society of Sahel, Monastir, Tunisia

Abstract. *Teucrium Polium L.* has been used since time immemorial as a medicinal plant with beneficial wound healing effects; this study aims to investigate the wound healing propriety of this plant. Two principal bioactive compounds were investigated in the hydroalcoholic extract of the aerial part of *Teucrium Polium L.*, tannin and flavonoids. The hydroalcoholic extract showed antioxidant activity with inhibition percentage more than 90% at a concentration of about 1 mg/mL and the antibacterial activity was carried out on *Staphylococcus aureus, Escherichia coli, Pseudomonas aeruginosa* and *Staphylococcus epidermidis*. The ability of *Teucrium Polium L.* extracts to inhibit the growth of bacteria, which is an indication of its broad-spectrum antimicrobial potential, which can be used in the management of microbial infections in the wound area.

Keywords: *Teucrium Polium L.* · Wound healing · Antibacterial · Medicinal plant

1 Introduction

The development of complex and advanced wound care represents one of the biggest challenges to health care systems worldwide. Textile materials are widely used as protection on skin wounds, to isolate and protect it from external contamination and friction or shock. The fibers used in wound care can be divided into natural and human-made fibers. The most important natural fibers are cotton, silk, and linen. Conventional human-made fibers, such as polyester, polyamide, polypropylene, polyurethane, polytetrafluoroethylene, etc., are used for wound-healing management as well (Voncina et al. 2016).

Recently, dressings that contain and release antimicrobial agents at the wound surface have entered the marketplace. These dressings usually provide a continuous or sustained release of the antiseptic agent at the wound surface to provide a long-lasting

S. Msahli and F. Debbabi (Eds.): CIRATM 2020, SPM 17, pp. 23–28, 2022.
https://doi.org/10.1007/978-3-031-08842-1_6

antimicrobial action in combination with maintenance of physiologically moist environment for healing. These agents are usually extracted from medicinal plants; it can be essential oil, vegetable oil, hydraulic extract or alcoholic extract. *Teucrium Polium L.* is used in folk medicine to treat wounds, it known as Gattabet Al-Ajrah or Jaada is a perennial and prostrate species with white to bright pink flowers, crenate leaves, and ramified indumentum (Boulila et al. 2008).

Present study attempts to investigate phyto-pharmacological and microbiological aspect of the hydroalcoholic extract of the aerial part of *Teucrium Polium L.*

2 Materials and Methods

2.1 Extraction Method

The aerial part of *Teucrium Polium L.* was collected in the city of Zaghouan, Tunisia, then identified by the botanist of the Faculty of Pharmacy of Monastir. The aerial parts of T. Polium. L (200 g) were air-dried in the shade and made into powder. Ethanol extract was provided using maceration method in 80% ethanol for 24 h, followed by evaporation by rotatory evaporator at 50 °C and then lyophilized in a freeze dryer (Esmaeili and Yazdanparast 2004).

2.2 Phytochemical Screening of the Hydroalcohol Extract of *Teucrium Polium L.*

2.2.1 Qualitative Screening

Test for tannins: 1 g of extract was diluted with 10 mL of water and tested with the following reagents:

$FeCl_3$ Reaction: 1 mL of ether, methanol or water extract was mixed with 10 mL of distilled water and filtered. Ferric chloride ($FeCl_3$) reagent (3 drops) was added to the filtrate. A blue-black or green precipitate confirmed the presence of Gallic tannins or catechol tannins, respectively (Karumi et al. 2004).

Stiasny reaction: 2 mL of aqueous solution add a few drops of Stiasny reagent and heat on a water-bath to boiling for 15 min. The appearance of a light brown precipitate indicates the presence of catechetical tannins.

Bate-Smith: In a test tube, to 1 mL of aqueous solution add a few drops of hydrochloric butanol reagent and bring to a water bath at 90 °C for 15 min. The formation of a red color indicates the presence of catechin tannins (Benzidia et al. 2018).

Test for flavonoïde (Shinoda reaction: Cyanidine): About 0.2 g of the extract was dissolved in 2 mL of methanol and heated. 2 mg of magnesium metal was added to the mixture followed by the addition of a few drops of concentrated HCl. The occurrence of a red or orange colouration was the indicative of the flavonoids which were further confirmed by thin layer chromatography (Aiyegoro and Okoh 2010).

2.2.2 Quantitative Analysis: Antioxidant Activity

Trapping free radical DPPH assay was conducted to evaluate the antioxidant activity. This test measures the ability of eliminating free radicals in Methanolic extract. DPPH is a molecule containing a stable free radical. In the presence of an antioxidant that can donate an electron to the DPPH, the purple color, typical free radical DPPH to disintegrate, the mixture was kept in the dark for 30 min and (OD) was recorded at 517 nm against a control sample (Brand-Williams et al. 1995). Different concentrations 1; 1/2; 1/4; 1/8; 1/16; 1/24 (mg/mL) were taken in separate test tubes. One millilitre of DPPH (2,2-diphenyl-1-picrylhydrazyl) solution was dissolved in 80% ethanol and added each to test tube and shaken vigorously. After the addition of DPPH solution, all of the test tubes were shaken gently and allowed to stand at 27 °C in a dark place for 45 min. The absorbance of the prepared samples was measured using UV spectroscopy at a wavelength of 517 nm. Radical scavenging activity of the tested crude extract samples was estimated as an inhibition percentage and was calculated by using the formula (1), (Alabri et al. 2014).

Measurement of radical scavenging activity (%):

$$\% \text{ inhibition} = \frac{Acontrol - Aextract}{Acontrol} 100 \quad (1)$$

where A control is the absorbance of DPPH + methanol; A extract is the absorbance of DPPH radical + sample.

3 Determination of Antimicrobial Activity

3.1 Qualitative Screening

Four bacterial species were used in this study, the gram-positive were *staphylococus aureus (S.aureus), Staphylococus epidermidis (S.epidermidis)* and Gram negative species bacteria were *Escherichia coli (E.coli), Pseudomonas aeruginosa (P. aeruginosa)* these species were obtained from the Laboratory of Microbiology of the University hospital of Monastir-Tunisia, These strains were obtained from bacteriological samples in hospitalized patients and/or consultants at the University Hospital of Monastir-Tunisia. Bacterial identification was performed using Gram staining, catalase test, tube coagulase, DNase agar, mannitol salt agar and API ID 20 STAPH galleries (bio-Mer-ieux, France) (Haddad et al. 2018).

3.2 Microdilution Method for the Determination of the MIC and MBC

The minimal inhibition concentration (MIC) and the minimal bactericidal concentration (MBC) values were determined for all bacteria tested in this study using the microdilution assay (Snoussi et al 2008). The inoculums of the bacterial strains were prepared from an overnight culture and suspensions were adjusted to 108 cells/mL ($DO\lambda = 600$ nm). Ethanolic extract of *Teucrium Polium* was dissolved in 10% dimethylsulfoxide (DMSO 10%). The highest concentration was 250 mg/mL, and then serial dilutions were prepared in concentrations ranging from 250 to 0.97 mg/mL. the tubes were incubated at 37 °C for 24 h. After incubation for 24 h at 37 °C, bacterial growth was evaluated by the presence of turbidity and a pellet on the well bottom. MIC was defined as the concentration that completely inhibited visible cell growth during a 24 h incubation period at 37 °C. To determine the minimum bactericidal concentration (MBC) values, 10 μL of each well medium with no visible growth was removed and pour plated with MHA. After 24 h of incubation at 37 °C, the number of surviving organisms was determined as cfu/mL. MBC was defined as the lowest concentration at which 99% of the bacteria were killed. Each experiment was repeated at least twice (Magina et al. 2009).

4 Results

The dry and fresh powder samples of *Teucrium Polium L.* were extracted with 80% ethanol solvent the maceration method. After the complete extraction, the ethanol solvent was evaporated using a rotary evaporator lyophilized in a freeze dryer producing solid mass crude extracts. The extract yield was 16% w/w.

4.1 Phytochemical Screening of the Hydroalcohol Extract of *Teucrium Polium L.*

4.1.1 Qualitative Screening

The phytochemical screening of the hydroalcoholic extract of the aerial part of *Teucrium Polium L.* (Table 1) showed the presence of showed the presence of flavonoids and tannins, these phytochemical compounds are known to support bioactive activities in medicinal plants and thus responsible for the antioxidant activities of this plant extract used in this study.

Table 1. Results of the phytochemical screening of hydroalcoholic extract of *Teucrium Polium. L.*

Phytochemical tests		*Teucrium Polium L.*
Tannins	$FeCl_3$ reaction	+++
	Stiasny reaction	+++
	Bate-Smith reaction	+
Flavonoid	Shinoda reaction	++

+++, High concentration; ++, moderate concentration; +, low concentration; -, absence.

4.1.2 Radical Scavenging (Antioxidant) Activity

The antioxidant activity of plant extract is determined by the free radical DPPH reduction method (Fig. 1).The results show that the percentage of inhibition of hydroalcoholic extracts of *Teucrium Polium* is greater than 90% at a concentration of the order of 1 mg/mL.

4.1.3 Determination of Antimicrobial Activity

The MIC of *Teucrium Polium* ethanolic extract against *S.aureus et S.epidermidis* strains was around 0.48 mg/mL, while *E. coli* and *P. aeruginosa* were more resistant (MIC = 7.81 mg/mL).

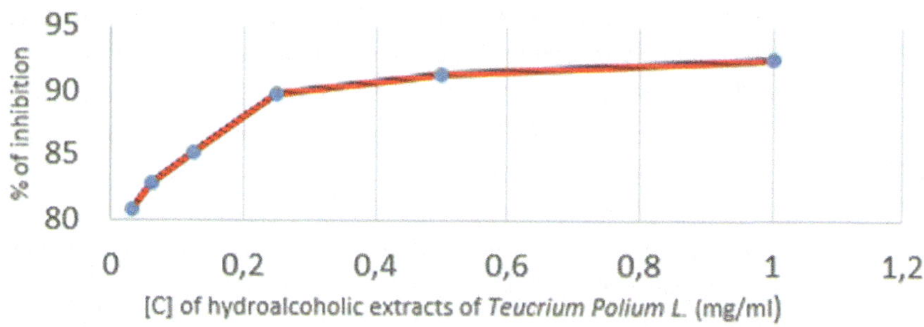

Fig. 1. DPPH radical scavenging activity of the hydroalcoholic extract of *Teucrium Polium L.*

The MBC values of *Teucrium Polium* ethanolic extract against *S.aureus et S.epidermidis* strains was around 0.97 mg/mL while the MBC *for E. coli* and *P. aeruginosa* were found at 15.62.

A concentration the *Teucrium Polium* ethanolic extract of 0.97 mg/mL and 15.62 mg/mL could kill 99% of gram negative bacteria and gram positive bacteria strains respectively tested in this study (Table 2).

Table 2. The CMI and CMB of the extract tested on the four selected microbial strains.

	E.c	P.a	S.a	S.e
CMI (mg/mL)	7.81	7.81	0.48	0.48
CMB (mg/mL)	15.62	15.62	0.97	0.97

5 Discussions

Tannins and flavonoids are phenolic compounds known to support bioactive activities in medicinal plants, these phytochemical compounds are likely to be responsible for the free radical scavenging effects observed (Bendif Hamdi 2017). Phenolic compounds are a major group of compounds that act as primary antioxidants or free radical scavengers (Potterat et al. 1997). Flavonoids have been shown to exhibit their actions through effects on membrane permeability, and by inhibition of membrane-bound enzymes such as the ATPase and phospholipase A2 (Potterat et al. 1997). Tannins and their derivatives are phenolic compounds considered to be primary antioxidants or free radical scavengers (Seka et al. 2012).

The results of present study showed antibacterial potential against *S.aureus, S.epidermidis, E. coli and P. Aeruginosa*, Ethanolic extract of *Teucrium Polium* aerial parts are effective against both gram-positive and gram-negative bacteria. In this work, we could verify that Gram-positive bacteria is susceptible to low extract concentration, On the other hand, the Gram-negative bacteria growth was only inhibited in higher extract concentrations. The phytoconstituents detected in the plant materials could be responsible for the antimicrobial activity of *Teucrium Polium L.* ethanolic extract. Flavonoids have also been reported to possess anti-bacterial activity, which could be attributed to their ability to form complex with extracellular, soluble proteins and bacterial cell walls (Tsuchiya et al. 1996).

References

Esmaeili, M., Yazdanparast, R.: Hypoglycaemic effect of Teucrium polium: studies with rat pancreatic islets. J. Ethnopharmacol. **95**(1), 27–30, 0378–874 (2004)

Voncina, B., Zemljič Fras, L., Ristic, T.: Active textile dressings for wound healing, advances in smart medical textiles. In: Lieva Langenhove, vol. 298, pp. 73–92. Elsevier, Woodhead Publishing (2016). ISBN: 9781782423799

Boulila, A., Bejaoui, A., Messaoud, C., Boussaid, M.: Variation of volatiles in Tunisian populations of Teucrium polium L. (Lamiaceae). Chem. Biodivers. **5**(7), 1389–1400 (2008). 11, 1612–1880 (2008)

Karumi, Y., Onyeyili, P., Ogugbuaja, V.: Identification of active principles of M. Balsamin. J. Med. Sci. **4**(3), 179–183 (2004). 3

Benzidia, B., et al.: Chemical composition and antioxidant activity of tannins extract from green rind of Aloe vera (L.) Burm. F. J. King Saud Univ. **31**(4), 1175–1181 (2018). 6, 1018–3647 (2018)

Aiyegoro, O., Okoh, A.: Preliminary phytochemical screening and In vitro antioxidant activities of the aqueous extract of helichrysum longifolium DC. BMC Complement. Altern. Med. **10**, 21 (2010). https://doi.org/10.1186/1472-6882-10-21. 8, 1–8, 1682–4474

Brand-Williams, W., Cuvelier, M.E., Berset, C.: Use of a free radical method to evaluate antioxidant activity. Food Sci. Technol. **28**(1), 25–30 (1995). 5, 1472–6882 (1995)

Alabri, T., Al Musalami, Ph., Hossain, M., Weli, A., Al-Riyami, Q.: Comparative study of phytochemical screening, antioxidant and antimicrobial capacities of fresh and dry leaves crude plant extracts of Datura metel L. J. King Saud Univ. **26**(3), 237–243 (2014). 6, 1018–3647 (2014)

Haddad, O., Merghni, A., Elargoubi, A., Rhim, H., Kadri, Y., Mastouri, M.: Comparative study of virulence factors among methicillin resistant Staphylococcus aureus clinical isolates. BMC Infect. Dis. **18**, 1–8 (2018). https://doi.org/10.1186/s12879-018-3457-2. 8, 1471–2334

Snoussi, M., et al.: In-vitro anti-Vibrio spp. activity and chemical composition of some Tunisian aromatic plants. World J. Microbiol. Biotechnol. 24, 3077 (2008). https://doi.org/10.1007/s11274-008-9848-6. 12, 1573–0972

Magina, M., et al.: Chemical composition and antibacterial activity of essential oils of Eugenia species. J. Nat. Med. **63**, 345–350. https://doi.org/10.1007/s11418-009-0329-5. 3(5) 1340–3443 (2009)

Potterat, O.: Antioxidants and free radical scavengers of natural origin. Curr. Org. Chem. **1**(5), 415–440 (1997). 25, 1385–2728

Li, H., Wang, Z., Liu, Y.: Review in the studies on tannins activity of cancer prevention and anticancer. Zhong Yao Cai, **26**(6), 444–448 (2003). 4, 1001–4454

Sekar, D., Kolanjinathan, K., Saranraj, P.: Screening of Phyllanthus amarus, Acalypha indica and Datura metel for its antimicrobial activity against selected pathogens. Int. J. Pharm. Biol. Arch. **6**(5), 1231–1235 (2012)

Hironori, T., et al.: Comparative study on the antibacterial activity of phytochemical flavanones against methicillin-resistant Staphylococcus aureus. J. Ethnopharmacol. **50**(1), 27–34 (1996). 8, 27–34, 0378-8741

Thermo-Mechanical Characterization of Post-consumer PP/Tunisian Organo-Clay Filaments

Kmais Zdiri[1,2,3](✉), Adel Elamri[2], Omar Harzallah[1], and Mohamed Hamdaoui[2]

[1] Université de Haute Alsace, ENSISA, LPMT EA 4365, 68100 Mulhouse, France
khmaiszdiri@gmail.com
[2] Université de Monastir, ENIM, MPTEX, UR17ES33, 5019 Monastir, Tunisie
[3] Université de Lille, ENSAIT, GEMTEX, 59000 Lille, France

Abstract. Extruded post-consumer PP (PCPP)/Tunisian clay blends were used to elaborate monofilament yarns using melt spinning method.

The scanning electron microscopy (SEM) images of cross section of filaments clearly showed a well dispersed system. We deduced from the differential scanning calorimetry (DSC) tests that the addition of clay increases the crystallization temperature (Tc) as compared to the raw PCPP filaments.

The stress-strain and bending analysis showed that the mechanical behaviors of the PCPP filaments were enhanced by the addition of clay fillers.

Keywords: Post-consumer · Polypropylene · Tunisian clay · Filament · Properties

1 Introduction

Polypropylene (PP) is a polymer widely used in several industrial sectors due to its low cost and density (Velásquez et al. 2019). This enormous use generates high quantities of PP residues, which considerably affects the environment (Zdiri et al. 2018). The recycling and reusing of waste PP plastics is the most successful solution to this environmental problem (Grigore et al. 2017).

The obtained post-consumer PP (PCPP) can be used for the elaboration of textile filaments. Nevertheless, the performance of these PCPP filaments sometimes proves to be insufficient during manufacturing, which reduces its thermo-mechanical properties (Adnan et al. 2018).

In this study, we use Tunisian clays as filler in order to ameliorate the performance of PCPP filaments. Also, we use the melt mixing technique in order to incorporate clay fillers into filaments, which takes the advantages of traditional polymer processing methods (Zdiri et al. 2018).

In this work, we used the melt spinning technique to elaborate Continuous monofilament yarn based on raw PCPP and PCPP/organo-clay composites. The objective was to study the impact of organo-clay on the morphological, thermal and mechanical properties of post-consumer PP filaments.

S. Msahli and F. Debbabi (Eds.): CIRATM 2020, SPM 17, pp. 29–35, 2022.
https://doi.org/10.1007/978-3-031-08842-1_7

2 Materials and Methods

2.1 Materials

The post-consumer PP investigated in this work was obtained from woven bags.

The organo-clay used in this study was obtained from Aleg formation located in the meridional Atlas of Tunisia (Zdiri et al. 2020).

2.2 Spinning of Filaments

Filament yarn of neat PCPP and PCPP/organo-clay composites were obtained using the method of melt spinning. The compositions and designations of the prepared filaments are listed in Table 1.

Table 1. Compositions and designations of the prepared filaments

	PCPP (%wt)	Organo-clay (%wt)	PP-g-MA (%wt)
PCPP	100	–	–
PCPPCC1	96	1	3
PCPPCC3	88	3	9
PCPPCC5	80	5	15

2.3 Structural Properties

A scanning electron microscope (Fig. 1) was used to study the surface morphology and cross-section of neat PCPP and PCPP/organo-clay filaments with different content.

2.4 Differential Scanning Calorimetry

The crystallization temperatures (Tc), the melting temperature (Tm) and the degree of crystallinity (Xc) of the prepared filaments were carried out using TA instruments.

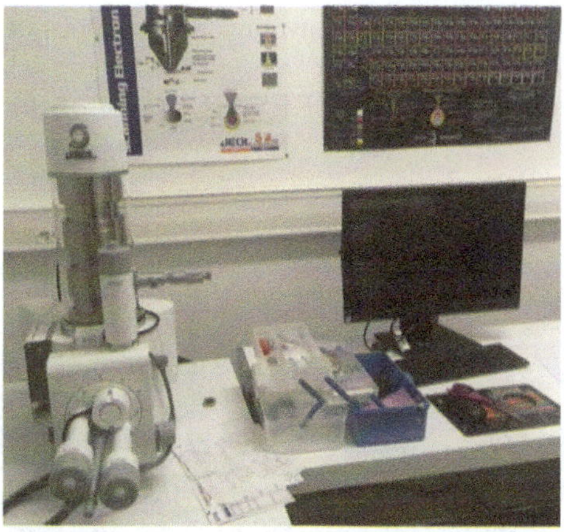

Fig. 1. Scanning electron microscope JEOL-JSM 100 LT

2.5 Tensile Test of Filaments

Tensile modulus and elongation at break were measured on an MTS/20 Instron dynamometer (Fig. 2). A constant length gauge of 20 mm was loaded at a constant crosshead speed of 20 mm.min^{-1} (NF EN ISO 5079).

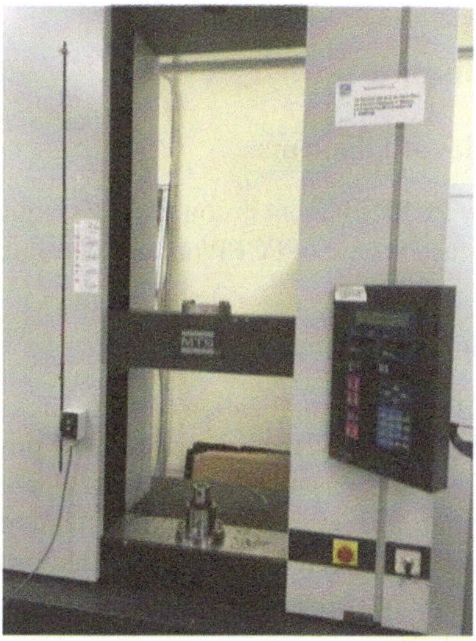

Fig. 2. Photograph tensile testing machine/ENSISA Mulhouse

2.6 Bending Test of Filaments

The bending hysteresis 2HB expressed in cN.cm/cm (gf.cm/cm) is obtained by averaging the two widths of the diagram at curves between -1 and $+1$ cm^{-1}. The bending rigidity

B expressed in cN.cm^2/cm (gf.cm^2/cm), average of the slopes of the diagram at curves between -1 and $+1$ cm^{-1}.

A bending measuring device, type KES-SH, was used for these measurements (Fig. 3).

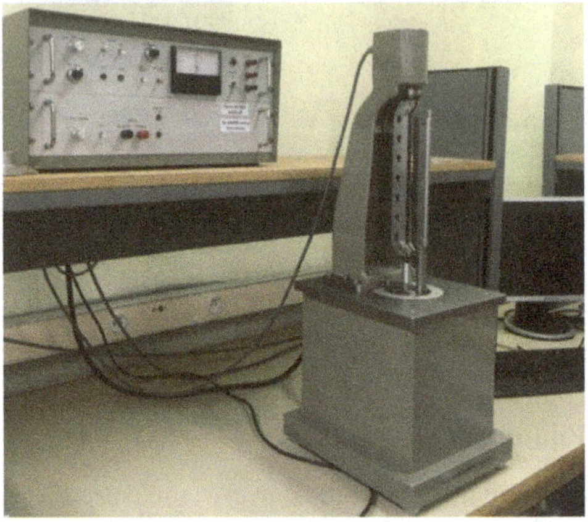

Fig. 3. Photo of the measuring device KES-SH/ENSISA Mulhouse

3 Results and Discussion

3.1 Structural Properties of Filaments

It can be observed that raw PCPP filament is completely smooth while a small aggregate appears in the filament surface in case of PCPP/organo-clay composite filaments (Fig. 4).

3.2 Differential Scanning Calorimetry

It can be seen from Table 2 that the crystallization temperatures of PCPP filament increase with increasing organo-clay content. This improve is most probably due to the effect of the organo-clay particles which act as nucleating agents (Lei et al. 2006).

On the other hand, the melting temperature of PCPP filament decreased after the addition of organo-clay particles. This decrease can be justified by the increased interactions between clay fillers and PCPP polymer (Pozsgay et al. 2002).

3.3 Tensile Behavior of Filaments

Table 3 shows the mechanical properties of the PCPP filaments with filler loading in various blend ratios. It can be seen that the mechanical behaviors of the pure PCPP filament are progressively improved by adding Tunisian organo-clay.

These enhancements may be attributed to the effect of clay platelets, which reduces the mobility of the polymer chains (Yin et al. 2016).

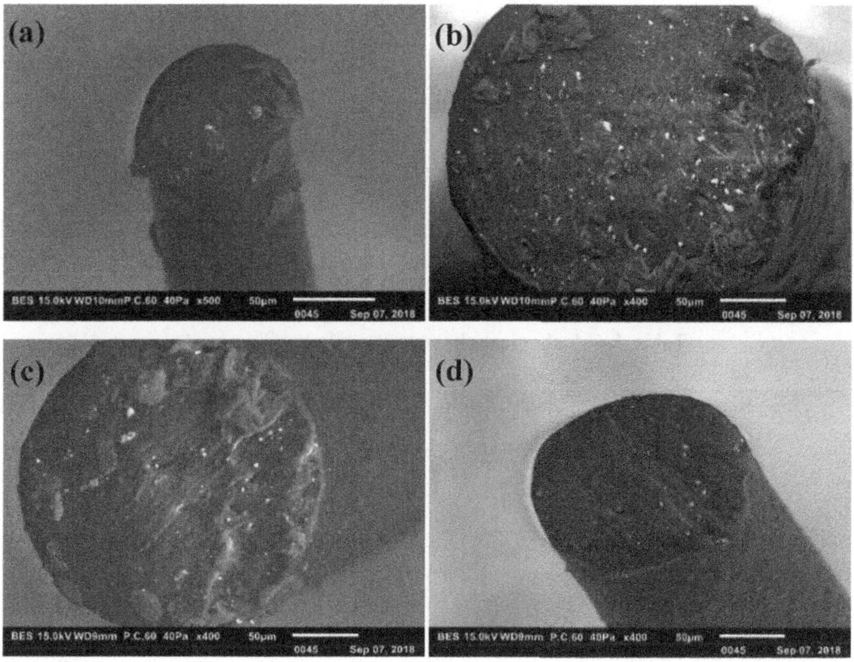

Fig. 4. Scanning electron micrographs of (a) PCPP, (b) PCPPCC1, (c) PCPPCC3 and (d) PCPPCC5.

Table 2. Differential scanning calorimetric analysis of the prepared filaments

	Tc (°C)	ΔHc (J/g)	Tm (°C)	ΔHm (J/g)	χ (%)
PCPP pur	119,36	93,44	163,37	82,7	44,7
PCPPCC1	120,84	94,14	162,19	75,78	45,04
PCPPCC3	122,45	95,76	161,96	75,24	45,81
PCPPCC5	123,51	96,17	161,72	74,88	46,01

Table 3. Mechanical properties of the prepared filaments

	E (GPa)	σs (MPa)	εb (%)
PCPP	7.20 ± 1,18	591 ± 42	708.86 ± 0,75
PCPPCC1	8.61 ± 0,75	735 ± 43	771.23 ± 3,30
PCPPCC3	10.22 ± 0,88	873 ± 44	798.89 ± 0,67
PCPPCC5	14.61 ± 0,06	940.2 ± 27	836.27 ± 1,73

3.4 Bending Tests of Filaments

Table 4 provides a summary of the averages obtained from the bending tests on the PCPP filaments. It is observed that the addition of the clay gives a greater flexural rigidity and flexural hysteresis values. Indeed, the specific rigidity R of PCPPCC5 filament is 0.21 $mN.mm^2/tex^2$, whereas it is 0.13 $mN.mm^2/tex^2$ for PCPPCC1 filament and only 0.09 $mN.mm^2/tex^2$ for the neat PCPP filament.

This effect can be attributed to the intercalation of polymer chains between clay sheets, which limits the mobility of the chains (De Bilbao et al. 2009).

Table 4. Mechanical characteristics in bending mode

	Bending rigidity (B) ($gf.cm^2/cm$)	Specific rigidity (R) ($mN.mm^2/tex^2$)	Bending hysteresis (2HB) (gf.cm/cm)	Specific hysteresis (H) ($mN.mm/tex^2$)
PCPP	0.035 ± 0,001	0.0919	0.0186 ± 0,00039	0.00488
PCPPCC1	0.0644 ± 0,002	0.132	0.0306 ± 0,0065	0.00631
PCPPCC3	0.088 ± 0,009	0.149	0.0524 ± 0,00748	0.00891
PCPPCC5	0.1597 ± 0,033	0.214	0.0707 ± 0,0134	0.00948

4 Conclusion

In this work, morphological and thermo-mechanical properties of post-consumer PP filaments filled with organo-clay at various concentrations (1, 3 and 5 wt.%) were investigated.

Surface morphology study of filaments shows a well dispersed system, which is the reason for the improvement in mechanical properties.

The results of the DSC analysis show an increase in crystallization indicating that organo-clay particles are acting as nucleating agents for PCPP crystallization. Additionally, a significant decrease of the melting temperature was observed.

The results obtained from stress-strain and bending tests indicate that mechanical properties of PCPP filaments are increased by the addition of organo-clay fillers.

References

Velásquez, E.J., et al.: Increasing the incorporation of recycled PET on polymeric blends through the reinforcement with commercial nanoclays. Appl. Clay Sci. **180**, 105185 (2019). https://doi.org/10.1016/j.clay.2019.105185

Zdiri, K., et al.: Reinforcement of recycled PP polymers by nanoparticles incorporation. Green Chem. Lett. Rev. **11**(3), 296–311 (2018). https://doi.org/10.1080/17518253.2018.1491645

Grigore, M.: Methods of recycling, properties and applications of recycled thermoplastic polymers. Recycling **2**(4), 24 (2017). https://doi.org/10.3390/recycling2040024

Adnan, M., et al.: In situ synthesis of hybrid inorganic–polymer nanocomposites. Polymers **10**(10), 1129 (2018). https://doi.org/10.3390/polym10101129

Zdiri, K., et al.: Rheological and thermal behavior of Tunisian clay reinforced recycled polypropylene composites. Adv. Poly. Technol. **37**, 3759–3768 (2018). https://doi.org/10.1002/adv.22159

Zdiri, K., et al.: Valorization of post-consumer PP by (Un)modified Tunisian Clay Nanoparticles Incorporation. Waste Biomass Valor **11**, 2285–2296 (2020). https://doi.org/10.1007/s12649-018-0427-2

Lei, S.G., et al.: Effect of clay types on the processing and properties of polypropylene nanocomposites. Composit. Sci. Technol. **66**(10), 1274–1279 (2006). https://doi.org/10.1016/j.compscitech.2005.09.012

Pozsgay, A., et al.: Nucleating effect of montmorillonite nanoparticles in polupropylene. J. Macromolec. Sci. Part B **41**(4–6), 1249–1265 (2002). https://doi.org/10.1081/mb-120013095

Yin, S., et al.: Comparative evaluation of virgin and recycled polypropylene fibre reinforced concrete. Construct. Build. Mater. **114**, 134–141 (2016). https://doi.org/10.1016/j.conbuildmat.2016.03.162

De Bilbao, E., et al.: Experimental study of bending behaviour of reinforcements. Exper. Mech. **50**(3), 333–351 (2009). https://doi.org/10.1007/s11340-009-9234-9

Optical Caracterisation of Commercial Photoluminescent Rare Earth Pigments Used in Textile Industry

Soumaya Sayeb[1(✉)], Faten Debbabi[2], and Karima Horchani-Naifer[1]

[1] Physical Chemistry Laboratory of Mineral Materials and Their Applications, National Center of Research in Materials Science, Soliman, Tunisia
sayebsoumaya@cnrsm.rnrt.tn

[2] Textile Engineering Laboratory, Monastir University, Monastir, Tunisia

Abstract. Nowadays, the development of textile finishing is able to create smart clothes witch have the ability to sense and react to external stimuli such us light, pressure, heat, ph.... Photo chromic pigments, as an example of smart textile materials, have an important role and potential in textile applications. In this paper two different commercial pigments mostly applied on textile fabric are investigated. A complete characterization (morphology and spectral characteristics) of these commercial luminescent pigments is realized to better understanding the applications of these pigments on textile fabrics. Different techniques are exploited: XRD and FTIR analysis, SEM coupled with EDX, and spectrofulometer measurements for emission and excitation data. Analysis reveals the presence of doped strontium aluminate in both pigment. The green pigment is doped with europium and dysprosium rare earth elements, for blue pigment, only europium was detected.

Keywords: Chromic materials · Textile · Luminescence · Pigments

1 Introduction

Photochromism is defined as light induced reversible changes in color which can caused by exposure to ultraviolet irradiation. New photochromic materials have been used for innovative textile in industry for the development of functional textiles. Recent researchers are rapidly increasing interest on the application of photochromic compounds in textiles (Robert 2013; Tawfik and Meram 2020). Thus, the use of photochromism in textiles can offer innovative opportunities to accomplish smart garments capable of blocking UV radiation, sensing environmental changes, security printing, sports or police clothing, fashion garments, fabric based electronic image displays, security barcodes, solar heat, sensor systems (Selestina et al. 2019; Tawfik et al. 2018a, b; Zhihua and Guibo 2010). Furthermore, photochromic effects can be applied for producing military clothing with camouflage effect (Hu 2008).

Photo luminescent pigments are synthetically processed crystalline compounds, which absorb energy, followed by emission of light with lower energy and longer wavelengths (Robert 2013). Luminescence in solids arises generally when an inorganic host

S. Msahli and F. Debbabi (Eds.): CIRATM 2020, SPM 17, pp. 36–42, 2022.
https://doi.org/10.1007/978-3-031-08842-1_8

material is doped with small amount of activator metal, which alters the electronic structure, resulting in a trap of charge carriers in metastable states upon excitation. Rare earth elements are among the most widely used activators within different hosts, generally, due to their high fluorescence efficiencies, when particles' size is reduces to the nanoscale (Zhihua and Guibo 2010).

The present paper focuses on the characterization of two types of commercial luminescent pigments applied on textile fabric.

A comprehensive characterization of used pigments was conducted using different techniques. Effectively, the XRD analysis, Fourrier Transform Infrared (FTIR) Spectroscopy data, Scanning Electron Microscope (SEM), spectrofluorometer for excitation and emission data were carried out in this study. The presented results provide various details (structure, excitation–emission spectra…) of luminescent pigments which will be helpful for textile manufactures for the development of new textile chromic materials.

2 Materials and Methods

Two photoluminescent pigments were analyzed with XRD and the diffractograms were performed on a Bruker X-ray *diffractometer*, type D8 with Cu K α radiation (λ = 0.15418 nm). FTIR measurements of these photoluminescent pigments were accomplished using FTIR System Spectrum NICOLET 560 Spectrometer using KBr pellets in the region of 4000–400 cm^{-1}. Scanning electron microscopy (SEM) coupled with EDX spectroscopy was employed to study the morphological properties and element composition of photoluminescent pigments.

Phosphorescence emission and excitation measurements were realized using spectrofluorometer (F980) equipped with phosphorescence accessory for phosphorescence life time measurements. The light source was a xenon arc lamp with slit band with 5nm for both of excitation and emission monochromators.

3 Results and Discussions

The obtained XRD diffractograms (Fig. 1) of the two luminescent pigments prove that the pigments have a good crystallinity. The observed peaks for the first and the second pigments were indexed according to the compounds $SrAl_2O_4$ (ICDD-PDF 010-0061) and $Sr_4Al_{14}O_{25}$ (ICDD-PDF 074-1810), respectively. As concluded, both pigments were identified as strontium aluminate. They are well-known metal aluminates, with phosphorescent properties, long afterglow and a broad spectrum distribution from blue to green region and emission maximum depending generally on the host (Selestina et al. 2019; Tawfik et al. 2018a, b; Tawfik and Meram 2020). Strontium aluminates can have different stoichiometries relying on molar ratio of SrO to Al_2O_3 during synthetic process. These aluminates are recognized as low cost, stable and ecofriendly hosts of rare earth luminescent materials (Duan 2003).

From vibrational properties of pigments assessed by FTIR spectroscopy in transmission mode (Fig. 2), bands in the range of 1000 to 500 cm^{-1} were observed and can be related to the stretching vibration of metal oxygen bonds (Al-O, Sr-O, and Sr-O-Al. Moreover, specific absorption bands at 840 cm^{-1} and 770 cm^{-1} are assigned to

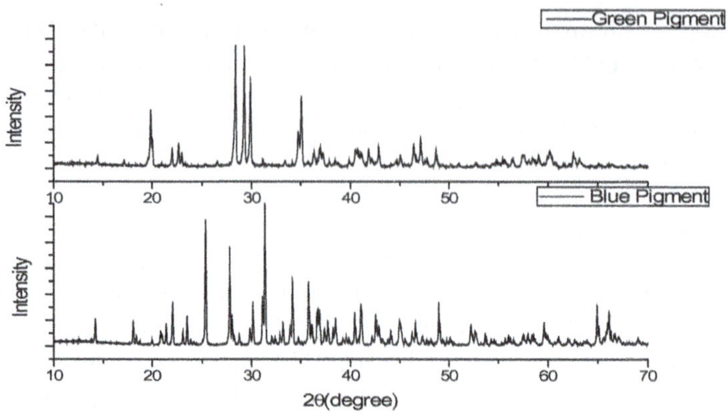

Fig. 1. X-ray diffraction (XRD) of photoluminescent pigments

the stretching vibration of $SrAl_2O_4$ (Yongliang et al. 2010). Bands at 690 cm^{-1} and 730 cm^{-1} are specific to $Sr4Al_{14}O_{25}$ (Sonika et al. 2020). Thus, the results with FTIR and XRD are corresponding.

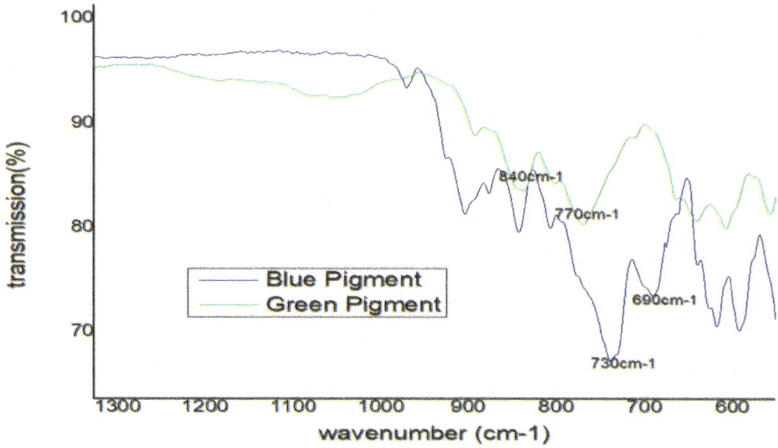

Fig. 2. Fourier Transform Infrared (FTIR) spectral lines of photoluminescent pigments

SEM coupled with EDX spectroscopy assessed the morphology of both pigment. Figure 3 shows remarquble heterogeneity for blue pigments and compacted particle structure: praticles are agglomerated and some fine powder is dispersed. For green pigments particles, they are less compacted and show a certain homogeneity, in deed shape particles are visually distinct. Particle size distributions measurements confirm that this herogenity is more pronounced within blue pigments figure with a size from 40–120 μm. However pigments have approximately a mean diameter of 57 μm for the blue pigment and 66 μm for the green one (Fig. 4).

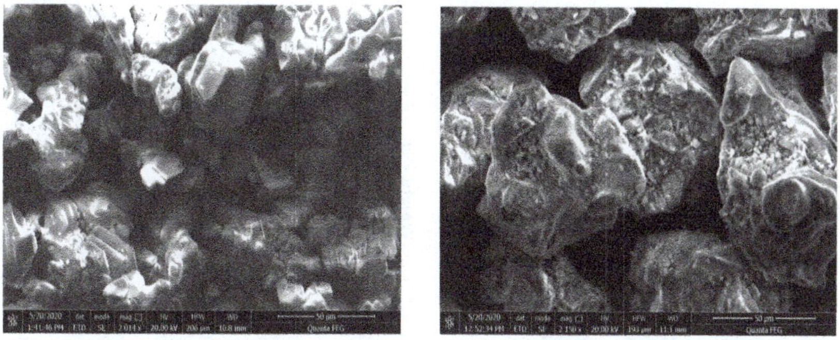

(a) (b)

Fig. 3. Scanning electron miscroscope (SEM) images of pigments: (a) blue pigments, (b) green pigments

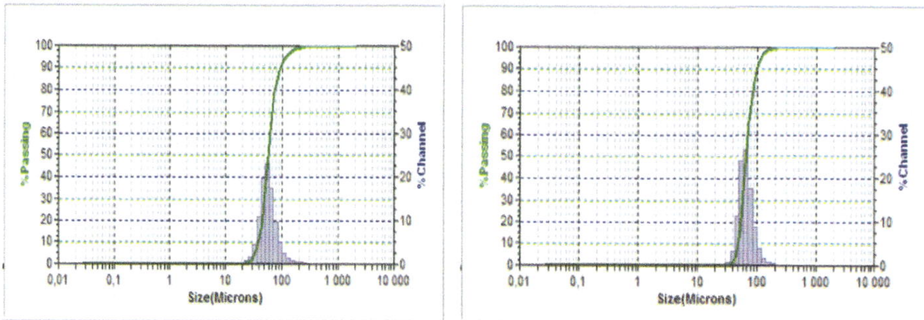

(a) (b)

Fig. 4. Size distributions of phosphorescent pigments: (a) blue pigment (b) green pigment

The EDX element distribution analyses were carried out with the intention to identify the presence of particular elements within pigments compounds, in order to identify elements assigned to the luminescence effect. Indeed, luminescent properties of pigments depend strongly on the chemical composition of the host material, as well as on the presence of particular dopant materials (Sonika et al. 2020). Accordingly, Fig. 5 reveals the presence of europium and dysprosium rare earth elements within the green pigment. For blue pigment, only europium was detected.

The excitation and emission spectra under UV excitation were inspected (Fig. 6). The main excitation peak at 250 nm was detected for both pigments. When, exciting the pigments at the corresponding maxima, the obtained emission spectra indicates respectively emission maxima at 522 nm for green pigment and 490 nm for blue one.

These finds corresponds with exciting data for $SrAl_2O_4$ doped with Eu^{2+} and Dy^{3+} rare earth ions, where 520 nm is the characteristic peak of the Eu^{2+} ion moving from the excitation state $4f^6 5d^1$ to the ground state $4f^7$ (Yanhong et al. 2012). However, blue pigment can be identified to $Sr4Al_{14}O_{25}$ doped with Eu^{2+}. The emission spectra

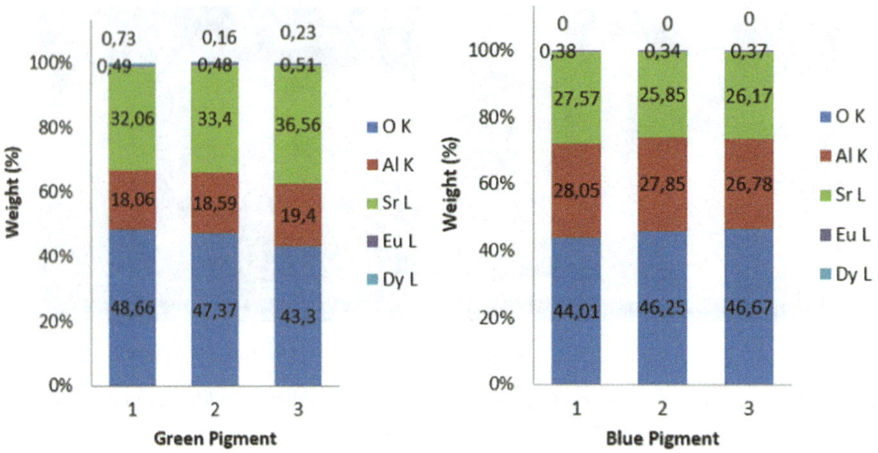

Fig. 5. EDX elements analysis extracted from 3 images of each pigment.

of $Sr_4Al_{14}O_{25}$:Eu^{2+}, shows the main emissive peak at 490 nm due to $4f^65d^1 \rightarrow 4f^7$ transition (Sonika et al. 2020).

The recorded chromatic characteristics of pigments were presented in Fig. 7. Their color coordinates are respectively $x_1 = 0.25708$ and $y_1 = 0.51180$ and $x_2 = 0.15526$ and $y_2 = 0.32433$.

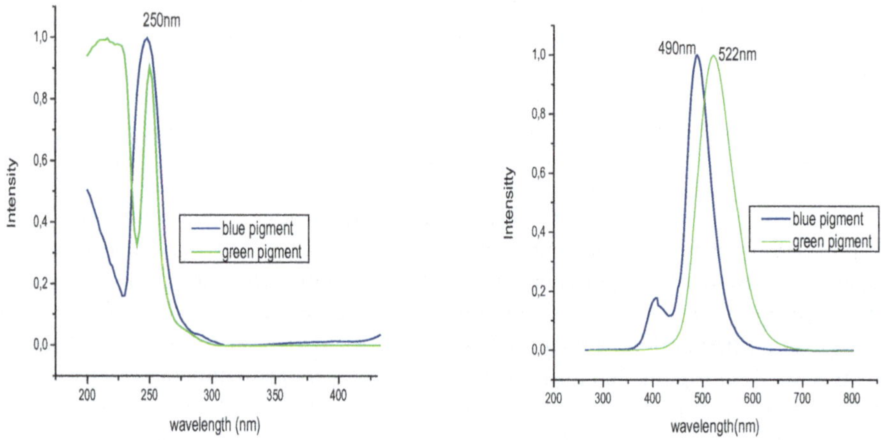

(a) (b)

Fig. 6. Excitation spectra (a) and emission spectra (b) at 250 nm of photoluminescent pigments

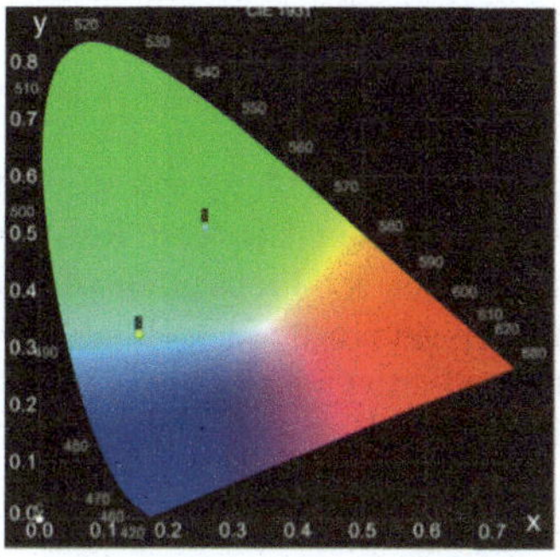

Fig. 7. CIE chromatic diagram showing the chromatic coordinates of pigments.

4 Conclusion

The photo luminescent characteristics of pigments were investigated in order to help textile manufactures to further exploitation of these pigments in technical application. The XRD analysis confirmed the presence of the homogenous single-phase monoclinic crystal structure for both pigments. The photo luminescence spectrum exhibits a hyper intense green emission peaking at 522 nm for the first pigment and also hyper intense blue emission peaking at 490 nm. Thus, the investigated pigments with high intense color emission ensure their potential applications in chromatic textile fabric. Further investigation including application techniques, pigments concentration can be conducted to obtain equal long afterglow intensity for fabrics as these photo luminescent pigments.

References

Aitasalo, T., Deren, P., Holsa, J., Jungner, H., Krupa, J.-C.: Persistent luminescence phenomena in materials doped with rare earth ions. J. Solid State Chem. **171**, 114–122 (2003)

Duan, X., Yi, L., Zhang, X., Huang, S.: Size-dependent optical properties of nanoscale and bulk long persistent phosphor SrAl2O4: Eu2+, Dy3+. J. Nanomater. **38**, 1–7 (2015)

Hu. J.: 3-D Fibrous Assemblies: Properties, Applications and Modeling of Three Dimensional Textile Structures, Elsevier (2008)

Christie, R.M.: Chromic Materials for Technical Applications, pp. 235–242. Woodhead Publishing Limites, UK (2013)

Gorgieva, S., Virant, N., Ojstrsek, A.: Complementary assessment of commercial photoluminescent pigments printed on cotton fabric. Polymers **11**(1216) (2019)

Kadyan, S., Singh, S., Sheoran, S., Samantilleke, A., Mari, B., Singh, D.: Synthesis, luminescent and structural characteristics of Sr4Al14O25:Eu2+ and Sr4Al14O25:Eu2+, RE3+ (RE = Nd, Dy) long persistent nanophosphors for solid state lighting. Optik- Int. J. Light Electron Opt. **204**, 164–159 (2020)

Khattab, T.A., Abdelrahman, M.S.: From smart materials to chromic textiles. Springer Nature Singapore Pte Ltd (2020)

Khattab, T.A., Rehan, M., Hamdy, Y., Shaheen, T.I.: Facile development of photoluminescent textile fabric via spray coating of Eu (II)-doped strontium aluminate. Indust. Eng. Chem. Res. **57**(34), 11483–11492 (2018)

Khattab, T.A., Rehan, M., Hamouda, T.:Smart textile framework: Photochromic and fluorescent cellulosic fabric printed by strontium aluminate pigment. Carbohyd. Polym. **195**(1143), 143–152 (2018)

Yan, Y., Ge, M., Li, Y., Kumar, D.N.T, Morphology and spectral characteristics of a luminous fiber containing a rare earth strontium aluminate. Text. Res. J. **82**(17), 1819–1826 (2012)

Yongliang Cheng, Y., Zhao, Y.Z., Cao, X.: Preparation of SrAl2O4:Eu2+, Dy3+ fibers by electrospinning combined with sol–gel process. J. Colloid Interface Sci. **344**, 321–326 (2010)

Chen, Z., Yin, G.: Suitability of Rare Earth Organic Light Conversion Agent of EI(III) Complex to improve Ultaviolet PRection Properties of Cotton Fabris. Text. Res. J. **80**(18), 1982–1989 (2010)

Thermogravimetric Analysis of a Double-Sided Knitted Fabric

Imene Ghezal[1,2(✉)], Ali Moussa[1,2], Imed Ben Marzoug[1,3], Ahmida El-Achari[4,5], Christine Campagne[4,5], and Faouzi Sakli[1,3]

[1] Textile Engineering Laboratory, University of Monastir, 5070 Ksar-Hellal, Tunisia
elghezalimene@hotmail.com
[2] National Engineering School of Monastir, University of Monastir, 5019 Monastir, Tunisia
[3] Higher Institute of Technological Studies of Ksar-Hellal, 5070 Ksar-Hellal, Tunisia
[4] Université Lille Nord de France, 59000 Lille, France
[5] ENSAIT, GEMTEX, 2 Allée Louise et Victor Champier, 59100 Roubaix, France

Abstract. The thermal degradation of a cotton/polyester double-sided knitted fabric was evaluated before and after undergoing a coating treatment. The applied coat which was a mixture of a fluorocarbon resin and an acrylic paste was applied to the polyester face of the fabric. The aim of the coating treatment was to impart water and windproofness to the used textile without restricting its breathability. Thermogravimetric analysis (TGA) was used in order to study the untreated fabric thermal degradation and to investigate the effect of the coating treatment on this fabric characteristic. Tests were done under ambient air and with a temperature ranging from 30 to 700 °C. For uncoated and coated samples degradation started in the range of 260 and 280 °C, respectively. The double-sided fabric showed a better resistance to thermal degradation after the coating treatment.

Keywords: Double-face fabric · Coating · Thermogravimetric analysis · Polyester/cotton fabric

1 Introduction

Over the forecasting horizon, the global market of waterproof breathable fabrics is projected to witness a compound annual growth rate (CAGR) of 5.7% (Grand View Research 2020; Mordor Intelligence 2020; Ghezal et al. 2019). There are many techniques for producing waterproof breathable fabrics such as coating fabrics, laminating textiles with membranes, producing closely woven structures, and developing smart textiles (Mukhopadhyay and Midha 2008).

Textile fabric can be made waterproof and breathable by coating their surfaces with chemical products (Ghezal et al. 2019, 2020). Coating is an increasingly growing technique. It is known for its importance as it imparts value to textile fabrics (Ghezal et al. 2019; Singha 2012).

Nowadays, studying textiles thermal stability and reducing fabric flammability have become a necessity (Mostashari and Mostashari 2009; Mostashari et al. 2009). Added

S. Msahli and F. Debbabi (Eds.): CIRATM 2020, SPM 17, pp. 43–48, 2022.
https://doi.org/10.1007/978-3-031-08842-1_9

to that, thermal decomposition is of great importance for some industries interested in recycling textiles. Thermal analysis (TA) is defined by the "International Confederation for Thermal Analysis and Colorimetry (ICTAC)" as a set of techniques that monitor changes in sample properties (a physical or a chemical property) as function of temperature and time in a controlled atmosphere and under a programmed temperature (Chartoff and Sircar 2004; Saadatkhah 2019). Thermogravimetric analysis (TGA) is widely used by researchers to study thermal properties of textile materials (Zhu et al. 2004; Xing et al. 2011).

In this research, a double sided knitted structure with a cotton inner-face and a polyester outer-face was coated with a fluorocarbon resin and an acrylic paste. The Thermal behavior of the double-sided fabric was evaluated before and after the coating treatment by using a thermogravimetric analyzer.

2 Materials and Methods

2.1 Coating Process

The tested fabric was a polyester/cotton double-sided knitting. The polyester face was treated by using the screen coating process. The applied coat was a mixture of two products; a fluorocarbon resin and an acrylic paste. Chemical formulas of both used products are represented by Fig. 1. Coated samples were then dried at 90 °C for 20 min and reticulated at 150 °C for 13.5 min.

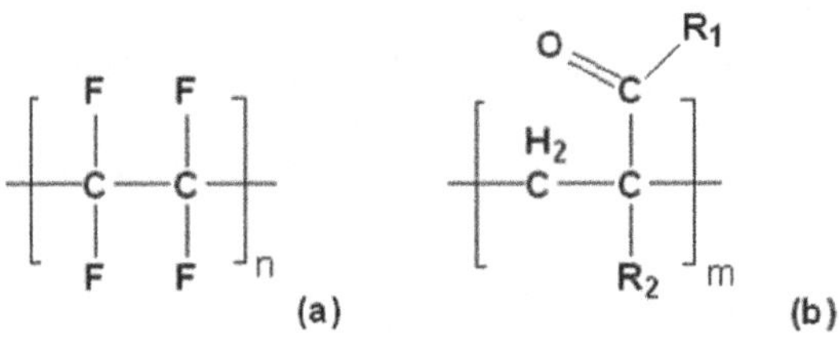

Fig. 1. Chemical formula: (a) fluorocarbon polymer, (b) acrylic polymer.

2.2 Thermogravimetry (TG)

The thermogravimetric analyzer consists of a furnace, a thermocouple, and a balance. The thermal decomposition of the sample can occur under an inert gas or ambient air. As well as temperature increases, the weight of the sample is recorded and weight loss is determined.

In this study, tests were done under ambient air and with a temperature ranging from 30 to 700 °C. Heating rate was of 10 °C/min. Weight of tested sample was equal to 10 mg. The weight loss of the sample as function of temperature was then plotted. Derivative thermogravimetry was also used to better visualize curves and picks obtained with the thermogravimetric analysis and to distinguish the different material degradation zones.

3 Results and Discussions

3.1 Thermogravimetry/Derivative Thermogravimetry Analysis

Thermogravimetry/derivative thermogravimetry (TG/DTG) experimental curves for the coated and uncoated double-face knitted fabric are represented by Fig. 2. For temperature values lower than 260 °C, a very small weight loss is shown by the thermogravimetry experimental curve of the uncoated sample. This is the result of the removal of physically absorbed water from the double-sided fabric. This small weight loss is accompanied with some changes in cotton fibers physical properties. The damage is mostly in the amorphous regions of the cotton fibers (Zhy et al. 2004). When temperature exceeds 260 °C, the thermogravimetry experimental curves can be divided into three zones. For a temperature between 260 and 370 °C (zone A), intermolecular dehydration of cotton takes place and cellulose is transformed to anhydrocellulose (Serbanescu 2010). For a temperature ranging from 370 to 460 °C (zone B), cellulose depolimirization takes place. Cellobiose decomposes into 1,4-α-D-glucopyranose which in turn is dehydrated to give 1,4 anhydro-α-D-glucopyranose (Serbanescu 2010). When temperature is higher than 460 °C (zone C), 1, 4 anhydro-α-D-glucopyranose is decomposed to give solid residues and volatile gases (Zhu et al. 2004; Serbanescu 2010).

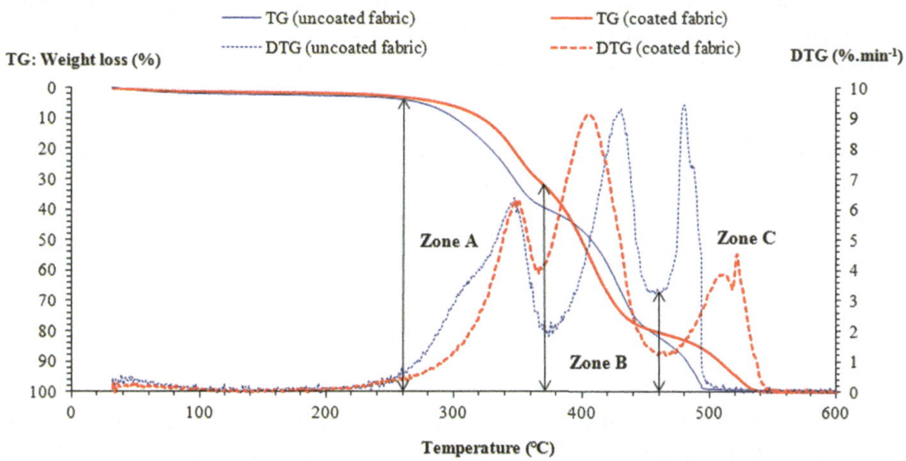

Fig. 2. TG (solid line)/DTG (dotted line) plot of polyester/cotton double-face knitted fabric before and after the coating treatment.

Based on previous researches, it was found that polyester multifilament weight loss starts when temperature reaches 350 °C (Zouai 2015; Carasalade 2009). The thermal degradation of polyester stars with polymeric chain scission after ester bounds breakage. The ethylene oxide and unstable radicals are formed (Zvanskii et al. 1997). From Fig. 2, DTG experimental curve for the uncoated fabric shows a pick in the weight loss when the temperature is equal to 430 °C. For higher temperatures, the ethylene oxide is transformed to acetaldehyde. As well as the temperature increases, a radical decomposition reaction takes place by a breakage on the C-C bounds producing benzene and releasing carbon dioxide and carbon monoxide gases. Ethylene gas can also be generated as the result of C-O bounds breakage (Zvanskii et al. 1997). The important weight loss is due to the volatile degradation. Polyester thermal degradation continues until the total polymer decomposition at 510 °C.

Based on thermogravimetric analysis curves, onset and offset temperatures for coated and uncoated samples were determined and recapitulated in Table 1. Onset and offset temperatures increased slightly after the coating treatment (Table 1). This is the result of the presence of fluorocarbon resin and acrylic paste on the polyester fabric outer-surface.

Table 1. Onset and Offset temperatures (°C) of coated and uncoated samples.

Samples	T_{onset} (°C)	T_{offset} (°C)
Uncoated knitting	260	510
Coated knitting	280	550

Weight losses (%) at onset and offset temperatures for coated and uncoated samples are shown in Table 2. For the untreated fabric, the thermal decomposition started at 260 °C with a weight loss equal to 3%. In the other hand, the decomposition temperature for the coated sample was noticed at 280 °C with almost the same weight loss obtained in the beginning of the uncoated sample thermal decomposition.

Table 2. Weight loss (%) of uncoated and coated samples at T_{onset} (260 °C) and T_{offset} (510 °C).

Samples	Weight loss at T_{onset} (260 °C) (%)	Weight loss at T_{offset} (510 °C) (%)
Uncoated knitting	3	98
Coated knitting	2	92

The increase in temperature leads to the polymer chain scission and gases release. The cotton is transformed to carbonaceous char with the formation of CO and CO_2 gases. The polyester is decomposed to carbonaceous char with the release of CO_2 gas (Muralidhara and Sreenivasan 2010; Devaux et al. 2002; Bautista et al. 2017). This explains the significant weight loss for high temperatures.

After the coating treatment, the obtained weight loss at 510 °C was equal to 92% which is lower than the value obtained with the uncoated sample (98%). The treated fabric showed a better resistance to thermal degradation. This difference in thermal degradation between uncoated and coated samples can be the result of the generation of not easily oxydizable and inert gases from the thermal degradation of the polymeric coat. These gases (H_2O, SO_2, and CO2) affect the sample's surrounded atmosphere by oxygen depletion or by preventing the oxygen access from the air (Mostashari et al. 2009).

4 Conclusion

Thermal behavior of a cotton/ polyester double-sided knitted fabric was studied before and after undergoing a coating treatment with a mixture of a fluorocarbon resin and an acrylic paste. Experiments were done by using a thermogravimetric analyzer under ambient air, a temperature ranging from 30 to 700 °C, and a heating rate of 10 °C/min. For low temperatures, it was found the weight loss shown by TG and DTG curves is not important. This was explained by the dehydration of cotton and polyester. The weight loss increased significantly when temperature exceeded 260 and 280 °C, respectively for uncoated and coated samples. For cotton and polyester, the thermal degradation process starts with polymeric chains scission. With the increase in temperature important weight loss was noticed for polyester and cotton. Cellulose and polyethylene terephthalate degradation gives char and volatile gases.

Before the coating treatment, the weight loss of the sample at 510 °C was equal to 98%. For the coated double-sided knitted structure, this value decreased to 92%. So, it can be deduced that the coating treatment increased slightly the resistance to thermal degradation of the cotton/ polyester double-sided knitted fabric.

References

Bautista, Y., et al.: Thermal degradation mechanism of thermostable polyester stabilized with an open-cage oligomeric silsesquioxane. Materials **11**(22) 1–13 (2017). ISSN 1996-1944

Carasalade, E.: Transitions et relaxations dans les assemblages polymères à base polyester à finalité ballons stratosphériques. University of Toulouse, France, Thesis (2009)

Devaux, E., et al.: Polyurethane/ Clay and polyurethane/ POSS nanocomposites as flame retarded coating for polyester and cotton fabrics. Fire Mater. **26**, 149–154 (2002), ISSN 1099-1018

Ghezal, I., et al.: Evaluating the mechanical properties of waterproof breathable fabric produced by a coating process. Cloth. Text. Res. J. **37**(4), 235–248 (2019). ISSN 1940-2473

Ghezal, I., et al.: Development and surface state characterization of a spacer waterproof breathable fabric. Fibers Polym. **21**(4), 910–920 (2020). ISSN 1875-0052

Grand View Research: Waterproof breathable textiles market size, share & trends analysis report by raw material (ePTFE, polyurethane, polyester), by fabric, by application, by region, and segment forecasts, 2020–2027, p. 120 (2020). Report ID 978-1-68038-316-4

Mordor Intelligence: Waterproof breathable textiles market-growth, trends, and forcast (2020–2025), p. 90 (2020). Report ID 4773608

Mostashari, S.M., et al.: Thermogravimetric analysis of cellulosic fabric treated with nickel sulfate hexahydrate as a flame-retardant. Cellulose Chem. Technol. **43**(1–3), 95–98 (2009). ISSN 2457-9459.

Mostashari, S.M., Mostashari, S.Z.: Thermogravimetric analysis of a cotton fabric incorporated by "Graham's salt" applied as a flame retardant. J. Thermal Anal. Calorimetry **95**(1), 187–192 (2009). ISSN 1588-2926

Mukhopadhyay, A., Midha, V.K.: A review on designing the waterproof breathable fabrics Part I: fundamental principles and designing aspects of breathable fabrics. J. Indust. Text. **37**(3), 225–262 (2008). ISSN 1530-8057

Muralidhara, K.S., Sreenivasan, S.: Thermal degradation kinetic data of polyester, cotton and polyester-cotton blended textile material. World Appl. Sci. J. **11**(2), 184–189 (2010). ISSN 1818-4952

Chartoff, R.P., Sircar, A.K.: Thermal analysis of polymers. Encyclopedia of Polymer Science and Technology. Wiley (2004)
Saadatkhah, N., et al.: Experimental methods in chemical engineering: Thermogravimetric analysis- TGA. Can. J. Chem. Eng. **98**(1), 34–43 (2019). ISSN 1939-019X
Serbanescu, C.: Étude et modélisation de la dégradation pyrolytique des mélanges complexes de composés organiques, Thesis, University of Toulouse, France (2010)
Singha, K.: A review on coating & lamination in textiles: processes and applications. Am. J. Polym. Sci. **2**(3), 39–49 (2012). ISSN 2163-1352
Xing, W., et al.: Flame retardancy and thermal degradation of cotton textiles based on UV-curable flame retardant coating. Thermochimica Acta **513**(1–2), 75–82 (2011). ISSN 0040-6031
Zhu, P., et al.: A study of pyrolysis and pyrolysis products of flame-retardant cotton fabrics by DSC, TGA, and PY-GC-MS. J. Anal. Appl. Pyrolysis **71**(2), 645–655 (2004). ISSN 0165-2370
Zouai, F.: Préparation et fabrication de nouveaux types de fibres et filaments polymères/Argile non traitée par la méthode RXR réversible crosslinking réaction et la technique extrusion réactive, Thesis, Ferhat ABBAS SETIF-1 University, Algérie (2015)
Zvanski, B.V., et al.: Mechanism of thermal degradation of polyester fibre in furnace pyrolyser. Fibre Chem. **29**(6), 363–366 (1997). ISSN 1573-8493

Usability of Electrostatic Charge Generated on Textiles

Juro Živičnjak(✉), Dubravko Rogale, and Antoneta Tomljenović

Faculty of Textile Technology, University of Zagreb, Prilaz baruna Filipovića 28a, 10000 Zagreb, Croatia

juro.zivicnjak@ttf.unizg.hr

Abstract. The generation of electrostatic charge in textile materials is commonly caused by contact charging described in the triboelectric series. The series only shows trends, but not the amount of charge generated. Therefore, with the purpose of determining the amount and the usability of generated electrostatic charge by rubbing two textile fabrics, in this paper, an adequate measuring instrument and method were developed. Triboelectric generator for storing static charge from clothing was used for testing five types of textile fabrics made of cotton, wool, silk, polyester and polyamide fibers. It was developed at the Department of Clothing Technology at the University of Zagreb Faculty of Textile Technology. Experimental results of generated electrostatic charge show a correlation with fabrics physical, sorption and electrostatic properties and that its amount could be utilized in further use.

Keywords: Textiles · Rubbing · Electrostatic charge · Triboelectric generator

1 Introduction

Textile materials as garments or technical textiles are prone to generation of electrostatic charge (Welker et al. 2006). There aren‘t many standard methods for determining properties of textile materials that lead up to their generation, due to a large number of factors they are affected by. The European standard related to testing of textile material electrostatic properties is the EN 1149, which provides the test methods for measurement of surface resistivity (part 1), vertical resistance (part 2) and charge decay (part 3). Standards are mainly used for personal protective equipment where any generation of electrostatic charge is reduced to a minimum, by the addition of antistatic materials that accelerate the process of charge dissipation. The majority of garment and textile products are not equipped with such addition and there are more prone to the generation of electrostatic charge. Triboelectric series of textile materials describes generated charge polarity, but does not provide information about generated charge amount (Nurmi et al. 2007). Even in small amounts, an electrostatic charge could prove as an alternate source of renewable energy for wearable electronics that are ever so present in day to day garment (Ünsal et al. 2020).

S. Msahli and F. Debbabi (Eds.): CIRATM 2020, SPM 17, pp. 49–54, 2022.
https://doi.org/10.1007/978-3-031-08842-1_10

The earlier proposed methods for determining the amount of electrostatic charge generated in textile materials were by rubbing: fiber against fiber, fiber against cylinder, yarn against cylinder, film against cylinder, fabric against cylinder/roller and plastic disk against disk (Liu et al. 2013). Results proved scattered and unreliable (Williams 2012; Mellouki et al. 2018) because the number of reasons, such as atmospheric conditions, material's high electrical conductivity and electrostatic discharge. The measurement method must contain clearly stated conditions for implementation, and a sufficient degree of automation to reduce human errors to a minimum. Furthermore, the instrument resolution should be in micro or nanoscale that would reduce the influence of textiles materials uneven surface (Morita et al. 2005; McCarty and Whitesides 2008).

Therefore, in this paper, an adequate method and newly developed measuring instrument were used to correlate physical, sorption and electrostatic properties of textile fabrics with generated amount of electrostatic charge and determine their usability.

2 Materials and Methods

Five types of textile fabrics made of cotton, wool, silk, polyester and polyamide were analyzed. Their physical, sorption and electrostatic properties were determined, with the purpose of better understanding how they affect generation of electrostatic charge, as follows:

- fabrics mass per unit area according to the ISO 3801. Fabrics are conditioned in a standard atmosphere (temperature of 20 ± 2 °C and relative humidity of 65 ± 4%), after which five samples of 100 cm^2 are cut and weighted with the precision of 10^{-3}g. The average value from five separate samples is calculated and given as mass per unit area in g/m^2.
- fabrics thickness according to the EN ISO 5084. Fabrics are conditioned in a standard atmosphere (temperature of 20 ± 2 °C and relative humidity of 65 ± 4%) and their thickness calculated as an average of ten measurements taken at ten different places on the fabric under the pressure of 1 kPa, using the presser foot of 2000 mm^2 and thickness gauge with an accuracy of 10^{-2}mm.
- fabric moisture regain according to the ASTM D 2654-89a. Fabric samples are conditioned in the standard atmosphere of air temperature of 20 ± 2 °C and relative humidity of 65 ± 4% for 24 h, after which they are weighted, dried in an oven at 105 °C and reweighed. The difference between the mass of conditioned and the mass of oven-dried samples is calculated as moisture regain and expressed in percentage.
- fabric surface resistivity according to the EN 1149-1. The method is intended for determining surface resistivity, by applying a potential of 100 ± 5 V on a fabric sample placed between the upper metal annular electrode and bottom insulating plate. Before testing samples are conditioned in the following atmosphere: air temperature of 23 ± 1 °C and relative humidity of 25 ± 5%. Ohmmeter used has an accuracy of ± 5% for resistances under 10^{12} Ω and ± 20% for resistances above 10^{12} Ω. The surface resistivity was calculated for each of the five values of resistance measured using the equation:

$$\rho = k \times R \quad (1)$$

- where: ρ is the calculated surface resistivity, in ohm (Ω); R is the measured resistance, in ohm (Ω); and k is the geometrical factor of the electrode (19.8). The geometrical mean value of surface resistivity was calculated and presented as a result for all fabric tested.
- fabric vertical resistance according to the EN 1149-2. Fabric samples are placed between two metal annular electrodes and subjected to a potential of 100 ± 5 V. Before testing samples are conditioned at the temperature of 23 ± 1 °C and relative humidity of 25 ± 5%. Instrument accuracy was ±5% for resistances under 10^{12} Ω and ±20% for resistances above 10^{12} Ω. Vertical resistance of the tested fabric, in ohm (Ω), is given as arithmetical mean of five measured values on five samples.

The generation and measurement of electrostatic charge, in this experiment, was conducted with the two-part *triboelectric generator for storing static charge from clothing* (Živičnjak and Rogale 2019) developed at the Department of Clothing Technology at the University of Zagreb Faculty of Textile Technology (Figs. 1a and 1b).

a) b)

Fig. 1. Triboelectric generator: a) mechanical rubbing device, b) measuring device for generated electrostatic charge (Q-meter)

The resolution of the *triboelectric generator for storing static charge from clothing* (Fig. 1b) can detect and store induced electrostatic charges in a range of 1 to 200 nC, separated in scopes of 1, 20 and 200 nC. The mechanical part of the triboelectric generator provides controlled manner for inducing electrostatic charge, by linear rubbing of two textile fabrics, at a rate of one cycle per second (Fig. 1a). With the mechanical part, can also be adjusted contact pressure between two textile fabrics by adding weight on top of the upper rubbing plate, as earlier researches have emphasize that contact pressure greatly affects the generation of the electrostatic charge (Castle 1997; Ohara et al. 2001). Conducted experiment method consisted of 10 rubbing cycles, under the contact pressure of 1960.8 Pa, with two textile fabrics placed face to face, on a contact area of 0.0051 m^2. Three samples of tested fabric were prepared and rubbed with all other types of fabrics selected for the experiment, including one of the same kind. Results are shown as a means of three measurements in nanocoulomb (nC) conducted at the room temperature of 23 °C and relative humidity of 50%.

3 Results and Discussion

The results obtained by the investigation are presented in Tables 1, 2 and 3. Values of physical (mass per unit area and thickness), sorption (moisture regain) and electrostatic properties (surface resistivity and vertical resistance) determined for all tested fabrics samples are shown in Table 1. Electrostatic charge polarity is the only unmeasured property in Tables 1 and 2, quoted from the triboelectric series (Nurmi et al. 2007) set from the positive (+) to negative (−) end.

Table 1. Properties of textile fabrics used

Triboelectric series	Mass per unit area (g/m^2)	Thickness (mm)	Moisture regain (%)	Surface resistivity (Ω)	Vertical resistance (Ω)
↑ (+) Wool	215.3	0.46	12.44	3.68 × 1012	884.2×10^7
Polyamide	180.7	0.55	3.17	3.36×10^{12}	529.0×10^7
Silk	99.3	0.21	10.41	19.38×10^{12}	702.0×10^7
Cotton	118.3	0.24	7.17	2.45×10^{10}	423.4×10^5
↓ (−) Polyester	147.8	0.29	0.47	11.49×10^{12}	308.4×10^9

The results presented in Table 2 shows the distribution of tested fabrics properties values - from the highest (+) to the lowest (−).

Table 2. Distribution of the textile fabrics by their property values

Triboelectric series	Mass per unit area	Thickness	Moisture regain	Surface resistivity	Vertical resistance
↑(+)Wool	↑(+)Wool	↑(+)Polyamide	↑(+)Wool	↑(+)Silk	↑(+)Polyester
Polyamide	Polyamide	Wool	Silk	Polyester	Wool
Silk	Polyester	Polyester	Cotton	Wool	Silk
Cotton	Cotton	Cotton	Polyamide	Polyamide	Polyamide
↓(−)Polyester	↓(−)Silk	↓(−)Silk	↓(−)Polyester	↓(−)Cotton	↓(−)Cotton

From the fabrics with lower thickness, values of mass per unit area and moisture regain, with higher surface resistivity and vertical resistance, the higher amount of electrostatic charge generation was expected. When we consider the type of fibers and their place in the triboelectric series, it can be seen from Table 2 that fabrics made from natural fibers (wool, silk and cotton) have the highest moisture regain values, but fabrics made of natural protein fibers (wool and silk) and polyester as well have high values of surface resistivity and vertical resistance. For cotton (natural cellulose fiber) a higher value of electrical conductivity was determined so it‘s less prone to generation of electrostatic charge.

When the materials are moved apart during the rubbing, especially if one of them is a weaker conductor, a significant amount of charge may be left on the two materials.

Table 3. Values of electrostatic charge generated after rubbing of each pair of fabrics (nC)

Fabric type	Wool	Polyamide	Silk	Cotton	Polyester
Wool	10	80	40	30	80
Polyamide		35	10	15	50
Silk			0	15	10
Cotton				0	0
Polyester					0

Results of generated electrostatic charge represented in Table 3, show that wool as one of the thicker fabrics, generates the most electrostatic charge after the ten rubbing cycles, especially when rubbed with fabrics made of manmade synthetic polyamide and polyester fibers. It has been found that rubbing of polyamide fabric with polyester fabric, of opposite triboelectric polarity, caused the lower amount of charge generated than when rubbed against wool (Table 3). Polyamide fabric rubbed against the same one caused the higher amount of generated electrostatic charges than other textile materials when rubbed with the material of the same type. When rubbed with cotton fabric, polyamide, wool and silk generates more charge than polyester fabric. The above indicates that: higher conductivity (lowest determined resistance) of cotton fabric cause faster charge dissipation; that smaller thickness and mass per unit area of the tested fabrics cause faster charge dissipation; that the mutual distance of the selected pair of fabrics in the Triboelectric series significantly affects the values of generated charge, and that the fiber composition, surface properties of yarns and fabrics have also a significant impact on the obtained results.

4 Conclusion

The conducted method proved to be adequate for testing of textile fabrics charge generation when rubbed with other textile fabrics. A measurement device could detect the amount of electrostatic charge generated. Amount of overall generated electrostatic charge, tested on small a surface area of 0.0051 m^2, that takes up only 0.3% of garment surface shows enough potential energy that could be used as an alternative, renewable source of electric energy for wearable electronics.

Acknowledgment and Funding. This paper has been supported by the Croatian Science Foundation through the project IP-2018-01-6363 Development and thermal properties of intelligent clothing – ThermIC.

References

ASTM D2654-89a: 1989 Test Methods for Moisture in Textiles

Castle, G.S.P.: Contact charging between insulators. J. Electrost. **40–41**, 13–20 (1997). ISSN: 0304-3886

EN 1149-2:2001: Protective clothing - electrostatic properties - Part 2: test method for measurement of the electrical resistance through a material (vertical resistance)

EN 1149-1:2007: Protective clothing - electrostatic properties - Part 1: test method for measurement of surface resistivity

EN ISO 5084:1996: Textiles - determination of thickness of textiles and textile products

ISO 3801:1977: Textiles - Woven fabrics - determination of mass per unit length and mass per unit area

Liu, L. et al.: Frictional electrification on polymeric flat surfaces. J. Eng. Fibers Fabrics **8**(1), 126–136 (2013). ISSN: 1558-9250

McCarty, L.S., Whitesides, G.M.: Electrostatic charging due to separation of ions at interfaces: contact electrification of ionic electrets. Angewandte Chemie **47**(12), 2188–2207 (2008). ISSN 1618-3258

Mellouki, H., et al.: Experimental modeling of the conformal-contact tribocharging of polymers, IEEE Trans. Dielect. Electric. Insul. **25**(1), 145–153(2018). ISSN 1070-9878

Morita, S., et al.: Microscale contact charging on a silicon oxide. In: Vilarinho, P.M., Rosenwaks, Y., Kingon, A. (eds.) Scanning Probe Microscopy: Characterization, Nanofabrication and Device Application of Functional Materials. NAII, vol. 186, pp. 289–308. Springer, Dordrecht (2005). https://doi.org/10.1007/1-4020-3019-3_13, ISBN 978-1-4020-3019-2, Algarve, Portugal, Dordrecht

Nurmi, S., et al.: Protection against electrostatic and electromagnetic phenomena. In: Duquesne, S., Magniez, C., Camino, G. (eds.) Multifunctional Barriers for Flexible Structure, Textile, Leather and Paper. SSMATERIALS, vol. 97, pp. 63–86. Springer, Heidelberg (2007). https://doi.org/10.1007/978-3-540-71920-5_4, ISBN 978-3-642-09101-8

Ohara, K., et al.: Frictional electrification between flat surfaces of polymers and of Langmuir-Blodgett layers, J. Electrost. **51–52**, 351–358 (2001). ISSN 0304-3886

Ünsal, Ö.F., et al.: Energy-generating textiles. In: Shahid-ul-Islam, Butola B.S. (ed.) Advances in Functional and Protective Textiles, 1st edn, pp. 415–456. Woodhead Publishing (2020). eBook ISBN: 9780128226766

Welker, R.W. et al.: Contamination and ESD Control in High-Technology Manufacturing. Wiley-IEEE Press (2006). ISBN: 978-0-471-41452-0

Williams, M.W.: Triboelectric charging of insulating polymers-some new perspectives. AIP Adv. **2**(1), 1–10 (2012). ISSN 2158-3226

Živičnjak, J., Rogale, D.: Triboelectric generator for storing static charge from clothing, Malaysia Technology Expo (MTE) International Exhibition of Innovations Kuala Lumpur, 21–23 February 2019, Gold medal, Silver medal and Special reward (2019)

Study on Triboelectricity Parameters

Sotiria F. Galata[1(✉)], Savvas Vassiliadis[1], Adrien Poujol[1,2], Aristeidis Repoulias[1], Dimitroula Matsouka[1], and George Priniotakis[3]

[1] Department of Electrical and Electronic Engineering, University of West Attica, Athens, Greece
sgalata@uniwa.gr

[2] Ecole Nationale Supérieure des Arts et Industries Textiles, Roubaix, France

[3] Department of Industrial Design and Production Engineering, University of West Attica, Athens, Greece

Abstract. Triboelectricity is a promising phenomenon towards the energy harvesting especially for the wearable systems and e-textiles. In this work we investigate this phenomenon on Polypropylene (PP) and Polyethylene (PE) rubbed on Teflon. Samples of different characteristics have been used and the related experiments were realized in order to obtain the electric measurements. A model was created to investigate the importance of various factors affecting their triboelectric behaviour.

Keywords: Triboelectricity · Triboelectric generator · Polypropylene (PP) · Polyethylene (PE) · Teflon

1 Introduction

Triboelectricity is defined as the phenomenon where electrical charges are generated between two materials when they are brought in contact, rubbed or separated (Lin et al. 2016). The intensity of the phenomenon is related to the electron affinities on their surfaces (Lin et al. 2016) and external factors (Molnar et al. 2018). The charges of the two materials are opposite in sign but equal in magnitude (Molnar et al. 2018). The two materials can be different or even seemingly chemically identical (Williams, 2012). Many efforts have been made over the last decades to harvest electrical energy based on the triboelectricity effect by building novel structures known as TEGs and TENGs. A Tribo Electric Generator (TEGs) is defined as the energy-harvesting mechanism which converts the external mechanical energy into electricity, based on the principles of triboelectricity and electrostatic induction, by using various set-up modes (Lin et al. 2016, Chacko et al. 2017, Somkuwar et al. 2020).

Energy harvesting using triboelectricity gained importance over the last years, because wearable electronics, sensors and e-textiles need power to operate (Fan et al. 2012). In addition, the basic requirements for the use of the textile fabrics (e.g. elasticity, flexibility, conductivity, breathability, washability), can be compatible with TEGs. Due to the fact that in textiles a large contact area is available too, their use for energy harvesting becomes challenging (Logothetis et al. 2017).

S. Msahli and F. Debbabi (Eds.): CIRATM 2020, SPM 17, pp. 55–61, 2022.
https://doi.org/10.1007/978-3-031-08842-1_11

2 Materials and Methods

The experimental setup is shown in Fig. 1. A carriage is made of wood, on top of it is placed a thin film of aluminium foil and both of them are surrounded either by Polypropylene (PP) or Polyethylene (PE) sheets. This structure is rubbed on a long (100 cm × 10 cm) Teflon plate. The sample is pulled by a light weight via a cable connected to a pulley. At the bottom of the Teflon plate, a same sized thin film of aluminium foil is placed. Electrical measurements were performed by an Agilent Technologies DSO3102A oscilloscope connected to the carriage and the aluminium foil (Fig. 1). Two wooden carriages of 4 cm × 5 cm and 8 cm × 5 cm and two Teflon thicknesses of 1 mm and 2 mm are used. The sliding distance for the rubbing is 30 cm or 60 cm. The pulling weight which is used is either 32 g or 64 g.

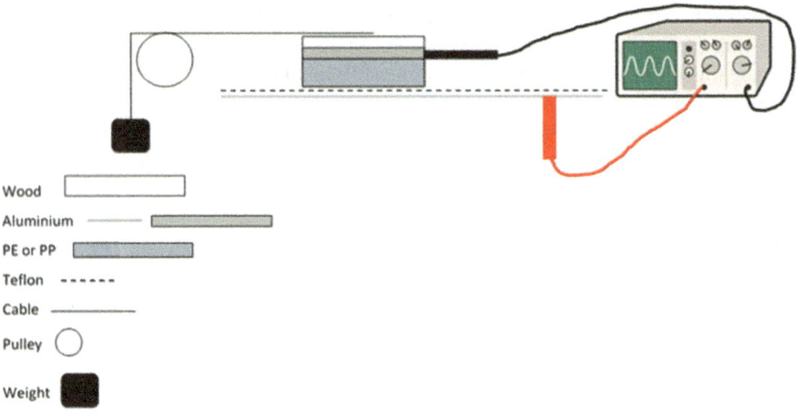

Fig. 1. Experimental setup

Table 1. List of parameters which were used on our experimental setup

	A	B	C	D	E	F	G
Strength of factor	**Material sourrounding the sample**	**Wooden carriage Size**	**Teflon Tickness**	**Distance of rubbing**	**Weight**	**Mode**	**Distance before other aluminium**
Unit	-	-	mm	cm	g	N	cm
-1	PE	Small	1	30	32	posed	0
1	PP	Big	2	60	64	dropped	10

All parameters which were used on our experiment setup are listed in Table 1 along with the notation for the "strength" for each factor. Each parameter was assigned to an arbitrary value either 1 or −1 in order to represent its importance according to factorial experiments/design method (Montgomery, 2017). Therefore, "1" corresponds to the highest value of the voltage output, whereas "−1" corresponds to the lowest value.

According to these 7 parameters, 2^7 different possible experiments can be created. However, we have chosen the following 16 experiments which provide an adequate spread through the values, as can be observed in Fig. 2. The measured voltage values for each experiment are also listed in Fig. 2.

	A	B	C	D	E	F	G	
	Material sourrounding the sample	**Wooden Carriage Size**	**Teflon Thickness**	**Distance of rubbing**	**Weight**	**Mode**	**Distance before other aluminum**	**Measured voltage values**
	A	B	C	D	E	F	G	V
1	1	1	1	1	1	1	1	25
2	1	1	-1	-1	-1	-1	1	4.8
3	-1	1	-1	-1	1	1	-1	6.4
4	-1	1	-1	1	1	-1	1	7.2
5	-1	-1	1	1	1	1	-1	21.2
6	-1	-1	-1	1	-1	1	1	6.8
7	-1	-1	-1	-1	-1	-1	-1	1.7
8	1	-1	-1	-1	1	1	1	8
9	-1	-1	1	-1	1	-1	1	10
10	-1	1	1	1	-1	-1	-1	5.2
11	1	1	-1	1	-1	1	-1	6.4
12	1	-1	1	1	-1	-1	1	5.4
13	1	-1	1	-1	-1	1	-1	4.2
14	-1	1	1	-1	-1	1	1	3
15	1	-1	-1	1	1	-1	-1	6
16	1	1	1	-1	1	-1	-1	15.6

Fig. 2. List of our sixteen experiments having a random spread of the arbitrary values 1 and −1 and the measured voltage values for each experiment

Therefore, as can be seen in Fig. 2, for example line 3 corresponds to an experiment where PE material was used having a big wooden carriage with a 1 mm Teflon thickness, 30 cm distance of rubbing, pulled by a 64 g weight, caused it to be dropped, at 0 cm distance before the second aluminium.

3 Results and Discussion

A typical electric measurement is illustrated in Fig. 3. A model was created basd on factorial experiments method (Montgomery 2017) and Plackett - Burman designs (Ledolter and Swersey 2007) in order to approach a triboelectric prediction procedure.

The aim of our research is to determine the precision of the model. Therefore, we calculated the value of each factor using the following equation:

$$(\text{Sum(arbitrary value of each factor} \times \text{measured Voltage}))/16 = \text{factor}(m, a, b, c, d, e, f, g) \quad (1)$$

Therefore, the theoretical values of the voltage could be calculated using the first-degree polynomial (Ledolter and Swersey 2007):

$$V_{th} = V * m + a * A + b * B + c * C + d * D + e * E + f * F + g * G \quad (2)$$

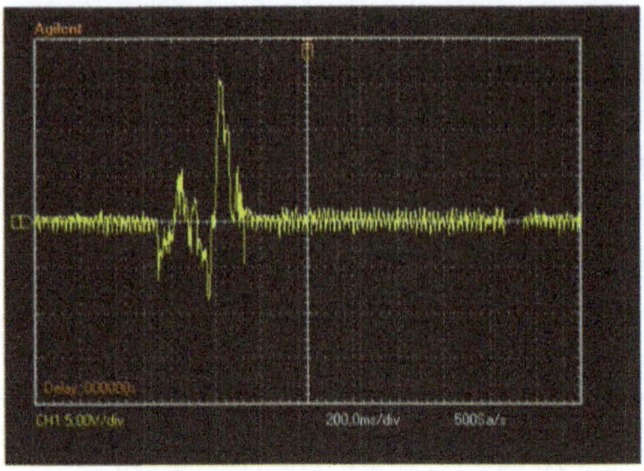

Fig. 3. Typical electrical pulse

where m, a, b.., g are the values of the factors, as have been calculated from Eq. (1) and V, A, B.., G are the values of experimental Voltage (V) and the arbitrary values for each experiment (A, B,.. G) respectively, as can be observed in Fig. 2. The residue is calculated as the difference between the experimental value of voltage minus the theoretical one for each experiment.

$$\text{Residue} = \text{V-V}_{\text{th}} \tag{3}$$

In addition, the sum of squares (SS) was calculated for the experimental voltage values (SSV), the theoretical voltage ones (SSV_{th}) and the residue values (SSR). The precision of the model R^2 (Devore 2011) and R^2 adjusted (Devore 2011, Yina and Fan 2001) was then calculated using the following equations:

$$\mathbf{R^2} = \frac{\mathbf{SSV_{th}}}{\mathbf{SSV}} * 100 \tag{4}$$

$$\mathbf{R^2 adjusted} = \left(1 - \frac{\frac{\mathbf{SSR}}{\mathbf{n-p-1}}}{\frac{\mathbf{SST}}{\mathbf{n-1}}}\right) * 100 \tag{5}$$

where n = 16 is the number of experiments and p = 7 is the number of parameters. For our model, R^2 was calculated as 73.13% and R^2 adjusted = 49.62%. Considering the logarithm of the experimental voltage values, the residue of the model became lower and the precision of our logarithmic model became better. So R^2 increased to 80.91% and R^2 adjusted to 64.20%, respectively. Figure 4 shows the effect of each factor using theoretical formula (1). In most of the cases the theoretical prediction matches the experimental intuition.

As it can be observed in Fig. 4, the effect of parameter A was not so important during the experiment, since PP and PE have similar values of electronegativity. The size of the sample (parameter B) didn't result in any extra electricity although it was expected that the bigger sample size would result in higher voltage values. In addition, the distance between the two aluminiums, (10 cm), (parameter G) is not adequate to create any extra triboelectricity. If the weight was dropped from 1 cm over the pulley, this extra force created more speed which caused extra triboelectricity (parameter F). The weight of the

	A	B	C	D	E	F	G
Name of factor	Material sourrrounding the sample	Wooden Carriage Size	Teflon Thickness	Distance of rubbing	Weight	Mode	Distance before other aluminum
Unit	-	-	mm	cm	g	N	cm

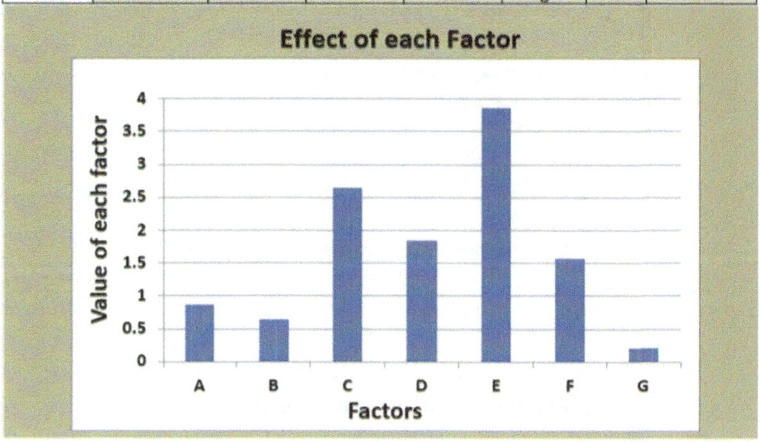

Fig. 4. Effect on the triboelectricity of each parameter

object used to pull the samples (parameter E) was quite important, since the heavier the object the bigger the force which was created between the samples and therefore a higher triboelectricity signal was produced. Also, Teflon thickness (parameter C) was also important, since the higher Teflon thickness resulted in higher output voltage values. Finally, the higher distance of rubbing (parameter D), as expected, produced extra voltage values. Therefore, all factors were either important or less important according to the prediction of our model. In order to examine the validity of our experimental values, it was important to compare them with the theoretical values, as have been predicted from our model. Figure 5 illustrates the experimental voltage values as have been extracted from our experiments over the theoretical ones which were predicted from our proposed model.

It can be concluded from Fig. 5 that there is a linear dependence between experimental values and theoretical ones. Therefore, our experimental values are good in terms of being quite close to the theoretical ones as predicted from our model. Using the logarithm of the experimental voltage values and performing the same prediction model, Fig. 6 was created.

From Fig. 6 we can deduce that the linear fit between the experimental and theoretical values has been improved, when used the logarithm of experimental voltage values. R^2 has increased to 80.91% and R^2 adjusted to 64.20% using Eqs. (4) and (5).

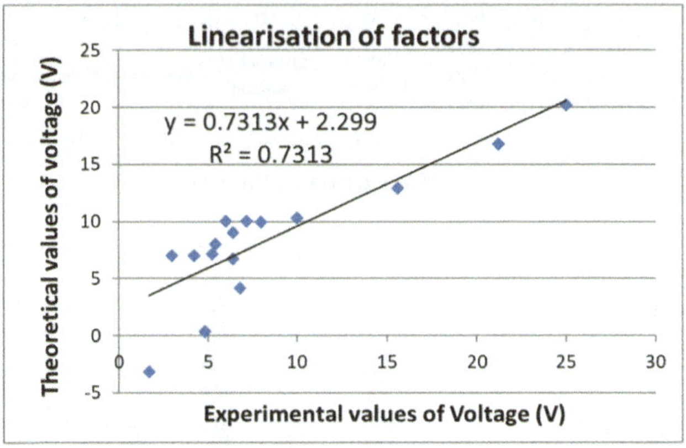

Fig. 5. Experimental voltage values over theoretical voltage values as predicted from our model

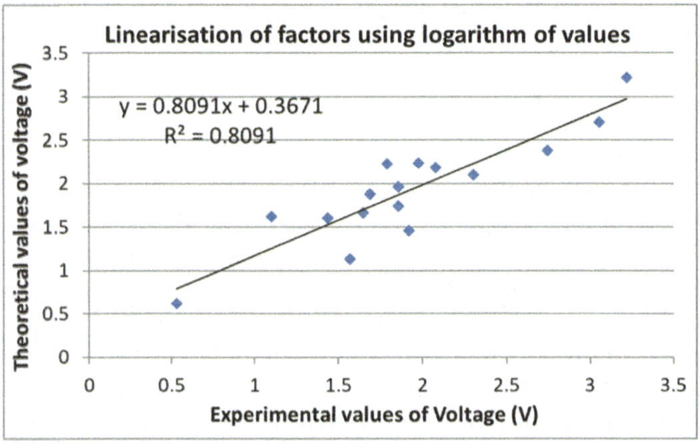

Fig. 6. Experimental voltage values over theoretical voltage values as predicted from our model using the logarithm of experimental voltage values

4 Conclusions

In this work the triboelectric phenomenon has been experimentally studied and the effect of various parameters has been thoroughly examined. A model based on the factorial experiments method was created to predict the effect of each parameter. It was found that our model can predict the experimental values quite well resulting to a R^2 value up to 81% and R^2 adjusted up to 64%. In addition, the importance of various factors was studied and it was concluded that the weight used to pull the samples as well as the distance of rubbing and the thickness of Teflon were the most important factors for creating higher values of triboelectricity, as expected. Further research has been organized in order to test triboelectricity in new materials.

References

Chacko, P., Kapildas, K.S., Jarin, T.: Nano generator intended for energy harvesting. Asian J. Appl. Sci. Technol. **1**(9), 88–91 (2017). ISSN: 2456-883X

Devore, J.L.: Probability and Statistics for Engineering and the Sciences, 8th edn. Brooks Cole Cengage Learning (2011). ISBN 978-0-538-73352-6

Fan, F.-R., Tian, Z.-Q., Lin Wang, Z.: Flexible triboelectric generator. Nano Energy **1**(2), 328–334 (2012). ISSN: 2211-2855

Ledolter, J., Swersey, A.J.: Testing 1–2–3: Experimental Design with Applications in Marketing and Service Operations. Stanford University Press (2007). ISBN: 978-0-8047-5612-9

Lin, Z., Chen, J., Yang, J.: Recent progress in triboelectric nanogenerators as a renewable and sustainable power source. J. Nanomater. 1–24 (2016). ISSN:1687-4110

Logothetis, I., Vassiliadis, S., Siores, E.: Triboelectric effect in energy harvesting. IOP Conf. Ser. Mater. Sci. Eng. **254**, 042021 (2017). https://doi.org/10.1088/1757-899X/254/4/042021

Molnar, A., Gerasimov, V., Kurytnik, P.I.: Triboelectricity and construction of power generators based on it, Przegląd Elektrotechniczny **1**, 167–171 (2018). ISSN: 0033-2097

Montgomery, D.C.: Design and Analysis of Experiments. 9th edn. Wiley (2017). ISBN: 9781119113478 (PBK)

Somkuwar, V.U., Pragya, A., Kumar, B.: Structurally engineered textile-based triboelectric nanogenerator for energy harvesting application. J. Mater. Sci. **55**(12), 5177–5189 (2020). ISSN: 1573-4803

Williams, M.W.: Triboelectric charging of insulating polymers–some new perspectives. AIP Adv. **2**(1), 010701–1–010701–9 (2012). https://doi.org/10.1063/1.3687233

Yin, P., Fan, X.: Estimating R^2 shrinkage in multiple regression: a comparison of different analytical methods. J. Exper. Educ. **69**(2), 203–224 (2001). https://doi.org/10.1080/00220970109600656

Effect of Through-the-Thickness Stitching on the Mechanical Performances of Textile Waste Reinforced Composites

Baccouch Wafa[1(✉)], Ghith Adel[2], Legrand Xavier[3], and Fayala Faten[1]

[1] National School of Engineering of Monastir, Laboratory of Thermal and Energetic Systems Studies, LESTE, University of Monastir, 5000 Monastir, Tunisia
wafaabaccouch@gmail.com

[2] National School of Engineering of Monastir, Textile Materials and Process Research, University of Monastir, 5000 Monastir, Tunisia

[3] National School of Arts and Textile Industries of Roubaix, GEMTEX, Research Laboratory of Textile Materials Engineering, Lille University, 59000 Lille, France

Abstract. The objective of this work is to study the effect of through the thickness stitching on the mechanical performances of textile waste reinforced composites. For this purpose, nonwoven mats were stitched using an industrial sewing machine. In the first part of the work, the effect of stitch row spacing on the mechanical behavior of cotton nonwoven waste was investigated. Results showed that stitching improved the tensile properties of the nonwoven and the space row of 10 mm presented the best Young modulus and tensile strength. The effect of through the thickness stitching on the tensile behaviour of polyester nonwoven was also studied and results showed an increase of Young modulus up to 5 times in 0°. In the second part of the work, stitched and unstitched nonwovens were used to manufacture composite panels via resin infusion method. Mechanical performances of composite materials were later tested using tensile test, three points bending test and charpy test. All tests results indicated that stitching improved significantly the mechanical behavior of all composites. The Young modulus showed an increase of 55% for cotton-based composite and 40% for polyester based composite. While the bending modulus of composite made of polyester nonwoven increased by 83% and the impact resistance increased by 10% .

Keywords: Composite material · Epoxy resin · Nonwoven waste · Recycling · Through the thickness stitching

1 Introduction

Textile waste recycling has become a necessity considering the augmentation of the amount of waste generated each year and the ecological problems that landfilling and burning can cause. Recycling textile waste reduces the need to landfill space, pollution as well as water and energy consumptions (Bureau of International Recycling 2018). Textile waste can be recycled into many different forms according to its composition and its final

S. Msahli and F. Debbabi (Eds.): CIRATM 2020, SPM 17, pp. 62–69, 2022.
https://doi.org/10.1007/978-3-031-08842-1_12

utilization. Using this waste as reinforcement to composite panels is a new recycling area that is being studied and many works has been conducted to study the performances of textile waste-based composites (Zou et al. 2011; Yalcin et al. 2013; Sadikoglu et al. 2003; Petrucci et al. 2015; Wei et al. 2014). Recycled fibers have the advantage of being environment friendly and low cost, however they present low mechanical performances compared to virgin fibers. Weaving, knitting, braiding and stitching have become major methods for 3D fiber structure to improve the performance of fiber reinforced composites (Ghazilan et al. 2017). Stitching is a cost-effective method that improves the interlaminar fracture toughness and impact damage tolerance (Mouritz et al. 1997). Stitched fiber composites have the advantage of improved through-the thickness mechanical properties, reduction of damage area and enhanced energy absorption capability (Ghazilan et al. 2017). Stitching has been used to produce advanced 3D composite material with notable success since the early 1980s even though it is found that it can cause degradation in some mechanical properties of composites such as tensile strength (Ravandi et al. 2016). In fact, a damage in the structure can occur during the manufacturing process or assembly phases where drilling operation can create holes or some local flaws in the structure that could degrade the mechanical behavior of composites (Barile et al. 2017). The objective of this work is improving the mechanical performances of composite materials made of textile waste as reinforcements using low cost process. For this purpose, through the thickness stitching was used as a method to improve the characteristics of the recycled textile waste in order to use it as reinforcement to composite material using resin infusion method. The performances of stitched and non-stitched reinforcements were tested using tensile testing machine, while the mechanical behavior of composite panels was evaluated using tensile test, three points bending test and Charpy test, based on the testing direction for both reinforcements and composite panels; machine and transverse direction.

2 Materials and Methods

2.1 Materials

Two different needle punched nonwoven wastes were used in this study as reinforcement materials. Wastes were supplied by textile recycling companies. The first waste (C) is made of 100% cotton reprocessed fibers that we transformed into nonwoven mat via carding and needle punching processes. The second waste (P) is mainly made of a blend of reprocessed fibers included mostly polyester fibers (90%) that were transformed into nonwoven mat using carding/overlapping/ needle-punching technology. Nonwoven images are presented in Fig. 1.

Epoxy resin (density = 1.175 ± 0.01 g/cm^3 at 20 °C) and its hardener (density = 1.01 ± 0.01 g/cm^3 at 20 °C) were used as matrix material and they were supplied by Sicomin company under the trade name SR 8200 and SD 7203. The mechanical characteristics of the resin are presented in Table 1.

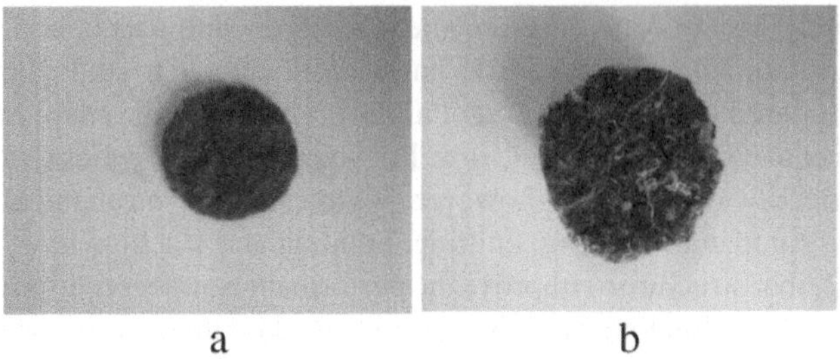

Fig. 1. Reinforcement materials: (a. cotton, b. polyester)

Table 1. Technical properties of the resin system (Sicomin epoxy systems 2017)

Tensile strenght (MPa)	39
Elongation (%)	1.5
Young modulus E (MPa)	2900
Bending strenght (MPa)	69
Elongation (%)	2
Bending Modulus (GPa)	3.44

2.2 Methods

Reinforcements Stitching

Nonwoven mats were stitched using an industrial sewing machine, the stitches were in straight rows in the nonwoven production direction (0°) and spaced evenly apart along the length or across the width of the reinforcements. Lock stitch was chosen for tests, with number of stitches equal to 4 stitches per centimeter. polyester sewing thread with fineness of 27 Tex was used in this study.The first set of experiments investigates the effect of stitch row spacing on the mechanical behavior of cotton nonwoven waste (C). Cotton reinforcements were stitched with three different stitching row spacing; 5 mm, 10 mm and 30 mm, and then five samples were tested using tensile test for each direction and results are presented in Fig. 2. Reinforcements thickness and areal densities are measured and presented in Table 2.

Composite Panels Manufacturing

Stitched and non-stitched nonwoven mats were then used as reinforcements to produce composite panels via resin infusion method. This method uses vacuum forces to infiltrate the resin inside the bagged reinforcement after eliminating the air using a vacuum pump. Materials are kept at ambient temperature for 24 h and then demolded. Reinforcement and matrix proportions were then determined after weighting resulting panels. Composites thickness and areal densities are measured and presented in Table 2.

Mechanical Characterization of Nonwoven Waste
Mechanical characterization of nonwoven was conducted according to ISO 9073-3 standard method, using the Instron universal testing machine with a load cell of 10 KN, and cross-head speed of 50 mm/min. Tests were performed in two directions: (0°) and (90°) according to nonwoven production directions. Five samples were tested for each direction and the arithmetic means of each lot were calculated.

Mechanical Characterization of Composite Panels
The mechanical properties of the composites were investigated using tensile test, three points bending test and Charpy tests. The tensile test is performed using an Instron universal testing machine with a load cell of 250 KN and a cross-head speed of 10 mm/min according to ISO 527-4 standard method. Three-point-bending test was carried out with the same tensile testing machine but configured for this purpose with a bending speed of 2 mm/min, according to ISO 14125. The Charpy impact test is done using the Instron Charpy test machine, according to EN ISO 179 standard method. Five samples were tested for each direction and the arithmetic means of each lot were calculated.

3 Results and Discussion

3.1 Physical Analysis

Thickness of non-stitched and stitched reinforcements and composites were measured while areal densities were calculated. Results are presented in Table 2. The results indicate that stitching reduces the thickness of reinforcements and composites by eliminating the space between fibers and increasing their cohesion. While adding sewing fibers to the structure didn't increase the surface mass of composites.

Table 2. Characteristics of stitched reinforcements and composite materials

	Thickness (mm)	Areal density (g/m^2)
R1	1,7	140
R2	9,6	914
SR1	1,5	145
SR2	7,9	922
CR1	0,72	665
CR2	4	3808
SCR1	0,7	670
SCR2	3,8	3815

3.2 Mechanical Characterization of Nonwoven Waste

In the first part of this work, the effect of stitch row spacing on the tensile properties of cotton nonwoven (C) was studied. Three different stitched nonwovens were produced according to the stitch row space, coded as C-S-5, C-S-10 and C-S-30 for row spacing of 5 mm, 10 mm and 30 mm respectively. Tensile test results are presented in Fig. 2. Results show an anisotropic behavior of stitched and non-stitched reinforcements where Young modulus and tensile strength of all samples are more important in direction 0°. This can be explained by production process that privileges the orientation of most of fibers in machine direction. Results show that stitching improves the mechanical performances of reinforcements in both directions. The space of 10 mm between rows has the best results where the Young modulus of SR10 coded sample is four times more important in both directions compared to non-stitched nonwoven. And the tensile strength increased by 13 times in 0° direction and 1.7 times in 90° direction. On the other hand, as expected, stitching decreased the nonwoven elongations to around the half, as stitches block the structure and improves the cohesion between fibers. This spacing is then used to test the mechanical characteristics of polyester nonwoven reinforcement in both directions, and the used to produce the composite panels. Tensile test results of unstitched and stitched polyester nonwoven coded R2 and SR2 respectively are presented in Fig. 3. The result is consistent with cotton nonwoven R1 where it shows a significant increase of Young modulus up to 5 times in 0° direction and around 3 times in 90° direction and an increase of tensile strength up to 36 times in 0° and 2.5 times in 90°.

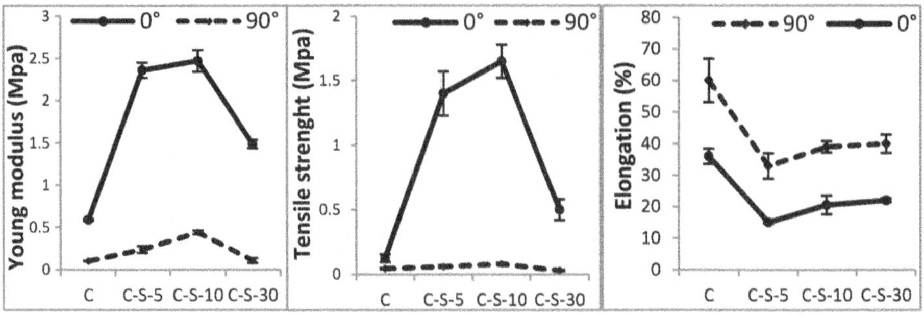

Fig. 2. Tensile test results for cotton nonwoven waste

3.3 Mechanical Characterization of Composite Panels

The Young modulus, tensile strength and elongation of non-stitched and stitched composite structures C-C and P-C are presented in Fig. 4 and Fig. 5, respectively. According to the results, all composite panels have a higher tensile strength in 0° direction compared to 90° direction. because of the higher orientation of textile waste fibers along the direction of nonwoven production. These results are consistent with reinforcements test results (Fig. 2 and 3). Stitching increased the Young modulus and the tensile strength of both composites. The Young modulus of composite material coded CSR1 is respectively

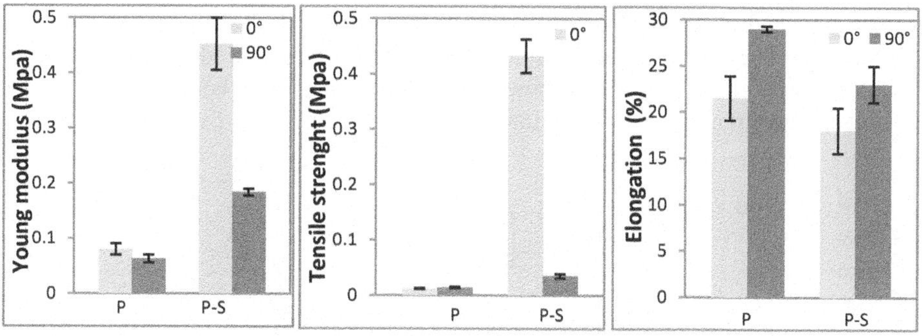

Fig. 3. Tensile test results for polyester nonwoven waste

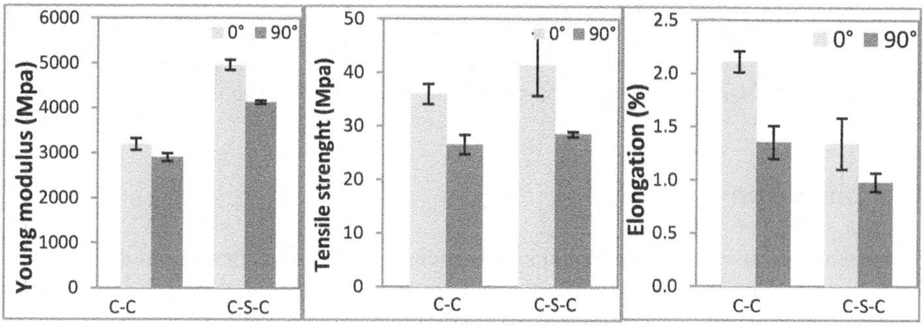

Fig. 4. Tensile test results of unstitched stitched and composite material (CR1)

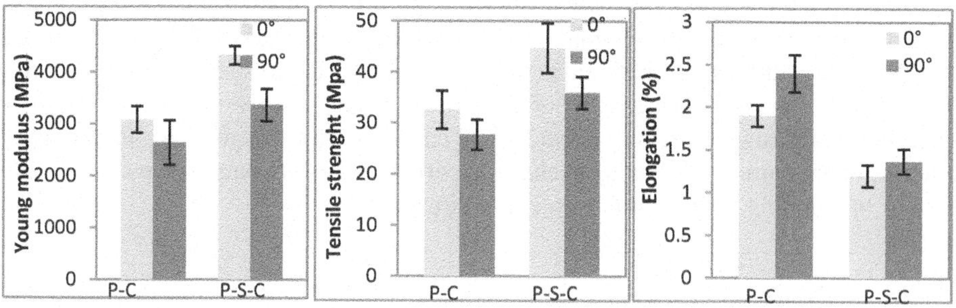

Fig. 5. Tensile test results of stitched and non-stitched composite material (CR2)

55% and 42% higher than unstitched samples (CR1) for 0° and 90° directions. While it is 40% and 27% higher for CSR2 coded composite panel.

The three points bending test results of composite panels are presented in Fig. 6. No breakage was noticed for CR1 composite during the test, while stitched composite SCR1 was broken, this is probably due to the creation of holes within the structure during stitching process, these holes can be considered as weakness points that can cause the specimen's fracture. In accordance with reinforcements and composites tensile tests, results indicated an anisotropic behavior of all samples. And through the thickness stitching improved the bending modulus and bending strength of all samples. Stitched polyester based composite SCR2 showed an increase of 83% in bending modulus and 8% in bending strength for 0° direction.

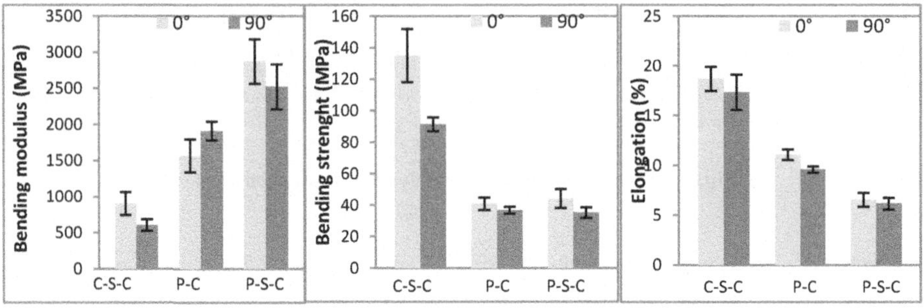

Fig. 6. Bending test results of stitched and non-stitched composite materials

4 Conclusion

In this study we used through the thickness stitching to improve the mechanical behavior of composite panels made of different reclaimed fibers, by enhancing the mechanical characteristics of the reinforcements. In the first place, three different row spacing were used and tested on cotton nonwoven reinforcements, results showed that the Young modulus of reinforcements stitched with 10 mm spacing is four times and a half more important that the non-stitched nonwoven in the machine direction and four times more important in the transverse direction. Nonwoven panels were made then by resin infusion method using stitched and non-stitched reinforcements. Results showed that stitching improved the mechanical behavior of composite panels. Results are more important in the stitching rows direction. Further tests can be performed to study the effect of other stitching parameters on the mechanical performances of composite panels such as density of stitching, type of stitch, stitching thread, etc.

References

Barile, C., et al.: The influence of stitching and unconventional fibres orientation on the tensile properties of CFRP laminates. Compos. B **110**, 248–254 (2017)

Bureau of International Recycling (2018). http://www.bir.org/industry/textiles

Ghazilan, A.L., et al.: Effect of stitching on the tensile mechanical property of empty fruit bunch oil palm fiber reinforced epoxy composites. IOP Conf. Ser. Mater. Sci. Eng. **184**, 012051 (2017)

Mouritz, A.P., et al.: A review of the effect of stitching on the in-plane mechanical properties of fibre-reinforced polymer composites. Compos. A **28A**, 979–991 (1997)

Petrucci, R., et al.: Tensile and fatigue characterisation of textile cotton waste/polypropylene laminates. Compos. B **81**, 84–90 (2015)

Ravandi, M., et al.: The effects of through-the-thickness stitching on the Mode I interlaminar fracture toughness of flax/epoxy composite laminates. Mater. Des. **109**, 659–669 (2016)

Sadikoglu, T.G., et al.: Usage of polyester textile wastes in composites. J. Sci. Ind. Res. **62**, 462–467 (2003)

Sicomin epoxy systems (2017). http://www.sicomin.com/fr

Wei, B., et al.: Fabrication and property of discarded denim fabric/polypropyelen composite. J. Ind. Text. **44**, 798–812 (2014)

Yalcin, I., et al.: Utilization of various non-woven waste forms as reinforcement in polymeric composites. Text. Res. J. **83**, 1551–1562 (2013)

Zou, Y., et al.: Reusing polyester/cotton blend fabrics for composites. Compos. Part B **42**, 763–770 (2011)

Advance in Manufacturing Process

Application and Comparison Between Exact and Evolutionary Algorithms for Color Recipe Prediction

Sabrine Chaouch(✉), Ali Moussa, Imed Ben Marzoug, and Neji Ladhari

Laboratory of Textile Engineering, University of Monastir, Monastir, Tunisia
chaouchsabrine@hotmail.fr

Abstract. This paper presents and compares evolutionary algorithms, such as ant colony and genetic algorithms, and exact approaches for color recipe prediction. The objective is to optimize the color formulation step in reproducing the desired shades by minimizing the color differences CMC (2:1) between the target color and the color obtained by the proposed recipe. Objective criteria were used to compare the proposed methods. Four direct dyes (CI Direct Red 227, CI Direct Orange 34, CI Direct Blue 85 and CI Direct Black 22) were used in this study for dyeing bleached woven fabrics (100% cotton).

Keywords: Color recipe formulation · Genetic algorithm · Ant colony · Color differences · Evolutionary algorithms

1 Introduction

In general, the most important problem in color recipe prediction is to determine the dyes to be applied and to find their respective correct concentrations to produce a match for a target shade.

The majority of color formulation methods are based on Kubelka-Munk theory. For a detailed description of the Kubelka-Munk theory, one can refer to literature (Kubelka and Munk 1931; Kubelka 1948, 1954). Only the equations used in this work are listed in this paper (Kubelka and Munk 1931; McGinnis 1967, Wright 1969):

$$\frac{K(\lambda)}{S(\lambda)} = \frac{[1 - R(\lambda)]^2}{2.R(\lambda)} \tag{1}$$

where K is the absorption coefficient, S is the scattering coefficient, and R is the reflectance factor at a specific wavelength λ.

$$\left(\frac{K}{S}\right)_{mix} = \left(\frac{K}{S}\right)_{sub} + C_i \times \sum_{i=1}^{n} \left(\frac{K}{S}\right)_i \tag{2}$$

where $\left(\frac{K}{S}\right)$ is the Kubelka-Munk function for a combination of n different dyes with concentrations of C_i at a specific wavelength, and mix and sub stand for mixture and substrate, respectively.

S. Msahli and F. Debbabi (Eds.): CIRATM 2020, SPM 17, pp. 73–81, 2022.
https://doi.org/10.1007/978-3-031-08842-1_13

The first equation presents a simple mathematical form of Kubelka-Munk theory, so it requires certain assumptions to formulate differential equations (Wyszecki and Stiles 2000):

- The absorption and scattering coefficients of individual pigments were directly proportional to its concentration and were additive in a mixture (Nobbs 1985).
- Each colorant in a mixture acts separately and the substrate is responsible for the majority of the scattering.

So, in using the Kubelka-Munk single-constant theory, we obtain the second equation. The linear relationship between the $\left(\frac{K}{S}\right)_i$ of each dye I and its concentration C_i allows the determination of the proportion of each dye concentration in any given mixture (Wyszecki and Stiles 2000).

Different methods and techniques, based on colorimetric and spectrophotometric algorithms (Agahian and Amirshahi 2008, Shams Nateri 2010, 2011) or on artificial intelligence techniques (Bishop et al. 1991, Ameri et al. 2005, Hai-Tao et al. 2007, Almodarresi et al. 2013, Jawahar et al. 2015), have been proposed and applied for the color recipe formulation.

In this work, we present and compare evolutionary algorithms, such as ant colony and genetic algorithm, and exact approaches to predict which colorant and what concentrations of colorant are required to obtain an exact color match. The comparison was made using two criteria: accuracy and calculation time of the optimal solution.

For evaluating the color differences between the target color and the color reproduced by the proposed recipe, we adopted the CMC(2:1) color differences formula in D65/10° (Clarke et al. 1984). It is currently the ISO standard for the textile industry (ISO 105-J03).

2 Experimental

2.1 Materials

Four direct dyes were used in this study: CI Direct Red 227, CI Direct Orange 34, CI Direct Blue 85 and CI Direct Black 22. These dyes were purchased from Huntsman (Basel, Switzerland), and used without further purification. For dyeing, bleached woven fabrics (100% cotton) were used and their colors were measured and evaluated.

2.2 Dyeing

Textile samples were dyed by the four direct dyes at 12 different concentrations (0.05, 0.1, 0.25, 0.5, 0.75, 1, 1.25, 1.5, 1.75, 2, 3 and 4%) thus forming a database of 36 different dyed samples. Dyeing parameters and processes are presented in Table 1 and Fig. 1, respectively. All dyeing were performed in a laboratory machine type DataColor AHIBA Nuance Speed (Datacolor International Company, USA) with a liquor ratio of 10:1.

Table 1. Recipe of dyeing with direct dyes

Levels	Products	Values
A	Direct dye Wetting agent Sodium carbonate	X%* 3mL/L Y**
B	Sodium sulfate	Z**

** X%: 0.05, 0.1, 0.25, 0.5, 0.75, 1, 1.25, 1.5, 1.75, 2, 3 and 4%.
**Y and Z are determined according the color shade from the manufacturer's instructions.

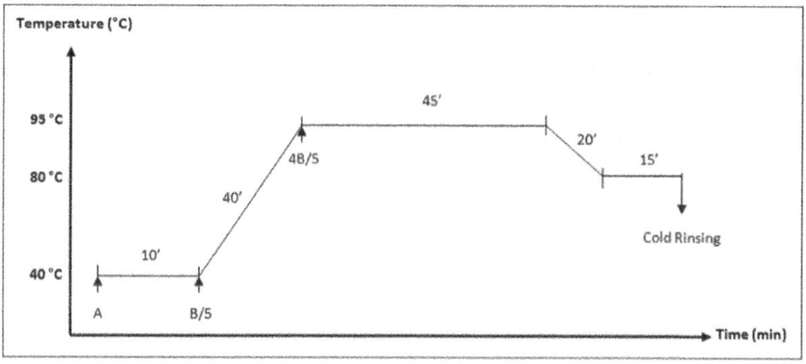

Fig. 1. Dyeing process.

2.3 Color Evaluation

The reflectance spectra of the dyed samples were measured using a spectrophotometer Spectraflash 600 Plus (Datacolor International Company, USA), with the following measuring conditions: spectra from 400 to 700 nm at 10 nm intervals, illuminant D65, and 10° standard observers.

3 Color Prediction

3.1 Exact Algorithm

The exact algorithm is a simplest approach to find exact solutions to an optimization problem, by scanning the search space exhaustively (Olaechea et al. 2014).

In the following we will propose an exact algorithm allowing the reproduction of a specific color based on the Kubelka-Munk theory while browsing all the possible combinations and choosing the best solution.

The different steps of this algorithm are as follows:

- **Step 1:** Measure and store the colorimetric coordinates of each target sample.

- **Step 2:** Browse all possible combinations of the dye mixture that can be used depending on the number of used dyes and the total number of dye's concentrations. We, calculate for each possible combination the Kubelka-Munk function (K/S) of the mixture by adding the spectra of the dyes that constitute it.
- **Step 3:** Calculate the CMC(2:1) color differences between the obtained spectrum and that of the target sample. If this difference is less than the tolerated threshold 1, the combination is considered acceptable and it will be memorized, otherwise we go on to the next combination.
- **Step 4:** Choose the optimal recipe that allows the minimum CMC(2:1) color difference, and display the corresponding color recipe.

3.2 Evolutionary Algorithms

Evolutionary algorithms are generally used to provide good approximate solutions to optimization problems that cannot be solved easily using other techniques (Pradnya 2016). They consist to work with a set of solutions simultaneously, which is gradually evolving. The use of several solutions simultaneously makes it possible to improve the exploration of the configuration space. In this work we have used ant colony and genetic algorithm to find approximate solutions for the color recipe formulation (Carlos 2005).

3.2.1 Ant Colony Optimization Algorithm (ACO)

ACO algorithm is inspired by the behavior of real ants during the colony searching for the shortest path (Dorigo and Stitzle 2004). When searching for food, ants initially explore the area surrounding their nest in a random manner. While moving, ants leave a chemical pheromone trail on the ground. Ants can smell pheromones. When choosing their way, they tend to choose, in probability, paths marked by strong pheromone concentrations. As soon as an ant finds a food source, it evaluates the quantity and the quality of the food and carries some of it back to the nest. During the return trip, the quantity of pheromones that an ant leaves on the ground may depend on the quantity and the quality of the food. The pheromone trails will guide other ants to the food source. It has been proved that the indirect communication between the ants via pheromone trails enables them to find shortest paths between their nest and food sources (Blum 2005).

To solve the problem of color recipe prediction we assume that paths followed by ants are the different concentrations of the various dyes.

The ACO algorithm process is as follows:

- **Step 1:** Set and initialization of algorithm parameters and pheromone trails;
- **Step 2:** Construction of ant solution;
- **Step 3:** Update pheromones;
- **Step 4:** Repeat step 2 and 3 until the termination condition is reached (All ants are chosen and the number of iterations is reached).

Initially, each ant selects randomly a starting concentration of the first dye of mixture. Then, he chooses randomly the next dye a concentration based on the probability values

P_{ij} defined as follows:

$$P_{ij}(t) = \frac{\tau_{ij}^{\alpha}(t).\eta_{ij}^{\beta}}{\sum_{l \in N_i^k} \tau_{il}^{\alpha}(t).\eta_{il}^{\beta}} \quad (3)$$

where $P_{ij}(t)$ is the probability to choose the concentration j of the next dye after choosing the concentration i of the current dye; τ_{ij} is the value of pheromones between the concentration i of the current dye and the concentration j of next dye; $\eta_{ij} = \frac{1}{\Delta E_{CMC(2:1)}}$ is the visibility between these two concentrations i and j; N_i^k is the set of concentrations of dye that the ant has not yet chosen at the moment t; α and β are the parameters controlling the relative importance between the rate of pheromones and visibility.

The updating of the value of pheromones traces is done by the following formula:

$$\tau_{ij}(t) = \rho.\tau_{ij}(t-1) + \Delta\tau_{ij} \quad (4)$$

where ρ is the evaporation parameter of pheromones and $\Delta\tau_{ij} = \frac{1}{\Delta E_{CMC(2:1)}}$.

3.2.2 Genetic Algorithm (GA)

GA is a search heuristic that is inspired of natural evolution. The concept of this algorithm was first introduced and used by John Holland in the early 1970s (Holland 1975).

This algorithm reflects the process of natural selection where the fittest individuals are selected for reproduction in order to produce offspring of the next generation (Abuiziah and Shakarneh 2013). The process of natural selection starts with the selection of fittest individuals from a population. They produce offspring which inherit the characteristics of the parents and will be added to the next generation. If parents have better fitness, their offspring will be better than parents and have a better chance at surviving. This process keeps on iterating and at the end, a generation with the fittest individuals will be found (Leardi 2009).

This notion was applied in color recipe prediction assuming that each individual in the population is composed of a single chromosome representing a color recipe prediction. Indeed, the number of genes in each chromosome is equal to the number of used dyes.

The GA process is as follows:

- **Step 1:** Generate randomly the initial population covering the entire range of possible solutions;
- **Step 2:** Roulette selection of parents;
- **Step 3:** A random crossover point is selected and the tails of its two parents are swapped to get new offsprings.
- **Step 4:** Creep mutation of children;
- **Step 5:** Construction of the new generation, by transmitting the best individual of the previous generation and supplementing the population by all the generated children.
- **Step 6:** Repeat step 2, 3, 4 and 5 until the termination condition is reached (The number of generations is reached).

The selection process is made to determine the set of chromosomes from the current population for reproduction according to their fitness *F(x)*:

$$F(x) = \frac{1}{\Delta E_{CMC(2:1)} + \varepsilon} \tag{5}$$

where $\Delta E_{CMC(2:1)}$ is the CMC(2:1) color differences value between the reproduced color and the desired target color and ε is a negligible constant set at 10^{-6} to avoid division by zero.

A pair of parent chromosomes is then chosen randomly from the current population to produce the new population based on these probabilities:

$$p(x_i) = \frac{F(x_i)}{\sum_{k=1}^{N} F(x_k)} \tag{6}$$

where $F(x_i)$ is the objective function of the individual x_i and N is The number of individuals in the population.

4 Results and Discussion

To evaluate and compare the effectiveness and the performance of the proposed algorithms, 20 target color samples were prepared using different binary and ternary mixtures of used dyes. For each target color sample, each proposed algorithm should predict the optimal recipe that minimizes CMC (2:1) color differences between the reproduced color and this target sample.

The parameters of evolutionary algorithms were fixed using full factorial design using MINITAB software (MINITAB® Release 14.1).

The parameters of ACO algorithm have been set at: The number of iterations = 2000 iterations; The number of ants = The number of concentrations of the used dyes; α = 0.5; β = 0.5 and ρ = 0.1.

The parameters of GA have been set at: The number of individuals = 1000 individuals and the number of generations = 10 generations.

Results obtained by applying the different algorithms to predict dyeing recipes of all the 20 target color samples are shown in Fig. 2.

We observed that the theoretical color difference between the optimal solutions proposed by the different developed algorithms and the target color was smaller than 1.

It is also observed that the values of color differences obtained by applying evolutionary algorithms were higher than the values obtained by exact algorithm but they remained below 1. In fact, exact algorithms gave the best solution of all possible solutions, while ACO and GA have proposed an approximate solution that is not necessarily the best solution but it remains acceptable (the average value of color differences of all testing data was lower than 1, which shows they are matched acceptably).

So, the exact algorithm gives better results, it is more precise and it allows to find the best solution. On the other hand, the time required by this algorithm to go through all the possible solutions and choose the optimal recipe is too long; it corresponds to 7–8 h for each target sample.

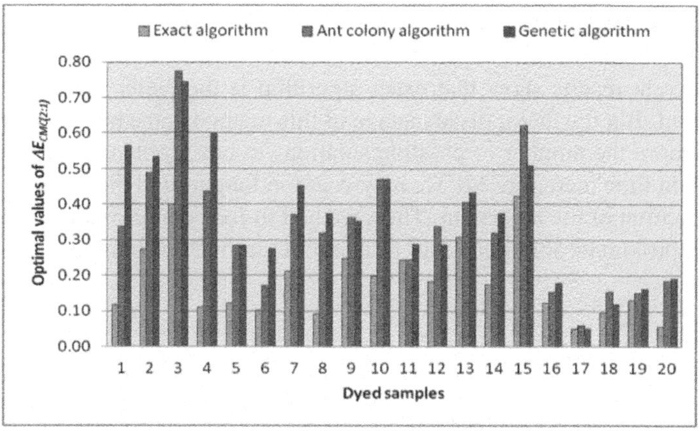

Fig. 2. Theoretical values of $\Delta E_{CMC(2:1)}$ between color samples proposed by the different proposed algorithms and the target colors.

The ACO algorithm converges to the optimal solution in 20–30 s, while the required time by GA to converges to the optimal solution correspond to 10–15 s.

It is difficult to conclude that one method is superior to another because of the wide variety of possible evaluating criteria. In this work, objective criteria such as accuracy and calculation time have been used. Table 2 presents a summary of the results obtained with the different proposed algorithm.

Table 2. Comparison of methods proposed.

Method		Precision	Calculation time
Exact algorithm		++	−−
Evolutionary algorithms	ACO	+	+
	GA	+	+

++: very good; +: good; -: poor; −−: very poor.

For accuracy, exact algorithm is better than evolutionary algorithms whether ACO algorithm or GA. For calculation time evolutionary algorithms was the best.

Generally, accuracy is considered the principal criteria to choose the good method.

But as the most accurate method requires a huge computation time, we have to make a compromise between accuracy and calculation time. Therefore, we have to focus on evolutionary algorithms to solve the problem of huge computing time.

5 Conclusion

This work presents and compares exact and evolutionary algorithms such as ACO algorithms and GA. The theoretical values of CMC (2:1) color differences between color samples proposed by all the proposed methods and the target colors were smaller than the textile threshold of 1.

The different results show that exact algorithm is the better from a purely accuracy viewpoint. But the major disadvantage of this method is the huge calculation time. In fact, the more the number of possible solutions to our problem increases, the more the calculation time increases. So, we may wait too long to find the optimal solution as we risk stagnation of the algorithm. Thus, we had to focus on evolutionary algorithms, to give an approximate solution to our problem in reasonable computing times.

References

Agahian, F., Amirshahi, S.H.: A New matching strategy: trial of the principal component coordinates. Color. Res. Appl. **33**(1), 10–18 (2008)

Almodarresi, E.S.Y., Mokhtari, J., Almodarresi, S.M.T., Nouri, M., Shams-Nateri, A.: A scanner based neural network technique for color matching of dyed cotton with reactive dye. Fibers Polym. **14**(7), 1196–1202 (2013)

Ameri, F., Moradian, S., Amani Tehran, M., Faez, K.: The use of fundamental color stimulus to improve the performance of artificial neural network color match prediction systems. Iranian J. Chem. Chem. Eng. Int. English Ed. **24**(4), 53–61 (2005)

Abuiziah, I., Shakarneh, N.: A review of genetic algorithm optimization: operations and applications to water pipeline systems, world academy of science, engineering and technology. Int. J. Math. Comput. Sci. **7**(12) (2013)

Bishop, J.M., Bushnell, M.J., Westland, S.: Application of neural network to computer recipe prediction. Color. Res. Appl. **16**(1), 3–9 (1991)

Blum, C.: Ant colony optimization: Introduction and recent trends. Phys. Life Rev. **2**, 353–373 (2005)

Carlos C.A.C.: An introduction to evolutionary algorithms and their applications. In: Ramos, F.F., Larios Rosillo, V., Unger, H. (eds.) ISSADS 2005. LNCS, vol. 3563, pp. 425–442. Springer, Heidelberg (2005). https://doi.org/10.1007/11533962_39

Clarke, F.J.J., McDonald, R., Rigg, B.: Modification to the JPC79 colour-difference formula. JSDC **100**, 128–132 (1984)

Dorigo, M., Stutzle, T.: Ant Colony Optimization. MIT Press, Cambridge, MA (2004)

Hai-tao, L., Ai-song, S., Bing-sen, Z.: A Dyeing Color Matching Method Combining RBF Neural Networks with Genetic Algorithms. IEEE Computer Society, pp.701–707 (2007)

Holland, J.: Adaptation in Natural and Artificial Systems, University of Michigan Press, Ann Arbor. (Technical Report ORA Projects 01252 and 08226). University of Michigan, Department of Computer and Communication Sciences, Ann Arbor (1975)

Jawahar, M., Babu, C., Kannan, N., Kondamudi-Manobhai, M.: Artificial neural networks for colour prediction in leather dyeing on the basis of a tristimulus system. Color. Technol. **131**(1), 48–57 (2015)

Kubelka, P., Munk, F.: Ein Beitrag zur Optik der Farbanstriche. Z. Tech. Phys. **12**, 593–601 (1931)

Kubelka, P.: New contributions to the optics of intensely light-scattering materials. Part I. J. Opt. Soc. Am. **38**, 448–457 (1948)

Kubelka, P.: New contributions to the optics of intensely light-scattering materials. Part II nonhomogeneous layers. J. Opt. Soc. Am. **44**, 330–334 (1954)

Leardi, R.: Genetic algorithms, chemical and biochemical data. Analysis **1**, 631–653 (2009)

McGinnis, P.H.: Spectrophotometric color matching with the least squares technique. Col. Engin. **5**, 22–27 (1967)

Nobbs, J.H.: Kubelka-Munk theory and the prediction of reflectance. Rev. Prog. Color. Relat. Top. **15**, 66–75 (1985)

Olaechea, R., Rayside, D., Guo, J., Czarnecki, K.: Comparison of exact and approximate multi-objective optimization for software product lines. In: Proceedings of the 18th International Software Product Line Conference, vol. 1, pp. 92–101 (2014)

Vikhar, P.A.: Evolutionary algorithms: a critical review and its future prospects. In: International Conference on Global Trends in Signal Processing, Information Computing and Communication (2016)

Shams-Nateri, A.: Prediction of dye concentrations in a three-component dye mixture solution by a PCA-derivative spectrophotometry technique. Color. Res. Appl. **35**(1), 29–33 (2010)

Shams-Nateri, A.: Dye concentrations determination in ternary mixture solution by using colorimetric algorithm. Iranian J. Chem. Chem. Eng. Int. English Edn. **30**(4), 51–61 (2011)

Wright, W.D.: The Measurement of Color, 4th edn. Hilger, London, U.K. (1969)

Wyszecki, G., Stiles, W.S.: Color Science: Concepts and Methods, Quantitative Data and Formulae, 2nd edn. Wiley, New York (2000)

An Eco-friendly Approach for Polyester Coloration Using Cochineal as Natural Dye

Najla Krifa[1(✉)], Wafa Miled[1,2], Riadh Zouari[1], Behary Nemeshwaree[3], Christine Campagne[3], and Morshed Cheikhrouhou[4]

[1] Textile Engineering Laboratory, Ksar Hellal, B.P 68, 5070 Ksar Hellal, Tunisia
krifa_najla@hotmail.com
[2] Department of Early Childhood, University College Taraba, Taif University, Taif, Saudi Arabia
[3] ENSAIT, GEMTEX, 59100 Roubaix, France
[4] Higher Institute of Arts and Crafts of Sfax, Sfax, Tunisia

Abstract. This study aims to provide an ecofreindly coloration process of polyester fabric with cochineal-based natural dye using the advanced waterless technology of plasma combined with padding techniques.

The impact of plasma treatment was characterized by water contact angle and capillarity measurements and the resulting dyeing performances were compared in terms of color strength (K/S) and fastness properties according to standard methods.

Experimental results revealed that surface activation mechanism increased successfully both wettability behavior and color strength of the polyester fabric. Indeed, up to a 40% enhancement of the wet pick-up rate of dye solutions on plasma treated substrate was observed as compared to untreated sample, whereas different fastness properties according to the selected padding conditions used, were noted.

Keywords: Eco-friendly · Colartion · Polyester · Cochineal · Plasma

1 Introduction

Eco-friendliness and energy conservation are crucial concerns in the textile industry (Hussain and Wahab 2018). Recently, there has been increased interest in the use of natural dyes for textile coloration, as alternatives to synthetic dyes that are toxic to a variety of organisms (Padhi 2012). Thus, reviving interest once again on natural colorants is being needed, as natural dyes are usually renewable and biodegradable. However, natural dyes have poor affinity for textiles and they require heavy-metal salts as mordants for fixation and color fastness. To overcome these limitations, plasma technology, consisting of randomly moving neutral and charged species, including electrons, free radicals and ions which can carry a large amount of internal energy, appears to be the most suitable alternative for textile pre-treatment and finishing unlike wet chemical treatment since plasma exposure of polymers enhances their surface attributes without alteration of bulk properties under energy efficiencies and lower chemical consumption, fitting appropriately the definition of eco-friendly textile manufacturing (Morent et al. 2008).

S. Msahli and F. Debbabi (Eds.): CIRATM 2020, SPM 17, pp. 82–87, 2022.
https://doi.org/10.1007/978-3-031-08842-1_14

Thus, the aim of this study is to investigate a green approach for polyester coloration using cochineal dye while combining plasma treatment and pad-dyeing technique in order to reduce excessive water and energy consumption used under exhaust dyeing processes (Mongkholrattanasit et al. 2010).

The influence of surface activation process on wetting behavior of polyester fabric was explored using specific wettability measurements, as well as the influence of the main dyeing conditions including temperature on the dyeing performance in terms of color strength and fastness properties were seeking.

2 Materials and Methods

2.1 Materials and Chemicals

A 100% polyester woven fabric (PET) of a density of 208 g/m^2 was used throughout this study. At first, in order to remove residual chemicals and auxiliaries, PET samples were sourced using Petroleum Ether and Ethanol in two different steps, then rinsed with deionized water for 20 min and finally dried for 24 h at ambient temperature (Figs. 1 and 2).

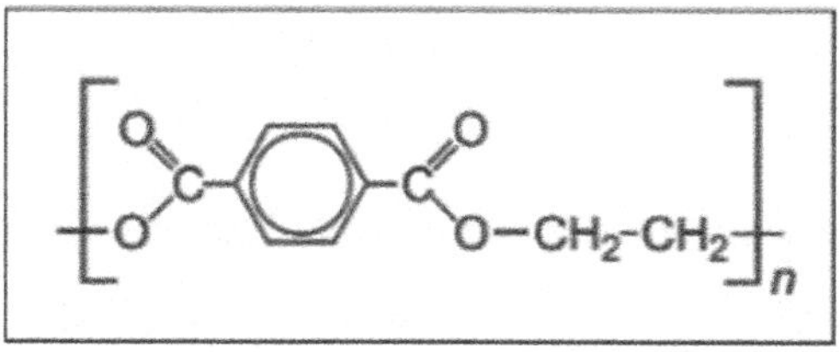

Fig. 1. Chemical structure of Poly (ethylene terephthalate). Red cochineal dye (carminic acid) was used as received from Couleurs de plantes (France).

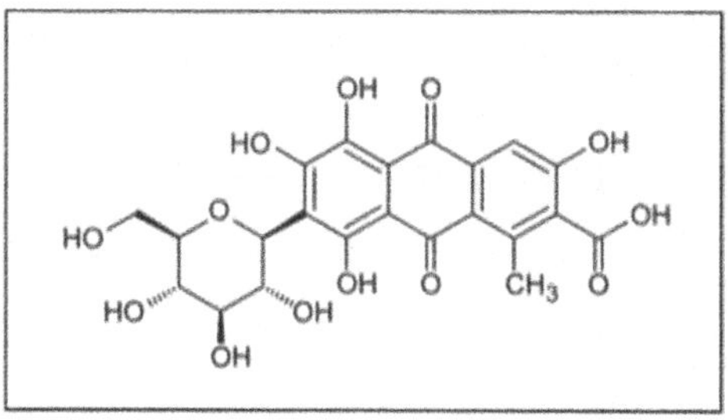

Fig. 2. Chemical structure of carminic acid

2.2 Plasma Treatment and Surface Characterization

The plasma treatment was carried out using air atmospheric plasma machine Coating Star (Ahlbrandt System, Germany) under an electrical power of 1 kW and a speed of

2 m/min. For the assessment of the effectiveness of surface activation process, water contact angle (°) and capillarity (%) measurements were carried out using a tensiometer "3S Balance" from GBX instruments according to the Wilhelmy principle method.

2.3 Pad-Dyeing Procedure

An aqueous cochineal dye solution (20 g/L) was prepared at pH 5 using acetic acid. PET fabrics were then padded with this solution using a laboratory-scaled padder system (Werner Mathis AG, Switzerland). The pressure was set to 2 bars and the rotation speed to 2.5 m/min. The wet pick-up rate was calculated from the weight uptake applying Eq. (1):

$$\text{Pick up rate}\ (\%) = \frac{W_f - W_i}{W_i} \times 100 \tag{1}$$

where W_i (g) and W_f (g) are the weights of the polyester samples before and after padding process, respectively.

The padded samples were then subjected to thermal treatment at different temperature (100–150–200 °C) for a duration of 2 min using a hot air dryer (Werner Mathis AG, Switzerland). Subsequently, the dried samples were washed in a commercial 5 g/L nonionic detergent bath with liquor ratio LR 50:1 (Labomat, Switzerland) at 40 °C for 20 min and then rinsed in soft water and left to air dry under ambient conditions.

2.4 Color Measurements of Dyed Samples and Fastness Properties

Spectral reflectance factors were measured using a Datacolor Spectraflash SF600 reflectance spectrophotometer. The wash fastness test was assessed using the standard ISO 105:C10 wash fastness test protocol. The rub fastness test was assessed using the standard ISO 105:X12. In addition to dry rub fastness, wet rub fastness was tested.

3 Materials and Methods

3.1 Contact Angle and Capillarity Measurement Results

Wettability measurement of polyester fabric before and after plasma treatment are presented in Table 1.

Table 1. Wettability measurement results

	Water contact angle (°)	Water capillarity absorption (%)
Untreated PET	77.3	1,683
Plasma treated PET	30.5	108,562

A significant decrease of the contact angle from 77.3° for untreated polyester fabric to 30.5° after plasma treatment was detected, along with a striking increase in the capillary rate up to 107 (%). The improvement of the hydrophilicity of plasma-treated fabrics may be attributed to the new chemical structure of the surface layer of the polymer material, which means that new active sites were created during the activation treatment conferring hydrophilic properties to the PET fabric (Leroux et al. 2006). In fact, the energetic electrons induced by plasma discharge using atmospheric air could cause several molecular chains scissions in the outer layers of the PET polymer especially on the ester bonds, which is the weakest bond in the polymer chain. These chain-scissions create a large amount of very reactive chain-ends which could then react easily with the oxygen atoms present in the atmospheric air plasma. Hence, formation of special chemical elements such as hydroxyl, carbonyl, and carboxyl groups takes place imparting hydrophilic character to the PET fabric surface due to their polar nature (Borcia et al. 2004).

3.2 Color Measurement Results

Figure 3 clearly indicates that plasma treatment of PET fabrics enhanced the coloration process by padding method. Indeed, the untreated PET surface had low affinity to the coloring substances since polyester fiber is highly hydrophobic with limited wetting behavior due to few polar interactions and is therefore difficult to dye, whereas plasma treated samples exhibited higher wettability reactivity, which was associated to the incorporation of numerous polar functional groups (OH, COOH, C = O). Consequently, the effective surface area increases after the plasma treatment, ensuring therefore a significant enhancement of the wet uptake rate of the dye solution from 62% for untreated PET fabric compared to 102% for plasma treated PET samples. Thus, the interaction and diffusion of more dye molecules are promoted forward.

However, air plasma treatment alone increased slightly the K/S value and it can be noticed that the dye yield depends closely on the temperature used; the higher it is, the higher was the color intensity, hence ensuring sufficient fastness properties as presented in Table 2.

In particular the dye diffused well into the PET when drying step was performed at 200 °C. In fact, temperature plays a very effective role, in the dyeing process. Rising drying temperature increase the mobility of polyester macromolecular chain which make the diffusion of logwood dye into the heart of the fiber more easily. Once inside the fiber, dye particles are trapped and they remain fixed into the fiber ensuring higher color strength. This process can be assimilated to thermosol dyeing which was developed and panted by du Pont in 1953 (Iqbal 2008).

3.3 Fastness Properties

Table 2 summarizes the color-fastness properties of the dyed fabrics.

When comparing the fastness rating of the samples dyed, it can be noticed that this coloration process produces different fastness grades. Table 2 shows that the wash fastness increase from poor (1–2) to moderate (3) with an increase in the drying temperature in PET fabric. The lower ratings can be explained by the dye accumulation on the fabric

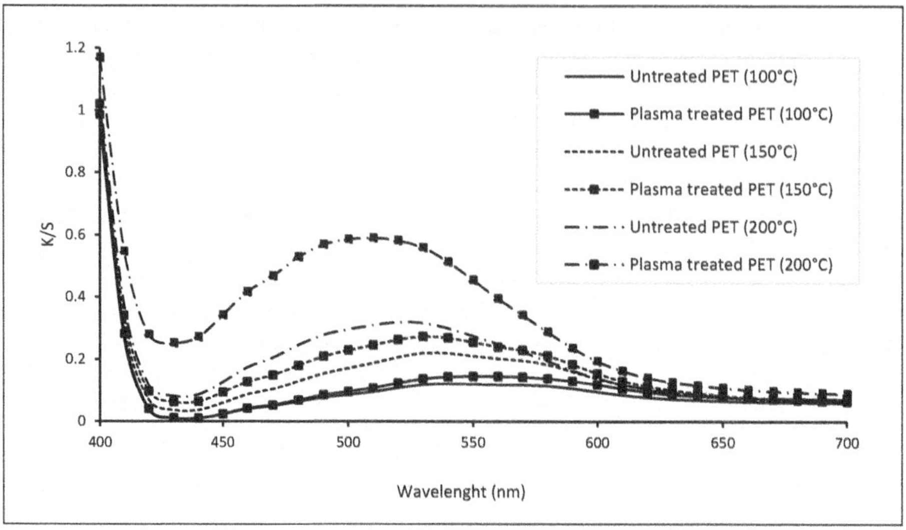

Fig. 3. Color yield results

Table 2. Washing (ISO 105:C10) and rubbing (ISO 105:X12) fastness properties

Thermal treatment temperature (°C)	Wash fastness		Rubbing fastness			
			Dry		Wet	
	U*	P.T*	U*	P.T*	U*	P.T*
100	1	1–2	2–3	3	1	1–2
150	2	2	2	3	2	2
200	3	3	3	4	2–3	3

U*: untreated sample, P.T*: Plasma treated sample

surface as the padding process relies only on surface deposition and affinity between the colorant and textile surface remain weak. Rising drying temperature on the polyester structure increase mobility of the macromolecular chain, and facilitates the relaxation and crystallization processes so better accessibility to carminic acid dye particles of the amorphous area of the fiber is assured. Once inside the fiber, dye particles stay fixed in the fine capillary spaces and can no longer be washed out.

Rubbing fastnesses was quietly greater in the case of the plasma-treated samples. This is mainly due to the modification of the surface morphology of polyester fabric after plasma treatment. In fact, plasma treatment was found to be a very effective tool to enhance the surface roughness of treated substrates (Yaman et al. 2012). Indeed, chain scission under plasma discharge can lead to low molecular weight fragments creation, which can be easily removed from surface. This material removal process is responsible for an etching effect and consequent appearance of cracks and holes. Such topography modification could provide new grooves to trap and hide dye molecules.

4 Conclusion

The investigation of a greener textile wet processing for the coloration of woven air-atmospheric plasma treated polyester fabric using carminc acid as natural dye was explored in this study through mordant free padding method.

Optimum dyeing results were reached while combining surface activation process with a thermal treatment at an elevated temperature (200 °C). Under these conditions, a greater technical reactivity between PET surface and more dye molecules is ensured, and the diffusion of the cochineal dye into the heart of the fiber is much more facilitated as macromolecular chain mobility increased under high temperature, achieving consequently acceptable wash and rub fastness properties.

References

Borcia, G., Anderson, C., Brown, N.: The surface oxidation of selected polymers using an atmospheric pressure air dielectric barrier discharge. Part I. Appl. Surf. Sci. **221**(1–4), 203–214 (2004). ISSN: 0169-4332

Hussain, T., Wahab, A.: A critical review of the current water conservation practices in textile wet processing. J. Clean. Prod. **198**, 806–819 (2018)

Iqbal, M.: Textile dyes. Rehbar, Karachi (2008)

Leroux, F., Perwuelz, A., Campagne, C., Behary, N.: Atmospheric air-plasma treatments of polyester textile structures. J. Adhes. Sci. Technol. **20**(9), 939–957 (2006)

Mongkholrattanasit, R., Kryštůfek, J., Wiener, J.: Dyeing and fastness properties of natural dyes extracted from eucalyptus leaves using padding techniques. Fibers Polym. **11**(3), 346–350 (2010). https://doi.org/10.1007/s12221-010-0346-8

Morent, R., De Geyter, N., Verschuren, J., De Clerck, K., Kiekens, P., Leys, C.: Non-thermal plasma treatment of textiles. Surf. Coat. Technol. **202**(14), 3427–3449 (2008)

Padhi, B.: Pollution due to synthetic dyes toxicity & carcinogenicity studies and remediation. Int. J. Environ. Sci. **3**(3), 940–955 (2012)

Yaman, N., Ozdogan, E., Seventekin, N.: Improvement fastnesses and color strength of pigment printed textile fabric. J. Eng. Fibers Fabr. **7**(2), 155892501200700207 (2012)

The Effect of Recycled Fibers on the Ecological Washing Performance of Denim Fabrics

Cheriaa Rim(✉) and Ben Marzoug Imed

Laboratory of Textile Engineering, University of Monastir, Monastir, Tunisia
cheriaar@yahoo.fr

Abstract. In this study, denim fabrics were produced with yarns obtained from cotton fibers (CO), recycled cotton fibers (r-CO), recycled Tencell fibers (r-Tencell) and recycled polyester (r-PET). Enzyme washing process was applied on these fabrics. A full factorial design was used to study simultaneously the effect of recycled fibers percentage in the fabric, enzyme type, enzyme concentration, treatment time and pumice stone dosage. The tear strength of treated fabrics was measured. The main goal of this work is to keep loss in fabric tear strength within tolerable limits (>2500 cN). The response surface methodology (RSM) and desirability functions were employed to obtain simultaneous optimum input parameters and to determine critical process for each composition.

Keywords: Recycled fibers · Denim fabrics · Enzyme washing · Tear strength properties · E-Flow technology

1 Introduction

Uncontrollable wastes generated by textile and apparel industry are a big load for ecology although improvement studies related to thermal efficiency, wastewater, and hazardous chemicals have been achieved (Abdurrahman et al. 2017). Sustainable procedure engaged in denim production facilitates innovative steps in denim washing named waterless washing. Enzymes itself comes from renewable energy and using a replacement of corrosive chemicals for chemical washing in one-bath enzymatic fading (Shibly et al. 2021). To address the environmental concerns, some other finishing techniques have been introduced as an alternative to the conventional wet processing. A new technology based on nano bubbles developed and patented by a Spanish company, Jeanologia, is known as E-Flow. The E-Flow ‘breaks up’ the surface of the garment, achieving soft hand feel and controlling shrinkage. A minimal quantity of water is needed and there is zero discharge from the process. Air from the atmosphere is introduced into an electro flow reactor and subjected to an electromechanical shock creating nano bubbles and a flow of wet air (Khalil 2016).

Pre-consumer and post-consumer garment wastes are an extra important problem (Dobilaite et al. 2017). Moreover, per capita fiber consumption in the world has been growing steadily depending on the increase in population and income level. The solution found or these problems is recycling applications. The recycling process starts by sorting

S. Msahli and F. Debbabi (Eds.): CIRATM 2020, SPM 17, pp. 88–95, 2022.
https://doi.org/10.1007/978-3-031-08842-1_15

on main fibre type and colour. The waste is then cut to 2–6 square-inch pieces. Subsequently, the waste is transported into the tearing machine through a take-in unit where the fabric is transformed into fibres by the mechanical action of a series of high-speed cylinders covered with, for example, saw wires or steel pins (Aronsson et al. 2020). The tearing machine generally consists of 3–6 rotating cylinders positioned one after another where the number of steel pins or saw wires usually increases with every cylinder within the machine. 8 The damage the materials suffer during the tearing process is severe. Compared to virgin fibres, the fibre length of the recycled staple fibres becomes significantly lower. According to (Gulich 2006), present technologies give between 25% and 55% of fibres longer than 10 mm in the tearing process. The negative features of yarns and fabrics produced from recycled fibers have been discussed in previous studies (Halimi et al. 2017; Hassani et al. 2010; Telli et al. 2015). It was known from these researches that recycled yarn has lower tensile strength, higher unevenness, and higher IPI values. There is a tendency to blend virgin material to compensate the losses (Esi et al. 2020).

According to (Abdurrahman et al. 2017), the negative yarn characteristics depending on recycled fiber properties can be ignored in denim sector owing to the working with higher yarn densities. Fabric appearance will not be affected negatively by the help of finishing process as long as denim fabrics produced from recycled yarns have acceptable physical and mechanical properties. Nevertheless, it is pointed out in the literature that washing process can deteriorate fabric mechanical features, especially strength and durability of the products. The negative effects of recycled fibers can be reemerged on the fabric in more abrasive washing processes in spite of the working with higher yarn weight in denim sector (Mezarcıöz et al. 2014).

In this study, 7 denim fabrics were produced with yarns obtained from recycled fibers (r-Co, r-Tencell, r-PES). And enzyme washing processes was applied systematically on these fabrics using E-Flow technology.

2 Material and Methods

2.1 Materials

The investigation has been carried out with 7 indigo dyed denim fabrics which mass area ranged from 314 g/m^2–462 g/m^2. Fabrics will be referred as Fabric 1–7 (Table 1).

3 Methods

3.1 Enzyme Treatment

Table 2 presents the enzyme washing recipe. Stone washing was combined with neutral cellulase enzyme to reduce the usage of pumice stone. A laboratory JEANOLOGIA machine (E-flow Lab), and NEF (cellulase enzyme, A); NSZ (cellulase enzyme, B), NCS (cellulase enzyme, B) were purchased.

Table 1. Fabric composition

Category 1		Category 2		Category 3		
1	2	3	4	5	6	7
90% Co/ **10% r-Co**	69% Co/ **30% r-Co/** 1% Elas.	13% Co/ **50% r-Co/ 36% r-Tencell/** 1% Elas.	16% Co/ **36% r-Co/ 48% r-Tencell**	77% Co/ **7% r-Co/ 14% r-PES/** 2% Elas.	67% Co/ **16% r-Co/ 16% r-PES/** 1% Elas.	66% Co/ **12% r-Co/ 20% R-PES/** 2% Elas.

Elas.: Elasthane, Co: Cotton

Table 2. Enzyme washing recipe.

Action	Liquor ration	Time, (min)	Temperature, (°C)	Chemical	Dosage
Rinsing	1/8	2	30	Water	--
Desize	1/8	5	40	Desizing agent	1.25 g/l
Rinsing	1/8	2	30	Water	--
Enzyme washing	**1/8**	**30-60 ♣**	**40**	**Enzyme Synthethic pumice stone**	**0.5% - 1% ♣ 100% - 200% ♣**
Rinsing	1/8	2	30	Water	-
Neutralisation	1/8	5	40	Na_2CO_3	1.5 g/l
Rinsing	1/8	2	30	Water	-
Drain	--	2	--	--	--
Drying	--	50	60	--	--

♣: See table 3

3.2 Testing and Analysis

Treated denim legs were conditioned for 24 h before testing. Tear strength was evaluated according to ASTM D 1424-1996.

3.3 Experimental Design

We established using MINITAB® 17.1.0 statistical software a mixed-level factorial design of experiments to study simultaneously the effects of 3 continuous factors and 2 categorical factors on the tear strengh loss. Factors from x (1) to x (5) with their correspondent levels were obtained based on primary tests, industrial and suppliers' recommendations, and enzymes data sheets (Table 3).

Table 3. Factors and levels related to enzymatic stone treatment for each denim category

Coded input		Uncoded input parameters	Levels		
			1	2	3
Categorical factors	x(1)	Recycled fiber percentage, %	P1♠	P2♠	-
	x(2)	Enzyme ID	A	B	C
Continuous factors	x(3)	Treatment time, min	30	60	-
	x(4)	Enzyme concentration, %	0.5	1	-
	x(5)	Pumice stone, %	100	200	-

♠ See details in table 1.

4 Results and Discussions

In biostoning finishing, cellulase is used to accelerate the abrasion; in this process the cellulase works by lessening the indigo dye on the denim surface. Cellulase tends to cause catalytic hydrolysis of 1, 4-beta glycoside bonds of cellulose molecules (Rahman et al. 2012).

Table 4 resumes ANOVA analysis used to determine the significance of each variable and interactions on the response.

Table 4. Summary of variance analysis results

Fabric category	Direction	Significant factors (p-value < 0,05)	Significant interactions (p-value < 0,05)	R-sq, (%)
1	Warp	- Recycled cotton % - Time - Enzyme concentration	–	90.32
	Weft	- Recycled cotton %	–	88.93
2	Warp	- Recycled Tencell % - Time Enzyme concentration	- Time * Enzyme type - Enzyme concentration * Pumice stone %	83.10
	Weft	- Recycled Tencell % - Enzyme type	- Recycled Tencell % * Enzyme type	62.29
3	Warp	- Recycled PET % - Time	- Enzyme concentration * Enzyme type	97.98
	Weft	- Recycled PET % - Enzyme type	- Enzyme type * Enzyme concentration - Time * Enzyme concentration	95.19

– No significant effect

From Table 4, the p-values of the variables: recycled fibers %; treatment time; enzyme concentration and enzyme type had statistically significant effects on tear strength loss (p-value < 0.05). Of the all two-way interactions, only interactions of enzyme type with treatment time, recycled fiber %, and enzyme concentration were statistically significant. The R-sq values are acceptable (>80%) expect for category 2 (62.29%), an indication that the models explained more than 80% of the variance in tear strength loss (%) after treatment.

Table 5. Summary of main effects plots

Category	Response: tear strengh loss, %	Factors				
		Recycled fibers, %	Enzyme type	Time, min	Enzyme concentration, %	Pumice stone %
1 (r-Co)	Warp	++	Ø . +	+	+	Ø
	Weft	++	− . +	+	+	Ø
2 (r-Tencell)	Warp	++	+ . +	+	+	+
	Weft	−−	++ . −	+	+	Ø
3 (r-PES)	Warp	++ . −−	Ø . Ø	Ø	Ø	Ø
	Weft	++ . −−	Ø . +	Ø	Ø	Ø

++ Highly positif significant effect, + positif significant effect, no significant effect,
−− Highly negatif significant effect, − negatif significant effect
Ø no significant effect

It can be seen from Table 5 that the recycled fibers percentage affects the fabric tear strength loss to a greater extent. Pumice stone dosage has a negligible effect on the fabric strength. Treatment time and enzyme concentration enhance tear strength loss of only fabrics of category 1 and 2. Tested enzymes had different effects on tested fabrics. Previous studies claims that tear strength of fabrics decreased after denim washing process (Mezarcıöz et al. 2014). However, tear strength results show increase with enzyme washing in some cases. In fact, tear strength is related to strengh and mobility of yarns within the fabric structure. Normally, denim fabrics are stiff, hard-wearing, and uncomfortable to wear without washing. The purpose of denim washing processes is to provide soft handle besides dyeing effects and desirable appearance. Washing treatments have been successful to improve fabric flexibility. In this study, higher tear strength was obtained thanks to the increased mobility of yarns within the fabric structure and the grouped yarns against tear depending on the more flexible and soft fabric structure.

The regression and variance analysis obtained using "stepwise option" show adequate regression models (high coefficient of determination $R^2 > 84\%$) and highly meaningful at 95% significance level.

The optimized process obtained using regression models and desirability functions to minimize tear strength loss in warp and in weft direction is the same for all fabrics: time: 30 min, enzyme 0,5%, and 100% pumice stone.

The critical process obtained from maximizing tear strength loss in warp and in weft direction is the same for all fabrics: time: 60 min, enzyme: 1%, and 100% pumice stone.

So, regardless of the enzyme used NEF, NEF or NCS, the tear strength of fabrics from category 1, 2 and 3 remains acceptable and > 2500 cN under optimal and critical recipe except for fabric 5 which have an initial tear strength equal to 2308 cN.

Figure 1 and 2 show validation results. It is clear that NEF is the best enzyme which allows the minimum tear strength loss under the critical washing recipe.

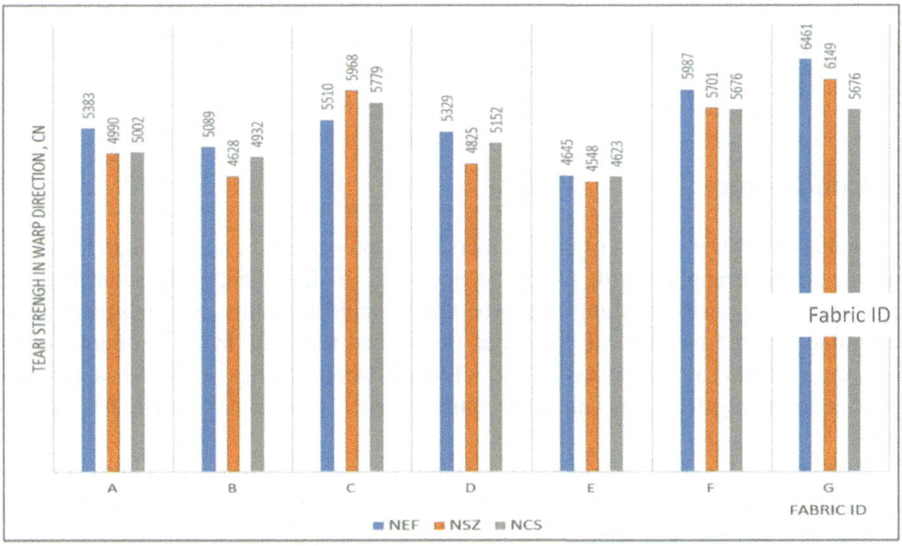

Fig. 1. Fabrics tear strength in warp direction under critical recipe

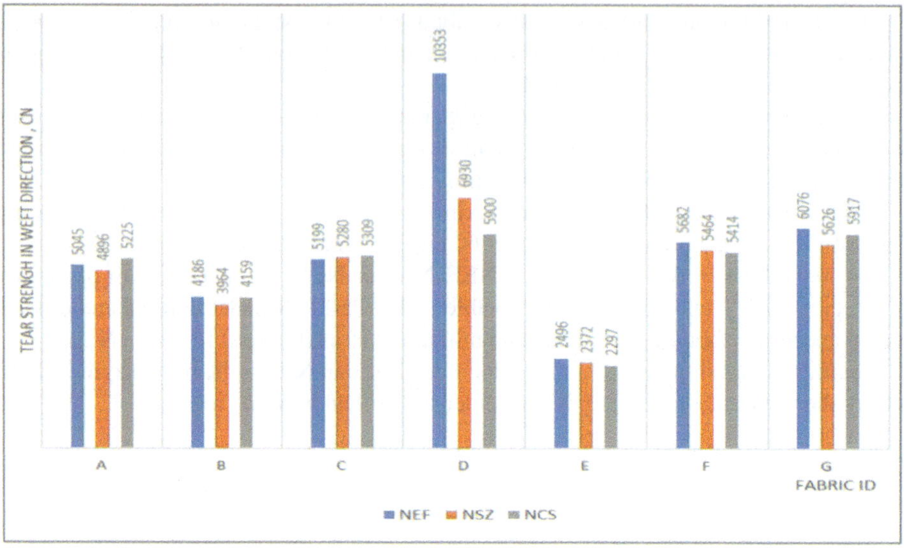

Fig. 2. Fabrics tear strength in weft direction under critical recipe

5 Conclusion

In conclusion, a clear benefit of r-PET fibers in the prevention of tear strength loss after enzyme washing applied with E-flow technolwogy is identified, r-PET contribute to resistance for agressive washing. It can be also concluded that the washing parameters had influenced the mechanical properties of finished recycled denim garments, especially denim fabrics made from r-Co and r-Tencell. Denim fabrics produced using yarns obtained from recycled fibers can show similar fabric performance to original denim. In fact:

Tear strength loss of denim fabrics containing r-Co fibers was able to achieve 35.45% and 24.46% (warp and weft direction respectively) under proposed critical recipe.

Tear strength loss of denim fabrics containing r-Tencell fibers was able to achieve 20.62% and 52.49% (warp and weft direction respectively) under proposed critical recipe.

Tear strength loss of denim fabrics containing r-PES fibers was able to achieve 41.35% and 27.56% (warp and weft direction respectively) under proposed critical recipe.

Tear strength loss of 100% cotton denim fabrics was able to achieve 29% and 21,8% (warp and weft direction respectively) under proposed critical recipe.

Tear strength loss of cotton/Elasthane denim fabrics was able to achieve 37% and 29% (warp and weft direction respectively) under proposed critical recipe.

References

Abdurrahman, T., Osman, B.: The effect of recycled fibers on the washing performance of denim fabrics. J. Text. Inst. **108**(5), 812–820 (2017)

Aronsson, J., Persson, A.: Tearing of post-consumer cotton T-shirts and jeans of varying degree of wear. J. Eng. Fibers Fabr. **15** (2020)

Dobilaite, V., Mileriene, G., Juciene, M., Saceviciene, V.: Investigation of current state of pre-consumer textile waste generated at Lithuanian enterprises. Int. J. Clothing Sci. Technol. **29**(4), 491–550 (2017)

Esi, B., Bayka, P.D.: Investigation of tensile strength and elongation properties of chenille upholstery fabrics including recycling polyester yarns. J. Eng. Fibers Fabr. **15**, 1–10 (2020)

Gulich, B.: Development of products made of reclaimed fibres. In: Wang, Y. (ed.) Recycling in Textiles, vol. 117, Woodhead Publishing, Cambridge (2006)

Halimi, M.T., Hassen, M.B., Azzouz, B., Sakli, F.: Effect of cotton waste and spinning parameters on rotor yarn quality. J. Text. Inst. **98**(5), 437–442 (2007)

Hasani, H., Semnani, D., Tabatabei, S.: Determining the optimum spinning conditions to produce the rotor yarns from cotton wastes. Ind. Text. **61**, 59–64 (2010)

Khalil, E.: Nano bubble technology: a new way to sustainable jeans finishing. In: 56th Convention of Institution of Engineers, Bangladesh (IEB) (2016)

Mezarcıöz, S., Toksöz, M.: Investigatıon of effect of special washing processes on denim fabrics'17 properties. Tekstil Ve Konfeksiyon **24**(1), 68–95 (2014)

Khan, M.M.R., Mondal, M.I.H., Uddin, M.Z.: Sustainable washing for denim garments by enzymatic treatment. J. Chem. Eng. **27**(1), 27–31 (2012)

Shibly, M.A.H., Hoque, M.M., Miah, S.: Development of eco-friendly denim fabric washing by natural resources. Int. J. Text. Sci. **10**(1), 1–6 (2021)

Telli, A., Özdil, N.: Effect of recycled PET fibers on the performance properties of knitted fabrics. J. Eng. Fibers Fabr. **10**, 47–60 (2015)

Effect of Finishing Products on the Mechanical Properties of Sewing Thread

Samar Mansouri[1,3(✉)], Chaabouni Yassin[2], and Cheikhrouhou Morched[2]

[1] Laboratoire de Chimie de l'environnement et des Procédés Propres, Faculté des Sciences, Université de Monastir, 5019 Monastir, Tunisia
Samar20481@yahoo.fr

[2] Higher Institute of Arts and Crafts of Sfax, 3069 Sfax, Tunisia

[3] Higher Institute of Arts and Crafts of Kasserine, 1200 Kasserine, Tunisia

Abstract. This research is an opportunity to highlight the influence of finishing products on sewing-thread's behavior. After sewing, the needle thread was removed then tensile properties (force and elongation at break) are determined. The results show a significant loss in mechanical properties values depending on finishing treatment and especially on products' concentration used. On the other hand, thread's diameter changes after sewing of woven fabrics finished with different concentration of various products.

Keywords: Finishing products · Sewing thread · Seaming · Tensile properties

1 Introduction

In the textile field, finishing process is occurred either to confer more pleasing appearance and handle to the fabric or to give it new properties for a particular end use. There are different ways to finish a textile fabric. In that, simple physical or mechanical treatments can change the appearance and properties of textile fabrics significantly. These are called dry treatments in which little or no water is used. On the other hand, chemical finishes are known as wet treatments. All of them have an important role in textile manufacturing. They are responsible for functionalizing textile fabrics which can create new conceptual textile systems for the 21st century (Nnamdi et al. 2019). So, finishing stages affect fibers, yarns and fabric's behavior (Sameii et al. 2008; Shakyawar et al. 2009). Also, they have an important influence on needle penetration force and seams mechanical behavior (Mansouri et al. 2014; Kordoghli et al. 2009).

Sewing process is used to transform bi-dimensional structure on three-dimensional one. It involves different elements. It is an assembly of woven fabric by sewing threads using the appropriate machine. So, all of these elements have an important effect on seamed product, which can be investigated. It is very important and complex, in that, it requires great efforts in order to achieve good quality to satisfy consumers' needs. Its importance can be recently justified by the large number of research studies. Indeed, several studies aim to investigate seaming operation and its environment (Evangelos et al. 1999; Hersh and Grady 1969; Sundaresan et al. 1998; Kawabata and Masako

S. Msahli and F. Debbabi (Eds.): CIRATM 2020, SPM 17, pp. 96–104, 2022.
https://doi.org/10.1007/978-3-031-08842-1_16

1998; Dorkin and Chamberlain 1963; Nikolić et al. 2003; Kordoghli et al. 2009; Gribaa et al. 2006; Mukhopadhyay et al. 2013).

Sewing thread as an important component of a sewn garment contributes significantly to the quality of clothes (Malek et al. 2020). An important consideration is given to thread's behavior during sewing steps (Mazari et al. 2016; Gurarda et al. 2005).

To the best of our knowledge, none of the past studies on sewing thread properties have considered the effects of finishing products. This study is devoted to highlight the modification occurred on the mechanical behavior of needle thread. In that, the lockstitch is formed by the interlacement of needle and bobbin threads. During seaming process, needle thread is subjected to repeated tensile stresses, heat, bending, pressure and torsion. Therefore, its mechanical properties change with variation of woven fabrics' properties. This paper focuses on the evaluation of mechanical properties of sewing thread, when changing the concentration of each finishing product.

2 Materials and Methods

We have adopted the following strategy: We carry out the assembly of two fabric inserts with a speed of 1000 trs/min and a point density equal to 5 points/cm. The sewn length is such that the length of the line makes it possible to carry out the tensile tests later. Then we destroy the bobbin thread to recover the needle one without exerting stress. After that, thread's properties were determined.

To measure the breaking strength and elongation of studied yarn we used the standard NFG 07-003. The distance between clamps is 500 mm, the thread's rupture must be after 20 s $\pm$ 3 s. The pre-load to be applied to the sample before tightening clamps is equal to 0.5 cN $\pm$ 0.1 cN per tex of linear density of the yarn.

Characteristics of Fabrics and Sewing Threads Used

To achieve the main objectives of the present investigation, DENIM fabric released in SITEX (Industrial Company of Textile) was used. The fabric is composed of 95% Cotton and 5% Elastane (C95E5). Its characteristics are presented in the following Table 1.

Table 1. Fabric's characteristics

Fabric code	Weave	Density (thread cm^{-1})		Thickness (mm)	Weight ($g\ m^{-2}$)
		Warp	Weft		
C95E5	Twill 4	26	17	0.720	294

Grey fabric was mercerized and then impregnated in a bath containing resins for stiffening, anti-wrinkling treatment and softening. Next, it was softened by fabric softener. These various treatments were carried out with three different concentrations of finishing products. Mercerizing and softening were applied in the Industrial Company of Textile "SITEX". Stiffening, anti-wrinkling treatment and softening using resin were done in the finishing factory in the same company. In addition, water-repellent treatment was carried out on non-treated fabric using an adhesive resin. To ensure the reaction of resin's polymerization, all last resin treatments were done at 170 °C for 1–2 min. Then samples were cooled at room temperature. Finally, the different tests were released. Each indicated value presents the average of 5 tests and the difference among values is less than 5% (Table 2).

Table 2. Different concentrations of used products

Treatments	Products	Concentrations (g. L-1)		
		C1	C2	C3
Mercerizing	Sodium hydroxide	140	160	180
Stiffening	*Rhenappret RBA (Thor SARL France)	40	60	80
Anti-Wrinkling treatment	*Quecodur LT (Thor SARL France)	5	10	20
Softening by resin	*Finistrol (Thor SARL France)	20	25	35
Softening by polishing substance	*Appreture 166 (Cognis)	10	15	20
Water repellent	Adhesive resin	20	35	50

* Industrial designation

Seaming operation was released using a needle thread presenting the following characteristics mentioned in the table below (Table 3).

Table 3. Characteristics of sewing thread used

Thread code	Fibres used	Linear density (tex)	Twisting (twist/m)	Tenacity (cN/tex)	Elongation (%)
S45	PES	45	438	47.13	27.13

3 Results and Discussions

In this part, we present the modifications of thread's properties in terms of resistance and elongation at breaking point, after assembly. This was done for the same fabric when varying concentrations of each product.

3.1 Influence of Mercerization

The fabric made from cotton wrinkles easily and is difficult to dye. It is, therefore, treated with sodium hydroxide to make it strong, lustrous and absorbent (Jordanov et al. 2010).

This treatment affects fabric's properties such as the resistance to needle penetration during seaming process (Mansouri et al. 2014).

First of all, we present the sudden variations noted in the thread's mechanical behavior. The Fig. 1 shows that the shape of the force/elongation curve remains unchanged. The change appears at the breaking point. Indeed, we notice a decrease in elongation. This resulting decrease is detected in the thread used to seam mercerized fabric with a NaOH concentration of 140 g/l, is 8% compared to a seamless yarn.

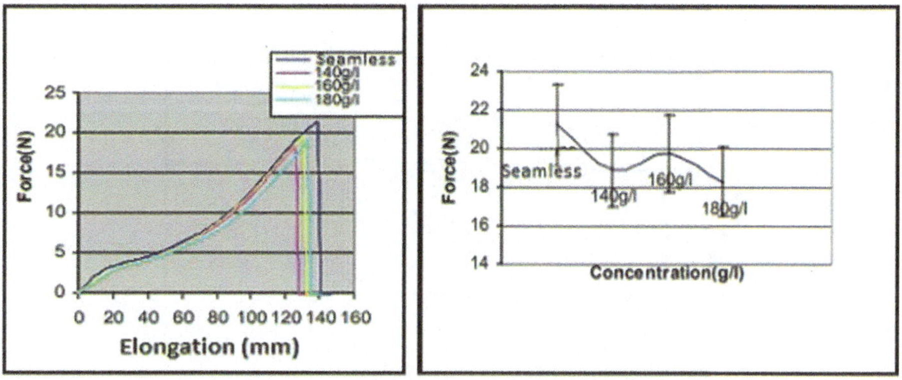

Fig. 1. Variation of the mechanical properties of the tested thread with NaOH concentration

By increasing the concentration of soda to 160 g/l and 180 g/l, we notice a low recovery at the level of elongation about 2.2%. As regards the variation of the breaking force, it follows the elongation's evolution. Indeed, the thread used to sew the mercerized fabric with 140 g/l of soda, loses 10.94% of its initial strength. At a concentration of 160 g/l, the strength improves by 4.5% compared to that of the assembly yarn used for fabric treated in a bath containing a concentration of soda equal to 140 g/l. The thread, frayed from the fabric treated in a bath containing a high quantity of soda, loses 13.7% of its initial strength. These results are justified by the fact that the swelling of cotton fibers, caused by soda even at low concentrations, causes the reduction of elongation and breaking force. In fact, significant friction with the fabric weakens the mechanical characteristics of the sewing thread. The improvement, even of small percentage, of the mechanical behavior results from the increase of the concentration of soda. Indeed, at high concentrations, the soda destroys the structure of the cotton fiber. So the structure of the warp and weft yarns will be modified. As a result, the interlacing of the sewing thread becomes easier. This keeps the mechanical properties after sewing. The decrease in strength and elongation of the thread used to assemble a mercerized fabric with 180 g/l of sodium hydroxide is linked to the narrowing of the mono filaments of elastane. Indeed, the higher the concentration of soda, the greater the shrinkage. This leads to the destruction of the mechanical properties of the sewing thread.

3.2 Influence of Stiffening

The purpose of the stiffening treatment is to improve the self-maintenance of the fabric to facilitate its handling during assembly operation. As a result, the load added to the fabric generates changes in thread's behavior inside the fabric. This generally affects the characteristics of the sewing thread. In our study, we used a flexible self-restretchable polyacrylic thermoplastic resin (Rhenapprêt). Figure 2 shows the evolution of the mechanical behavior (force/elongation) of the sewing thread according to different concentrations of the resin acquiring stiffness to the treated fabric.

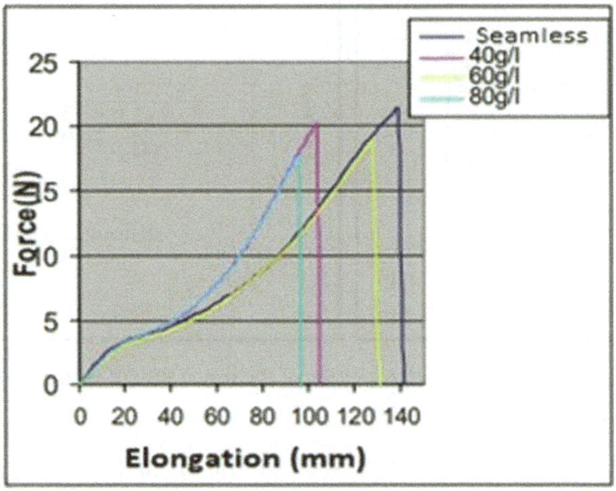

Fig. 2. Evolution of the thread's mechanical behavior at different concentration of Rhenappret RBA

We note that the shape of the force/elongation curves of the different concentrations is the same in the elastic zone. Therefore, with low tensile stresses, the concentration of the resin has no influence on the mechanical behavior of the sewing thread. At rupture's zone, we observe variations in strength as well as elongation of the assembling yarn. We notice a loss of the mechanical properties of the sewing thread. This is proven when comparing the values found for a yarn used in a stiffened fabric assembly by 40 g/l of the resin with the results of the seamless yarn. The highest reduction in elongation is noticed for the concentration of 80 g/l. The percentage decrease is equal to 32%. For a concentration of 60 g/l, the loss in elongation is minimal. The variation in resistance is linear. This is given by Fig. 3. Indeed, the breaking force of the sewing thread is inversely proportional to the concentration. The percentage decrease is of about 6%.

This variation is produced by the amount of resin. Indeed, the latter is self-restretchable, so its polycondensation causes the cross linking of the cellulosic chains of cotton fibers. This reduces the deformability of warp and weft yarns. As a result, the insertion of the sewing thread becomes more difficult due to the rigidity of the fabric threads. Thus, friction increases, which destroys the sewing thread's structure.

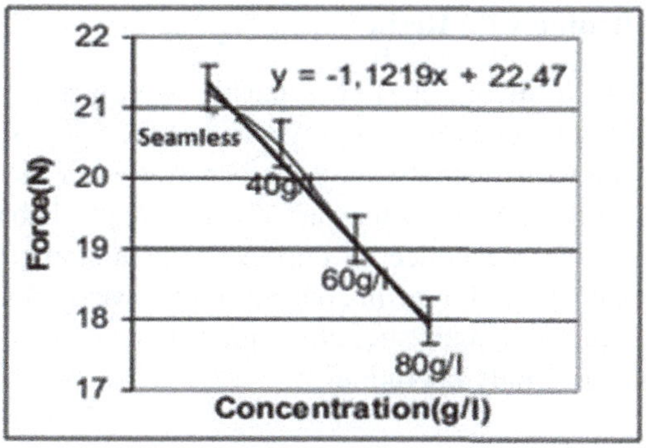

Fig. 3. Variation of the breaking force of the thread

3.3 Influence of Wrinkle-Free Treatment

We used a «wash and wear» treatment. We used the most usable resin in the textile field. It is characterized by its smoothing and irreversible efficiency Michallet (1991). Figure 4 shows the evolution of the mechanical behavior of yarn took of the treated fabric. We notice a loss in its strength and elongation of break. The percentage of decrease in the strength is 14.3%. In the same way, the breaking elongation change. These results are justified by the very principle of smoothing treatment. Indeed, the product used is a reactive resin that is fixed by a chemical bond thanks to the OH groups of cellulose. As a result, the internal interstices of the cellulosic structure will be filled. Subsequently, the cotton fibers swell. This favors the smoothness of the textile material. But in the other hand, the interlacing of the sewing thread through this fabric causes its destruction.

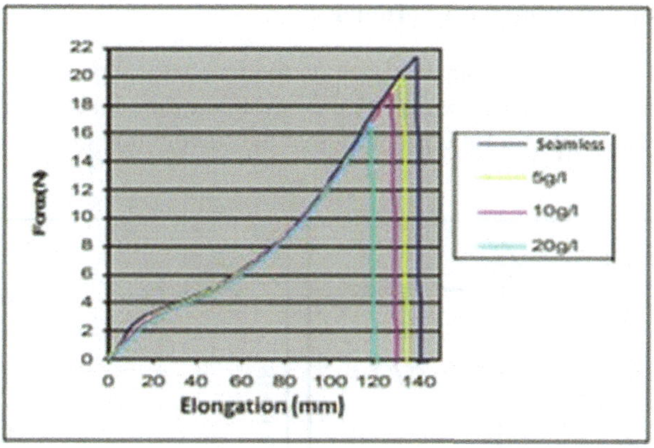

Fig. 4. Variation of the mechanical properties of the sewing thread after "wash and wear" treatment

3.4 Influence of Softening with Resin

This treatment is done by a silicone elastomer that allows easy penetration into the fiber. This product allows a soft touch and an improvement of the fiber's elasticity. Therefore, the deformability of the yarns increases. So the penetration of the sewing thread into the fabric will be easy. Figure 5 shows the variation of thread's behavior with softening resin's concentration. Thus, we notice an improvement in strength and elongation at high resin concentration. With high resin concentrations, we detected 10% of decrease by comparing the pre- and post-seam strength of the softened fabric with 20 g/l of resin. This may be due to the assembly operation.

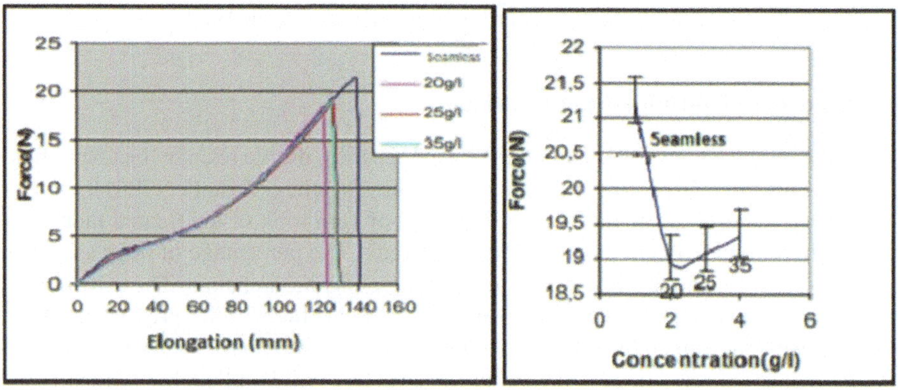

Fig. 5. Variation of the mechanical properties of sewing thread with the resin concentration

3.5 Influence of Softening with Softener

The influence of the softener concentration on the mechanical properties of the sewing thread is given in Fig. 6. We notice that by increasing the amount of softener in the bath, we will have a conservation of the properties of the sewing thread.

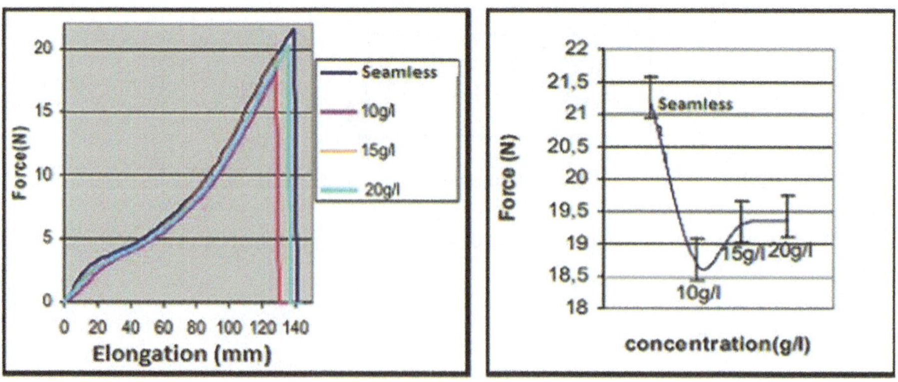

Fig. 6. Variation of the thread's mechanical behavior with the softener concentration

This result is justified by the role of the lubricant that the softener plays. Indeed, the decrease in the coefficient of friction between the sewing thread and the warp and weft threads guarantees us the protection of the structure of the sewing thread after interlacing through the fabric.

3.6 Influence of Water Repellent Treatment

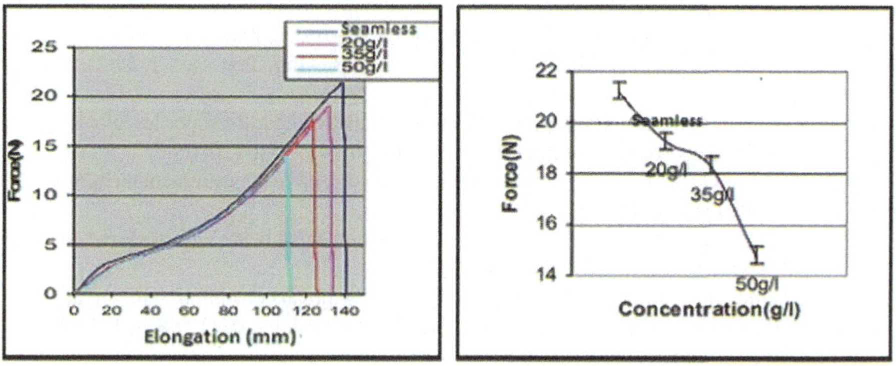

Fig. 7. Variation of thread's behavior after water-repellency treatment

Figure 7 shows the evolution of sewing thread's behavior when varying the concentration of the water repellent product after seaming operation. We find that when the concentration increases, we notice a loss in strength that can reach 30% of the initial thread's strength before sewing. This results from the blocking of the yarns inside the fabric by the resin. Therefore, their rigidity increases. As a result, there is a high friction coefficient, which destroys the thread's structure and alters its mechanical characteristics.

4 Conclusion

After different tests of extracted sewing thread from treated fabrics, it is clear that according to the properties wanted (softness, stiffness, non-wrinkling…) the finishing products used affect the sewing thread's behavior differently. So that stiffening, waterproofing and anti-wrinkling treatment which increase the fabric's resistance to needle penetration because of resins used when making film on the surface (Mansouri et al. 2014) also have an effect on sewing thread's characteristics. They weaken its mechanical resistance. But, concerning the treatment of softening, the change in thread's properties is lower; this is due to lubrication produced by products responsible for this treatment either resin or polishing substance. On the other hand, all finishing products must be applied with optimum quantities.

References

Dorkin, C.M.C., Chamberlain, N.H.: The facts about needle heating. Technological report no. 13, Clothing institute (1963)

Evangelos, L., Ruxu, D., Dan, S., Jasmina, B., Frank, L.: An experimental study of needle heating in sewing heavy materials using infrared radiometry. Int. J. Clothing Sci. Technol. **11**(5), 300–314 (1999)

Gribaa, S., Amar, S.B., Dogui, A.: Influence of sewing parameters upon the tensile behavior of textile assembly. Int. J. Clothing Sci. Technol. **18**(4), 235–246 (2006)

Gurarda, A., Meric, B.: Sewing needle penetration forces and elastane fiber damage during sewing of cotton/elastane woven fabrics. Text. Res. J. **75**(8), 628–633 (2005)

Hersh, S.P., Grady, P.L.: Needle heating during high speed sewing. Text. Res. J. **39**(2), 101–120 (1969)

Jordanov, I., Mangovska, B., Taveer, P.F.: Mechanical and structural properties of mercerized cotton yarns, bio-scoured with pectinases. Tekstil **10**(10), 439–446 (2010)

Kawabata, S., Masako, N.: Clothing engineering based on objective measurement technology. Int. J. Clothing Sci. Technol. **10**(3/4), 263–272 (1998)

Kordoghli, B., Cheikhrouhou, M., Kacem Saidene, C.: Mechanical behaviour of seams on treated fabrics. AUTEX Res. J. **9**(3), 88–92 (2009)

Malek, S., Khedher, F., Adolphe, D.C., Jaouachi, B.: Sewing thread consumption for chain stitches of class 400 using geometrical and multilinear regression models. AUTEX Res. J. **20**(4), 415–425 (2020)

Mansouri, S., Chaabouni, Y., Cheikhrouhou, M.: Influence of finishing products on sewing needle penetration force. Int. J. Appl. Res. Text. **2**(2), 33–41 (2014)

Mazari, A., Akagun, E., Havelka, A., Mazari, F.B., Kejzlar, P.: Effect of sewing speed on the physical properties of firefighter sewing threads. Int. J. Mater. Metall. Eng. **10**(3), 312–315 (2016)

Michallet, G.: Les apprêts chimiques. Ind. Text. (1218), 66–73 (1991)

Mukhopadhyay, A., Midha, V.K.: The quality and performance of sewn seams. In: Jones, I., Stylios, G.K. (eds.) Joining Textiles Principles and Applications, pp. 175–207. Woodhead Publishing, United Kingdom (2013). ISBN: 978-1-84569-627-6

Nikolić, G., Šomođi, Z., Šarić, D.F.: Mechanical properties of sewing stitch performed in frozen state. Int. J. Clothing Sci. Technol. **15**(3/4), 198–203 (2003)

Iheaturu, N.C., Aharanwa, B.C., Chike, K.O., Ezeamaku, U.L., Nnorom, O.O., Chima, C.C.: Advancements in textile finishing. J. Polym. Text. Eng. **6**(5), 23–31 (2019). p-ISSN: 2348-0181

Sameii, N., Mortazavi, S.M., Rashidi, A.S., Sheikhzadah-Najar, S.: An investigation on the effect of hot mercerization on cotton fabrics made up of open-end yarns. J. Appl. Sci. **8**(22), 4204–4209 (2008)

Shakyawar, D.B., Behera, B.K.: Influence of softening treatments on hand value of woven fabrics produced from Indian wool and their blends. Indian J. Fibre Text. Res. **34**, 76–81 (2009)

Sundaresan, G., Salhotra, K.R., Hari, P.K.: Strength reduction in sewing threads during high speed sewing in industrial lockstitch machine, part II: effect of thread and fabric properties. Int. J. Clothing Sci. Technol. **10**(1), 64–79 (1998)

Decolorization of Reactive Black 5 by Laccase

Echhida Sayahi(✉) and Neji Ladhari

Textile Engineering Laboratory, University of Monastir, Monastir, Tunisia
chahida.sayahi@yahoo.fr

Abstract. The decolorization of Reactive black 5 (RB5), an azo dye extensively used in textile industry was studied using purified laccase from a white rot fungus *Trametes trogii.* We observed that the presence of 1-hydroxybenzotriazole (HBT) is important and enhance the decolorization of RB5. We have studied the effects of three variables dye (25; 50; 100 mg/L), enzyme (0.1; 0.5; 1 U/mL) and redox mediator (0.1; 0.5; 1 mM) concentrations. The contour plots methodology was used as an efficient experimental strategy to determine the optimal conditions and the correlation between the effects of these variables on the decolorization of RB5. The results clearly indicated that the HBT concentration was the main factor influencing the RB5 decolorization yield. The selected optimal conditions were enzyme concentration 1U/ml, mediator concentration 0.7 mM, and dye concentration 25 mg/l. These conditions allowed more than 92% of RB5 decolorization.

Keywords: Laccase · Reactive black 5 · HBT concentration · Enzyme concentration · Dye concentration

1 Introduction

Large numbers of chemically different dyes are used for various industrial applications and significant proportion appears in the form of wastewater and is spilled into the environment.

The majority of physicochemical dye removal methods have many limitations. Biological degradation is an environmental-friendly and cost-competitive alternative to the physicochemical decomposition process.

White rot fungi was the most used in the dye biodegradation. Its dye degrading ability is due to its lignolytic enzyme system consisting of lignin peroxidase, manganese-dependent peroxidase and laccase.

Many studies have interested to the capacity of laccase to decolorize textile dyes (Zille et al. 2003; Abadulla et al. 2000; Soares 2001). This enzyme is not able to oxidize all dyes because some of them are too large to penetrate in the enzyme active site or presented a high redox potential. As a solution, redox mediators have been used to enhance the dyes degradation by laccase. It has been shown that the efficiency of the dye decolorization by laccase depends on many factors such as the reaction time, the concentration of the enzyme and the concentration of the dye, and the redox mediator (Murugesan et al. 2007; Tavares et al. 2009).

S. Msahli and F. Debbabi (Eds.): CIRATM 2020, SPM 17, pp. 105–109, 2022.
https://doi.org/10.1007/978-3-031-08842-1_17

2 Materials and Method

2.1 Materials

Reactive Black 5 "RB5" (Primazin Noir BN) was supplied by the Higher Institute of Technological Studies (I.S.E.T.) Ksar-Hellal – Tunisia. 1-hydroxybenzotriazole (HBT) and laccase of 'Trametes trogii' were provided by Engineers National School, Sfax, Tunisia.

2.2 Methods

The reaction mixture for RB5 decolorization experiments contained laccase (0.1; 0.5; 1 U/ml), HBT (0.1; 0.5; 1 mM) and RB5 dye (25; 50; 100 mg/l). All the reactions were conducted at pH 5 and ambient temperature and the residual dye concentration was determined by monitoring the decrease in absorbance at 596 nm using a UV–Vis spectrophotometer (Shimadzu UV-256). Dye decolorization was expressed in terms of percentage.

Contour plots methodology was used to determinate the optimum HBT, enzyme and dye concentration in order to obtain maximum RB5 decolorization.

3 Results and Discussion

The reactive dye RB5 is widely used for textile dyeing process which is resistant to biodegradation. In the present study, we have tested the ability of purified laccase from *Trametes trogii* to decolorize the azo dye RB5. We have obtained a partial decolorization. Only 49% of RB5 was removed after 24 h. However, in the presence of 0.4 mM HBT, a decolorization percentage of about 85% of RB5 was obtained in 24 h. The absorbance spectrum of the RB5 and its decolourization by laccase is shown in Fig. 1 which revealed that HBT is essential for the decolourization of RB5. This result shows consistency with the results of other researchers reported for RB5 decolourization by laccase (Ramsay and Nguyen 2002; Mazmanci and Unyayar 2005).

Although the important role of laccase-mediator systems on dye degradation, the mediator inhibits the laccase activity over certain concentration (Garcia et al. 2003). Therefore optimization of HBT concentration is essential for the successful decolorization. In the present study we have controlled the combined effects of enzyme, dye, and HBT concentrations at various levels for the decolorization of RB5 using contour plots.

The effect of the studied variables on RB5 removal was observed through Fig. 2, 3 and 4. Figure 2 shows the interactions between concentrations of enzyme and HBT on dye decolorization at 100 mg/l of RB5 using the contour plots. The presented figure demonstrates an increase of the decolorization with increasing the HBT concentration even at low enzyme concentration. However a slight augmentation was observed with increasing enzyme concentration and keeping all other factors constant. The contour plot shows the best decolorization (>90%) obtained with an enzyme concentration more than 0.9 U/ml and a HBT concentration in the range of 0.75–1 mM. Murugesan (2007) had also reported that increasing HBT concentration from 0.5 to 1.5 mM increased the rate reactive black 5 decolorization.

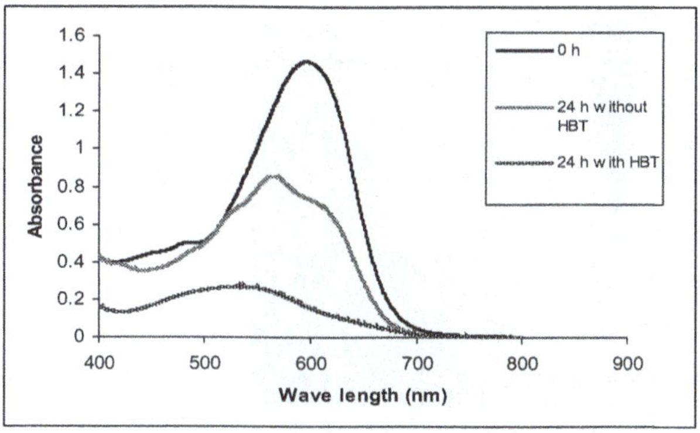

Fig. 1. UV-vis absorption spectrum of RB-5 treated by purified laccase with and without 0.4 mM HBT incubated for 24 h.

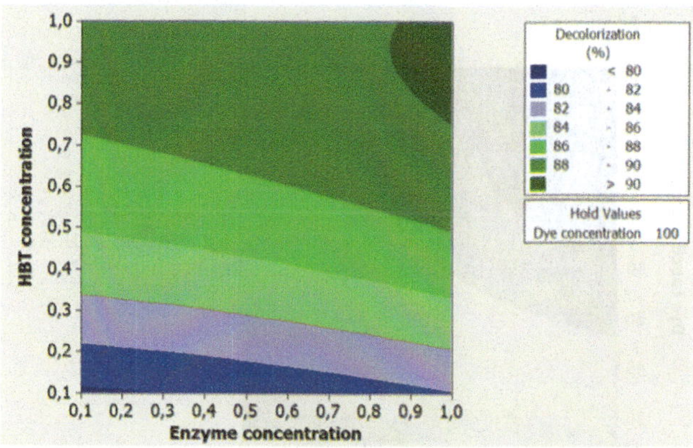

Fig. 2. Contour plots for the effect of enzyme and HBT concentrations at constant dye concentration (100 mg/l) on the decolorization of RB5.

Figure 3 represents the effect of varying dye and HBT concentration at fixed concentration of enzyme. The contour plots of Fig. 3 also support the important role of HBT concentration. It can be seen from Fig. 3 that increasing the dye concentration, keeping all other factors constant, decreases the decolorization. Experimentations conducted with a mediator concentration more than 0.7 mM and dye concentration low than 30 mg/l led to relatively high decolorization yields (>92%) as shown in Fig. 3.

Figure 4 represent the interaction between the dye and enzyme concentration. The positive effect of enzyme concentration was also demonstrated in this contour plots. However, the dye concentration presents a negative effect on RB5 degradation. At a HBT concentration of 1 mM, high decolorization yields (>91.5%) can be reached when using a relatively low dye concentration (<30 mg/l) and an enzyme concentration more than 0.8 U/ml.

The results presented above showed that the concentration of the HBT mediator was the more relevant factor for the RB5 decolorization.

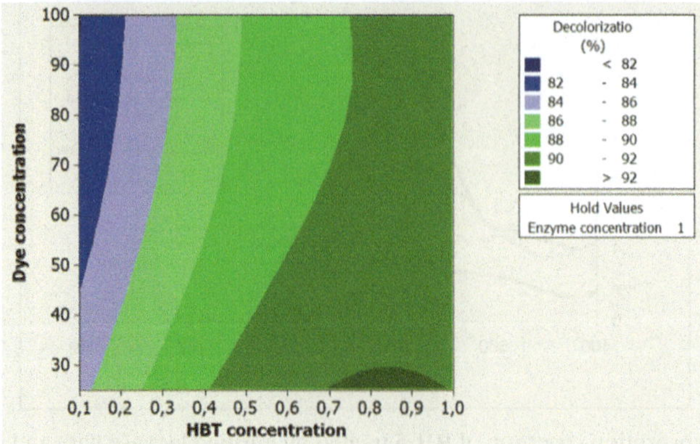

Fig. 3. Contour plots for the effect of dye and HBT concentrations at constant enzyme concentration (1 U/ml) on the decolorization of RB5.

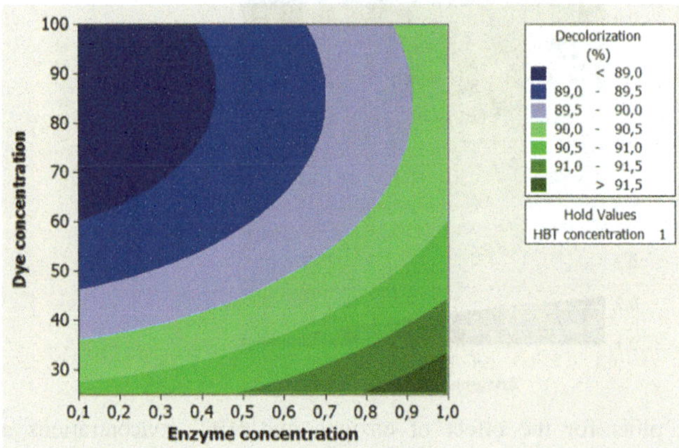

Fig. 4. Contour plots for the effect of enzyme and dye concentrations at constant HBT concentration (1 mM) on the decolorization of RB5.

The optimum concentrations of dye, enzyme and HBT were found to be 25 mg/l, 1 U/ml and 0.7 mM, respectively, for maximum decolorization of RB5 more than 92%

4 Conclusion

This work revealed that the contour plots were a useful tool to determine the optimal experimental conditions for the decolorization of the textile dye, Reactive Black 5. The presence of HBT mediator enhanced the decolorization of RB5 by *Trametes trogii* laccase.

The concentration of the HBT proved to be the principal factor that affected the yield of the dye decolorization.

The selected optimal conditions (enzyme concentration 1 Um/l, mediator concentration 0.7 μM, and dye concentration 25 mg/l) were checked. Maximum decolorization obtained at these conditions was more than 92%.

References

Abadulla, E., et al.: Decolorization and detoxification of textile dyes with a laccase from Trametes hirsuta. Appl. Environ. Microbiol. **66**, 3357–3362 (2000)

Boaventura, R.A.R., Macedo, E.A.: Application of statistical experimental methodology to optimize reactive dye decolorization by commercial laccase. J. Hazard. Mater. **162**, 1255–1260 (2009)

Garcia, O., et al.: Optimization of a laccase-mediator stage for TCF bleaching of flax pulp. Holzforschung **57**, 513–519 (2003)

Mazmanci, M.A., Unyayar, A.: Decolorisation of Reactive Black 5 by Funalia trogii immobilised on Luffa cylindrica sponge. Process Biochem. **40**, 337–342 (2005)

Murugesan, K., et al.: Decolorization of reactive black 5 by laccase: optimization by response surface methodology. Dyes Pigm. **75**, 176–184 (2007)

Soares, G.M.B., et al.: Use of laccase together with redox mediators to decolorize Remazol Brilliant Blue R. J Biotechnol. **89**, 123–129 (2001)

Tavares, A.P.M., et al.: Decoloration of textile dyes by Trametes versicolor and its effect on dye toxicity. Biotechnol. Lett. **24**, 1757–1761 (2002)

Zille, A., et al.: Immobilized laccase for decolorization of Reactive Black 5 dyeing effluent. Biotechnol. Lett. **25**, 1473–1477 (2003)

Development of Airlaid Non-woven Panels for Building's Thermal Insulation

Melek Ayadi[1(✉)], Riadh Zouari[1], Cesar Segovia[2], Ayda Baffoun[3], Slah Msahli[1], and Nicolas Brosse[4]

[1] Textile Engineering Laboratory, 5070 Monastir, Tunisia
melek.ayadi@univ-lorraine.fr
[2] Centre d'Essais Textile Lorrain, 88000 Epinal, France
[3] Textile Materials and Process Research Unit, 5035 Monastir, Tunisia
[4] Laboratoire d'Etude et de Recherche sur le Matériau Bois, BP 70239, Nancy, France

Abstract. This study aims to explore a textile technology using Airlaid process to develop non-woven fabrics made of natural fibers extracted from *Posidonia Oceanica*'s waste for assessing their suitability for insulation products in the construction field. The prepared panels are analyzed for their thermal conductivity which were close to commonly used thermal insulation materials, ranging between 0.03515 W/m.K and 0.03957 W/m.K. The second part of this work aims to determinate the panel's resistance to five common mold types in buildings. In fact, at high moisture content, molds are likely to develop on cellulosic materials affecting indoor air quality and eventually causing a variety of health risks to occupants. However, optic microscope results showed no growth of molds on the Posidonia samples which allows conceiving reliable thermal insulation materials.

Keywords: Thermal insulation · Airlaid · Mold · Posidonia Oceanica

1 Introduction

In recent years, increased consumption of energy resources accompanied with environmental pollution, have significantly contributed to reconsideration of green alternatives to develop sustainable eco-friendly materials. Daily consumption of burning fossil fuels, which provide most of the energy needs, have obviously caused long-lasting negative impacts on public health, ecosystems, and the global climate. The International Institute of Refrigerating in Paris (IIF/IIR) has estimated that 20% of the overall electricity used worldwide is employed for refrigeration and air-conditioning processes. In this regard, in the construction sector, reliable thermal insulation materials play a crucial role in acquiring buildings' energy efficiency (Chastas et al. 2016), contributing to the reduction of the CO_2 emissions, being the first part of the energy saving strategies in buildings life cycle (Asdrubali et al. 2016). The choice of the integrated fibers depends on different parameters including their length, diameter, porosity, bulk density, tortuosity and their orientation within the non-woven structure. Wang et al. reported that

S. Msahli and F. Debbabi (Eds.): CIRATM 2020, SPM 17, pp. 110–117, 2022.
https://doi.org/10.1007/978-3-031-08842-1_18

the increasing in fiber length leads to an increase of thermal conductivity to a sufficient fiber length (Gao et al. 2007). Gibson and Lee showed that as the fibers get finer, thermal resistance increases (Gibson et al. 2007). In this work, we used Mediterranean Posidonia (*Posidonia Oceanica*) fibers to make thermal insulation non-woven panels for buildings. That is a marine flowering plant which is endemic of Mediterranean Sea. *Posidonia Oceanica* covers a quasi-continuous surface of Mediterranean coasts (between 30000 and 40000 km^2) locally interrupted at the estuaries and ports (Green et al. 2003). Dead leaves, lost most often in autumn are transported by storms and can thus be found along sandy coasts. This waste in form of balls, from a few centimeters to several meters of thickness is designated "aegagropiles" (Chessa et al. 2000). The valorization of this renewable and low-price biomass for producing more environmentally friendly industrial products presents an economical and ecological challenge (Dural et al. 2011). Over the last years, some researches were conducted to explore the use of *Posidonia Oceanica* as reinforcement of polymeric matrix to elaborate natural composites based in a mixture of *Posidonia Oceanica* and pine wood particles in polyurethane matrix with a view to investigate their performances as structural panels (Macia et al. 2016). Other authors evaluated the minimum bulk density for the use of Posidonia as a building solution on a flat roof. It was showed that to guarantee a thermal conductivity of 0.044 W/m.K, a layer of thermal insulation made of *Posidonia Oceanica* fibers should have a density of 185 kg/m^3 (Cristian et al. 2018). Another study (Korjenic et al. 2011) proposed natural fiber insulation materials developed from flax, hemp and jute with the addition of binder and shives. It showed that thermal conductivity ranged from 0.039 W/m.K to 0.048 W/m.K for density ranging from 26 kg/m^3 to 82 kg/m^3. In the present work, we propose to present the development process and thermal characterization of insulating panels made of *Posidonia Oceanica* waste manufactured through Air-laid technology. Furthermore, when moisture exceeds the tolerance of materials, it can lead to the growth of potentially harmful organisms and the decline of air quality. This paper evaluates the panel's resistance to five common mold strains in buildings.

2 Materials and Methods

2.1 Development of the Non-woven Panels

The balls have been mechanically opened to separate the fibers (Fig. 1), then sun-dried for 48 h. Through Airlaid process, Posidonia fibers (5–45 mm) and a proportion of 10% of thermoplastic bicomponent polyethylene terephthalate (PET) fibers were mixed with air flow to form a uniform air-fiber mixture which is then deposited on a moving-air permeable conveyor belt (Fig. 2). This technique offers the feature to develop isotropic non woven structures by orienting randomly the fibers on the fabric surface. The fiber web is then passed in an oven between heated calendar rollers at a temperature of 170 °C which is the melting temperature of the constituent thermoplastic fibers to consolidate the fabric, followed by cooling to 70 °C to solidify the bonding areas (Fig. 3). The thickness of the fabric is adjusted to 35 mm in the oven through a calendar with a hot metal roll. At the exit of the oven machine, the non-woven fabric was cut into seven panels of 300 mm in length and 300 mm in width (Fig. 4) and termed P1, P2, P3, P4, P5, P6 and P7.

Fig. 1. Posidonia Oceanica plant (a), balls (b) and fibers (c)

Fig. 2. Posidonia Oceanica/PET bicomponent Airlaid web structure

Fig. 3. Non-woven structure after thermal bonding at 170 °C

Fig. 4. General view of developed thermal insulation non-woven panels (300 mm × 300 mm)

2.2 Thermal Conductivity

Thermal conductivity factor at steady state is measured by means of hot/cold plate apparatus (EN 12667 2001). Thermal conductivity is measured at respective cold and hot plate temperatures of 10 °C and 30 °C then at 0 °C and 20 °C. Measurements are made on the samples dried in an oven until a relative humidity close to 0% and a mass variation of almost zero. Thermal conductivity (k) in one dimension can be described by the simplified form of the Fourier's law:

$$q = -k\,A(dT/dx) \tag{1}$$

where q is the steady-state heat flux in direction x (wm^{-2}), A is the cross-sectional area of the sample (m^2), - dT/dx is the temperature gradient (K/m) (McADAMS 1942).

2.3 Mold Growth

The test was carried out on Posidonia non-woven samples, chosen as pieces of 50 × 50 mm^2. The mold strains used in this study are *Aspergillus niger, Penicilum funiculosum, Trichoderma viride, Chaetomium globosum, Paecilomyces varioti*. Mold inoculums were maintained by storing them on malt-oats-agar (20 g of oatmeal, 10 g of malt extract and 20 g of agar in 1000 ml of water) (EN 15101-1 2013). Three filter paper control samples immersed in the suspension were placed as inoculum on agar in Petri dishes. The Four Posidonia samples were preconditioned for five hours at temperature and moisture conditions of respectively 28 °C ± 2 °C and RH 95% ± 4%. Thus, they were added onto the inoculated Petri dishes and let for incubation under the same temperature and moisture conditions favorable for mold growth for 28 days. Mold growth was then qualitatively determined by both visual inspection and under the Electronic Microscope.

3 Results and Discussions

3.1 Thermal Conductivity

The thermal conductivity from the seven panels (P1, P2, P3, P4, P5, P6 and P7) was plotted as a linear function of panel density (Table 1), as shown in Fig. 5. The results showed that at the same temperature, higher density values lead to higher thermal conductivity coefficients. In fact, *Posidonia Oceanica* fibers are porous structures having cellular cavities (voids) filled by air which serve as scattering centers of phonons (Fig. 6). The heat flow transfers through solid substance and void, and while the thermal conductivity of air within the voids is much lower than that of solid substance, that leads to a lower thermal conductivity of the whole material (Chessa et al. 2000).

Table 1. Density of fibers panels

Panel's reference	Panel's density (kg/m^3)
P1	43
P2	44
P3	46
P4	49
P5	85
P6	89
P7	103

It can be also noticed that the thermal conductivity values measured at mean temperature of 20 °C were higher than those measured at mean temperature of 10 °C (Fig. 7). Thus, operating thermal range is a determining factor in the thermal performance of the insulating panels. These variations can indicate the thermal conductivity estimated for the winter periods (internal and external temperatures of the building respectively of 20 °C and 0 °C) and summer periods (internal and external temperatures of the building respectively of 10 °C and 30 °C).

The Table 2 compares the thermal conductivity of the non-woven panels and commonly-used thermal insulation materials. The results showed that the Posidonia panels present a thermal conductivity ranging between 0.03515 W/m.K and 0.03957 W/m.K

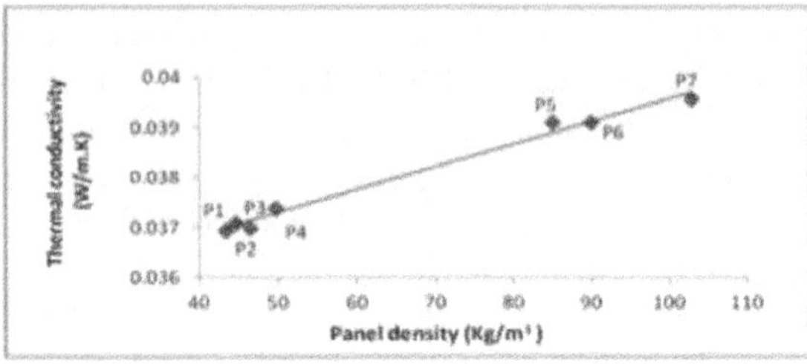

Fig. 5. Thermal conductivity of the panels at mean temperature of 20 °C as function of panel's density.

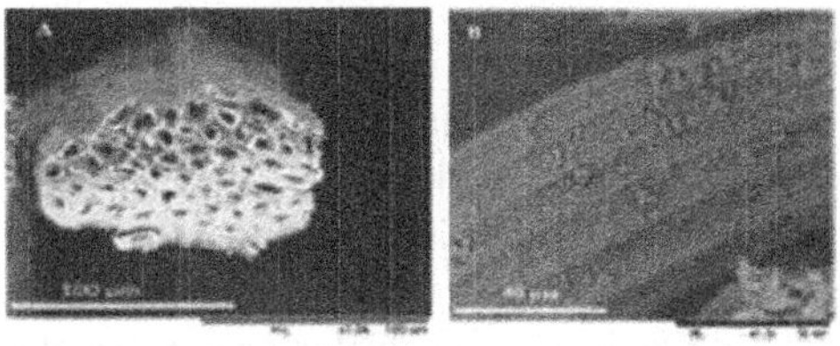

Fig. 6. Scanning Electron Microscopy (SEM) images of Posidonia Oceanica fibers. (A) transverse cross-section, (B) longitudinal section of the fiber. Scale bar equal to: (A) 100 μm and (B) 50 μm.

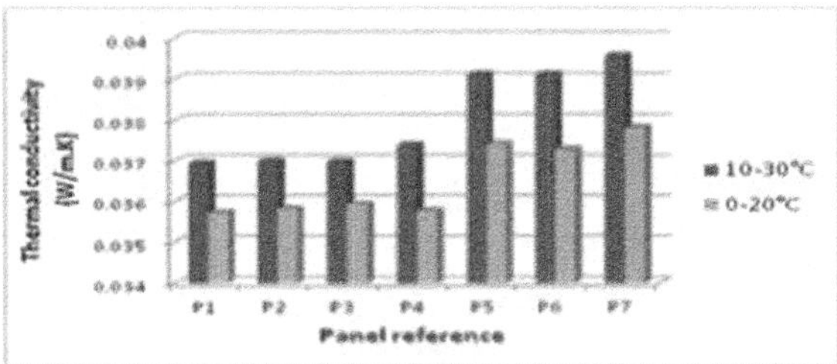

Fig. 7. Comparison of the thermal conductivity of the panels at cold and hot plate temperatures of 0 °C and 20 °C vs 10 °C and 30 °C of the Flowmeter apparatus.

which is revealed in the same range of fiberglass within the same density range and slightly lower than that of others cellular materials such as expanded perlite and vermiculite.

3.2 Mold Resistance

Mold presence was investigated by macroscopic observation (A) and (B) by Optical Microscopic observation at 40× as recommended in the NF EN 15101-1. The test was valid since the control filter papers were covered with mold on their entire surface after 5 days of incubation with mold. After four weeks of incubation, pine specimens were 50% colonized in the agar plate test, visible to the naked eye (Fig. 8). However, there was no mold growth observed on the samples (Fig. 9). Thus the non- woven samples correspond to the rating 0 according to NF EN 15101-1. The main finding of this part

Table 2. Thermal conductivity of various thermal insulation materials

Materials	Density (kg/m^3)	Thermal conductivity (W/m.K)	Reference
Posidonia panel wood (pine, lauan)	43–103	0.035–0.039	
	450–630	0.151	(Suleiman et al. 1999)
Fiberglass	24–120	0.034–0.047	(ASTM Standard C 1997)
Expanded perlite	78–224	0.047–0.061	(ASTM Standard C 1997)
Rockwool (natural rocks)	40–200	0.037	(ASTM Standard C 1997)
Vermiculite	80–200	0.047–0.07	(ASTM Standard C 1997)
Kenaf binderless board	150–200	0.051–0.058	(Suleiman et al. 1999)

was that the non-woven materials, present a natural resistance to the five mold strains in question which may allow sparing additional costs of chemical treatments.

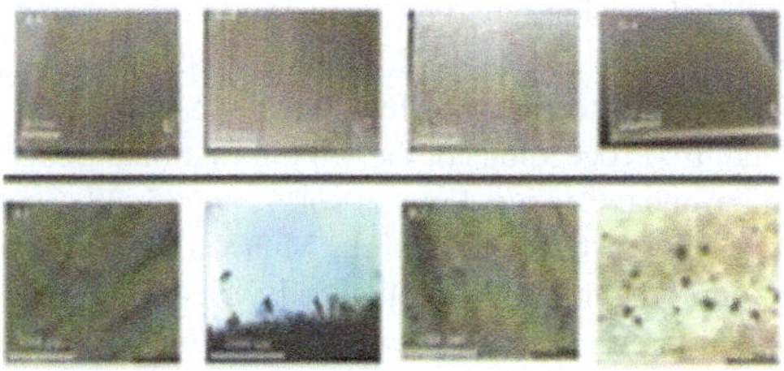

Fig. 8. Evaluation of the mold growth in four pine sapwood comparison materials (50 × 50 mm) by (A) macroscopic observation, (B) Optical Microscopic observation. Scale bars represent 10 mm in (A) and 1000 μm in (B).

Fig. 9. Evaluation of the mold growth in four specimens of the nonwoven material by (A) macroscopic observation, (B) Optical Microscopic observation. Scale bars represent 10 mm in (A) and 1000 μm in (B).

4 Conclusion

This paper presents a method of development and thermal characterization of thermal insulation panels made of *Posidonia Oceanica* fibers for buildings. Posidonia fibers were transformed into non-woven panels through Airlaid process and thermal consolidation with bicomponent Polyethylene terephthalate fibers. Thermal conductivity of manufactured panels was then investigated. The results show that the non-woven panels may conquer commonly-used thermal insulation materials with a thermal conductivity ranging between 0.03515 W/m.K and 0.03957 W/m.K for densities ranging between 43 kg/m^3 and 103 kg/m^3, which include the panels in the standard of thermal insulation materials according to the Heating Regulation RT 2012. It was found that an increase in bulk density resulted in lower insulation. The study of the mold resistance of the panels showed no infection of the nonwoven with mold after 28 days of incubation. This study offers an alternative to develop reliable thermal insulation materials from *Posidonia Oceanica*'s waste that can be integrated into various applications particularly for the huge field of buildings.

References

Asdrubali, F., et al.: Experimental and numerical characterization of innovative cardboard based panels: thermal and acoustic performance analysis and life cycle assessment. Build. Environ. **95**, 145–159 (2016)

Chastas, P., Theodosiou, T., Bikas, D.: Embodied energy in residential buildings towards the nearly zero energy building: a literature review. Build. Environ **105**, 267–282 (2016)

Chessa, L.A., et al.: Contribution to the knowledge of 'Banquettes' of Posidonia oceanic (L.) Delile in Sardinia Island. In: Proceedings of the 4th International Seagrass Biology Workshop, Biologia Marina Mediterranea, Corsica, France, vol. 7, pp. 35–38 (2000)

Cristian, C., Horrach, G., Olivier, C., Forteza, F.J., Munoz, J.: Posidonia oceanica as thermal insulation: determination of the minimum bulk density, according to project specifications, for its use as a building solution on a flat roof. Revista de la Construcción **17**, 250–257 (2018)

Dural, M.U., Cavas, L., Papageoriou, S.K., Katsaros, F.K.: Methylene blue absorption on activated carbon prepared from Posidonia oceanica (L.) dead leaves: kinetics and equilibrium studies. Chem. Eng. J. **168**, 77–85 (2011)

EN 12667: Thermal performance of building materials and products - determination of thermal resistance by means of guarded hot plate and heat flow meter methods - products of high and medium thermal resistance (2001)

EN 15101-1: Thermal insulation products for buildings – in-situ formed loose fill cellulose (LFCI) products–part 1: specification for the products before installation (2013)

Gao, J., Yu, W., Pan, N.: Structures and properties of the goose down as a material for thermal insulation. Text. Res. J. **77**, 617–626 (2007)

Gibson, P., Lee, C., Ko, F., Reneker, D.: Application of nanofiber technology to nonwoven thermal insulation. J. Eng. Fibers Fabr. **2**, 32–40 (2007)

Green, E.P., Short, F.T.: World atlas of seagrasses: present status and future conservation (2003)

Korjenic, A., Petranek, V., Zach, J., Hroudova, J.: Development and performance evaluation of natural thermal-insulation materials composed of renewable resources. Energy Build. **43**, 2159–2168 (2011)

Macia, A., Baeza, F.J., Saval, J.M., Ivora, S.: Mechanical properties of boards made in biocomposites reinforced with wood and Posidonia oceanica fibers. Compos. B **104**, 1–8 (2016)

McAdams, W.H.: Heat Transmission. 2nd ed., McGraw-Hill (1942)
Suleiman, B.M., Larfeldt, J., Leckner, B., Gustavsson, M.: Thermal conductivity and diffusivity of wood. Wood Sci. Technol. **33**, 465–473 (1999). https://doi.org/10.1007/s002260050130
Terminology relating to thermal insulating materials. ASTM Standard C, pp. 97–168 (1997)

Usage Quality Assessment of Weft Knitted Fabrics Made of Modal Fibers

Antoneta Tomljenović(✉), Juro Živičnjak, Zlatko Vrljičak, and Veronika Stamać

Faculty of Textile Technology, University of Zagreb, Prilaz baruna Filipovića 28a, 10000 Zagreb, Croatia

antoneta.tomljenovic@ttf.unizg.hr

Abstract. Knitwear that are worn in direct contact with the skin, are often made of man-made artificial fibers from cellulose which provide silky touch and exceptional contact comfort. In this paper three different circular weft double jersey knitted fabrics made of single modal ring, rotor and air-jet spun yarns of the same linear density, all made of bright staple modal fibers of linear density of 1.3 dtex and length of 38/40 mm, were used. With the purpose of analyzing the influence of differently spun modal yarns and the knitted fabrics processing level on their properties, along with the basic characterization of yarns, the usage quality of raw and finished knitted fabrics were evaluated and their applicability assessed.

Keywords: Modal knitwear · Double jersey · Ring, rotor and air-jet spun yarn · Textile testing

1 Introduction

Knitwear that are worn in direct contact with the skin, are often made of man-made artificial fibers from cellulose (e.g. viscose, modal or lyocell) which provide silky touch, better hydrophilicity and exceptional contact comfort (Kreze et al. 2003). Knitted fabrics for lingerie were usually made of single spun yarns produced by conventional ring spinning system. More recently in application were also spun yarns made by unconventional rotor and air-jet spinning systems. Different spinning systems provide spun yarns of different structure and properties (Skenderi et al. 2019). As the number of European standards related to testing and characterization of knitted fabrics are low (Tomljenović et al. 2016), it is necessary to expand the research in the field of their quality assessment.

Therefore, in this paper three different circular weft double jersey knitted fabrics made of single modal ring, rotor and air-jet spun yarns of the same linear density, all made of the same bright staple modal fibers were used. With the purpose of analyzing the influence of differently spun modal yarns and the knitted fabrics processing level on their properties, the usage quality of raw and finished knitted fabrics were evaluated and their applicability assessed. Along with the basic characterization of yarns, testing of mass per unit area, thickness and number of wales and courses per unit length of knitted fabrics were carried out. Knitwear usage quality were evaluated by determination of breaking strength and elongation, dimensional change after laundering, permeability of fabrics to air, their propensity to surface piling and abrasion resistance, all according to the standardized test methods.

S. Msahli and F. Debbabi (Eds.): CIRATM 2020, SPM 17, pp. 118–123, 2022.
https://doi.org/10.1007/978-3-031-08842-1_19

2 Materials and Methods

Three differently spun modal yarns of the same linear density (20 tex) chosen for the knitting purpose were used: standard single ring spun yarn as a reference, single rotor spun and air-jet yarns, all made of bright staple modal fibers of linear density of 1.3 dtex and length of 38/40 mm. Tensile properties of the modal yarns used are shown in Table 1.

Table 1. Tensile properties of modal yarns used

Yarn type	Breaking strength (cN)	Breaking elongation (%)	Tenacity (cN/tex)
1 Ring modal	487	10.2	24.3
2 Rotor modal	325	7.2	16.3
3 Air-jet modal	406	9.0	20.3

Three samples of plain double jersey weft knitted fabrics were made using circular double-bed knitting machine with E17 gauge and needle bed diameter of 200 mm (8 in.) under the same conditions (Table 2).

Table 2. Construction characteristic of double-bed circular knitting machine used

Machine gauge, E	Needle bed diameter	Number of knitting systems, S	Number of needles, N_i	Working speed of the cylinder
17	200 mm (8 in.)	8	432 × 2	60 r/min

All dry relaxed knitted fabrics (raw samples) were finished in the production plant: firstly washed thoroughly at 40 °C, further treated with the addition of detergent, bleaching and stabilization agent at 95 °C; rinsed, cold washed with neutralization and softening; and dried at 150 °C with a passage rate of 0.15 m/s. Mass per unit area, thickness and total number of wales and courses per unit length of raw and finished modal knitted fabrics determined after the conditioning (at temperature: 20 ± 2 °C and air relative humidity: 65 ± 4%) are shown in Table 3. The usage quality of conditioned row and finished knitted fabrics were evaluated by determination of:

- breaking strength and breaking elongation according to the EN ISO 13934-1 using the strip method. Arithmetic mean of the five measurements in the length and width directions and coefficient of variation were calculated;
- dimensional change in the length and width directions of knitted fabric in tubular form, after one washing and drying cycle according to the procedure 4M at 40 °C (mild agitation during heating, washing and rinsing) of EN ISO 6330 with non-phosphate ECE reference detergent (without optical brightener) and open-air drying (procedure A, line dry). The percentage change in the length and width of the knitwear was

calculated, and the state of whether the dimension has decreased (shrinkage) was expressed my means of minus;

- permeability of fabrics to air according to the EN ISO 9237 using test surface area of 5 cm^2 and pressure drop of 100 Pa. Arithmetic mean of the 10 individual readings and coefficient of variation were calculated;
- propensity to surface fuzzing and pilling according to the Martindale method (EN ISO 12945-2). The knitted fabrics were rubbed with wool abradant fabric and visually assessed (by grades: 1–5) after 125, 500, 1000, 2000, 5000 and 7000 rubbing cycles;
- abrasion resistance by determination of specimen breakdown using the Martindale abrasion tester according to the EN ISO 12947-2. When using this method the specimens moves according to the Lissajous curve, and standard woven wool fabric is abraded over the entire surface.

Table 3. Characterisation of raw and finished modal knitted fabrics

Knitted fabrics sample	Mass (g/m^2)	Thickness (mm)	No. of wales/cm	No. of courses/cm
1 Ring modal/raw	162	0.84	20	12
1 Ring modal/finished	145	0.72	19.5	13
2 Rotor modal/raw	144	0.79	18.5	13
2 Rotor modal/finished	160	0.70	20.5	13
3 Air-jet modal/raw	150	0.85	19	12.5
3 Air-jet modal/finished	150	0.71	18.5	12.5

3 Results and Discussion

The results obtained by the investigation are presented in Tables 3, 4, 5, 6 and 7.

After finishing the thickness of all modal knitted fabrics samples is reduced (Table 3). Changing of mass per unit area additionally affects the dimensional change of knitwear (confirmed by changing of total number of wales and courses per unit length). That is also confirmed with results of air permeability of raw and finished modal knitted fabrics presented in Table 7. Permeability to air and the breaking strength (Table 4) of finished knitted fabrics are changed primarily because of their dimensional and structural changes.

Breaking strength of knitted fabrics (Table 4) is conditioned by breaking strength of the yarn used for knitting (shown in Table 1). Rotor spun yarn have lower tenacity when compared with ring and air-jet modal yarns. In the length direction of all knitwear samples it has been found higher values of breaking strength and lower breaking elongation, and in the width direction the lower breaking strength and higher breaking elongation. After finishing the breaking strength and breaking elongation of all knits in both directions is reduced.

Table 4. Breaking strength and breaking elongation of raw and finished modal knitted fabrics

Knitted fabrics sample	Breaking strength (N) with coefficient of variation (%)		Breaking elongation (%) with coefficient of variation (%)	
	In length	In width	In length	In width
1 Ring modal/raw	313.8/12.28	93.7/4.00	44.4/6.30	150.0/2.77
1 Ring modal/finished	209.8/11.22	84.7/2.15	40.9/3.00	119.7/1.52
2 Rotor modal/raw	261.1/3.66	79.1/1.93	41.3/2.50	182.2/4.18
2 Rotor modal/finished	222.8/7.15	65.1/1.23	40.9/5.90	135.5/2.37
3 Air-jet modal/raw	292.4/5.97	82.7/3.02	42.0/2.79	175.6/0.97
3 Air-jet modal/finished	214.8/12.42	62.2/6.27	40.5/1.20	124.1/3.24

Fabric shrinkage is a serious problem for knitwear, originating from dimensional changes in the fabric, particularly stiches. The on-balanced condition of modal knitwear is quite easy changing when knitwear is laundered and dried. Therefore, the geometrical dimensions of knitwear are changing (Hashimoto et al. 2018; Mikučionienė and Laureckienė 2009). All raw knitted fabrics show high deformability after laundering (Table 5). Dimensional stability after laundering are significantly improved in all finished knitted fabrics.

Table 5. Changes in dimensions in the length and width directions of raw and finished modal knitted fabrics after one washing and drying cycle

Knitted fabrics sample	Dimensional stability in length (%)		Dimensional stability in width (%)	
	Raw	Finished	Raw	Finished
1 Ring modal	−18.8	−6.9	6.7	−6.7
2 Rotor modal	−16.9	−8.8	−8.3	−1.7
3 Air-jet modal	−20.5	−8.8	−2.0	−1.4

With increasing the number of abrasion cycles, in all knitwear samples increases the propensity to surface fuzzing and pilling (Table 6). Best rated knitwear are those made of modal air-jet spun yarns because of their lower hairiness and specific regular and tightly structure when compared with ring and rotor spun yarns. After 7000 pilling rubs, slightly lower final grades were found in all tested finished fabric samples.

Table 6. Visually assessed propensity to surface pilling of raw and finished modal knitted fabrics

Knitted fabrics sample	Number of pilling rubs					
	125	500	1000	2000	5000	7000
1 Ring modal/raw	4	3/4	3	3	2	2
1 Ring modal/finished	4	3	2/3	2/3	2	1/2
2 Rotor modal/raw	4	4	3/4	3/4	3	3
2 Rotor modal/finished	4	3/4	3	2/3	2	2
3 Air-jet modal/raw	5	4/5	4	4	3/4	3/4
3 Air-jet modal/finished	4/5	4/5	4	3/4	3	3

Table 7. Abrasion resistance and permeability to air of raw and finished modal knitted fabrics

Knitted fabrics sample	Abrasion resistance (number of rubs to reach endpoint*)		Air permeability (mm/s) with coefficient of variation (%)	
	Raw	Finished	Raw	Finished
1 Ring modal	40 000	45 000	2476.3/2.86	2730.4/5.22
2 Rotor modal	30 000	30 000	3610.2/3.23	2434.1/6.62
3 Air-jet modal	30 000	30 000	2996.5/4.23	2379.5/6.78

*Determination of specimen breakdown

Yarn structure, linear density, twist and hairiness are the main properties which affect abrasion resistance of textile materials. The yarn spinning method has also an influence on the abrasion resistance (Özdil et al. 2012). Knitted fabrics from ring spun yarns have better abrasion resistance than knitted fabrics from rotor and air-jet spun yarns. Ring spun yarns are hairier but more compactly structured than rotor yarns, this well aligned compact structure doesn't promote easy fibre wear of (Table 7). By finished knitwear samples made of single ring modal yarns, specimen breakdown occurs by 40 000 and 45 000 abrasion rubs. This can be connected also with the fact that the ring spun yarns have higher tenacity (Table 1), and modal fibres are highly twisted on the surface of the yarn.

4 Conclusion

On the basis of the results obtained, it was concluded that for selection of the modal spun yarn for knitted fabrics production is necessary to consider their structure and the characteristics, but also the fact that yarn spinning technique, as well as the process of knitted fabric finishing significantly influence knitwear usage quality.

Acknowledgment and Funding. This paper is funding by the Croatian science foundation within the Project IP-2016-06-5278 (Comfort and antimicrobial properties of textiles and footwear, principal investigator: Prof. Zenun Skenderi, PhD).

References

EN ISO 12947-2:2016: Textiles - determination of the abrasion resistance of fabrics by the Martindale method - part 2: determination of specimen breakdown

EN ISO 12945-2:2020: Textiles - determination of fabric propensity to surface fuzzing and to pilling - part 2: modified Martindale method

EN ISO 13934-1:2013: Textiles - tensile properties of fabrics - part 1: determination of maximum force and elongation at maximum force using the strip method

EN ISO 9237:1995: Textiles - determination of the permeability of fabrics to air

Hashimoto, Y., et al.: Effect of washing and drying conditions on dimensional change in various articles of knitted clothing. J. Fiber Bioeng. Inform. **11**(4), 227–240 (2018). ISSN: 1940-8676

ISO 6330:2012: Textiles - domestic washing and drying procedures for textile testing

Kreze, T., Malej, S.: Structural characteristics of new and conventional regenerated cellulosic fibres. Text. Res. J. **73**(8), 675–684 (2003). ISSN: 0040-5175

Mikučionienė, D., Laureckienė, G.: The influence of drying conditions on dimensional stability of cotton weft knitted fabrics. Mater. Sci. **15**(1), 64–68 (2009). ISSN: 1392-1320

Özdil, N., et al.: Analysis of abrasion characteristic in textiles. In: Adamik, M. (ed.) Abrasion Resistance of Materials, pp. 119–146. InTech, Rijeka (2012). ISBN: 978-953-51-0300-4

Skenderi, Z., et al.: Study on physical-mechanical parameters of ring, rotor and air-jet spun modal and micro modal yarns. Tekstilec **62**(1), 42–53 (2019). ISSN: 0351-3386

Tomljenović, A., et al.: Durability assessment of functionally printed knitted fabrics for T-shirts. Fibers Text. Eastern Europe **24**(4), 129–138 (2016). ISSN: 1230-3666

Nanoclay Incorporation for Improved Dielectric Properties of Polyester Fabric

Ezzine Sawssen(✉), Abid Khaled, and Laadhari Neji

Textile Engineering Laboratory, University of Monastir, Monastir, Tunisia
Sawssenezzine88@gmail.com

Abstract. In this study, nano-scaled particles namely nano clays were incorporated in the structure of polyester fabrics via coating techniques to develop functionalized nanocomposite fabric with improved dielectric characteristics and to prove that the resulting resistivity can only be increased in comparing with the fabric resistivity.

The nanocomposite coating materials were fabricated using nanoclay into polyacrylate and polyurethane resin. The concentration of nanoclay was 0, 5 and 10 wt %. The nanoclay was used without any further modifications. For the morphology and the surface chemical composition analysis of the coated polyester fabric sample and uncoated ones, a scanning electron microscope (SEM) was used. And a new measuring device has been developed to determine the surface resistivity of different samples.

Keywords: Polyester fabric · Nano clay · Nanocomposite · Dielectric properties

1 Introduction

Textile industry has been looking for new technologies to light consumer demands. Nowadays, nanomaterial applications during manufacturing, finishing and coating processes to produce nanocomposite structures have come into prominence for obtaining multifunctionality or special functions for textiles. The most important parameter of the textile materials is its dielectric resistivity. It can be declined by using conductive composites (Gaurav et al. 2013; Mago 2014) or enhanced by adding a non conductive coating. This last can be a mixture of nanoclay and organic resin, and subsequently the textile fabric will have hybrid dielectric characteristics: one of the fabrics and the other of the constant thin film. Electrical conductivity is the reciprocal of electrical resistivity. The characteristic of resistivity and dielectric constant is the important thing for investigating the dielectric properties of the nanocomposite.

So, in this study, nano-scaled particles namely nano clays were incorporated in the structure of polyester fabrics via coating techniques to develop functionalized nanocomposite fabric with improved dielectric characteristics and to prove that the resulting resistivity can only be increased in comparing with the fabric resistivity. A new measuring device has been developed to determine the surface resistivity and electric permittivity.

S. Msahli and F. Debbabi (Eds.): CIRATM 2020, SPM 17, pp. 124–128, 2022.
https://doi.org/10.1007/978-3-031-08842-1_20

2 Materiels and Methods

The nanocomposite coating materials were fabricated using nanoclay into polyacrylate and polyurethane resin. The concentration of nanoclay was 0, 5 and 10 wt %. The nanoclay was used without any further modifications. The samples were submitted to regular coating. A pressure-controlled rake was used for this purpose. The nanocomposite formation was confirmed in our previous work (Ezzine 2018). After a 5 min drying operation permitting for water to evaporate, the polymerisation of these fabric coatings was done at 150 °C for a period of 5 min (Abid 2010).

The composition of the different samples was presented in the next table (Table 1).

Table 1. The composition of different samples

N°	PU (%)	PAC (%)	Nanoclay (%)
1	0	0	0
2	95	0	5
3	90	0	10
4	0	95	5
5	0	90	10

For the morphology and the surface chemical composition analysis of the uncoated raw polyester fabric sample and uncoated ones, Thermo Scientific™ Apreo™ scanning electron microscope (SEM) was used.

The used electric circuit for measuring the surface resistivity of the fabrics is shown in the next figure (Fig. 1).

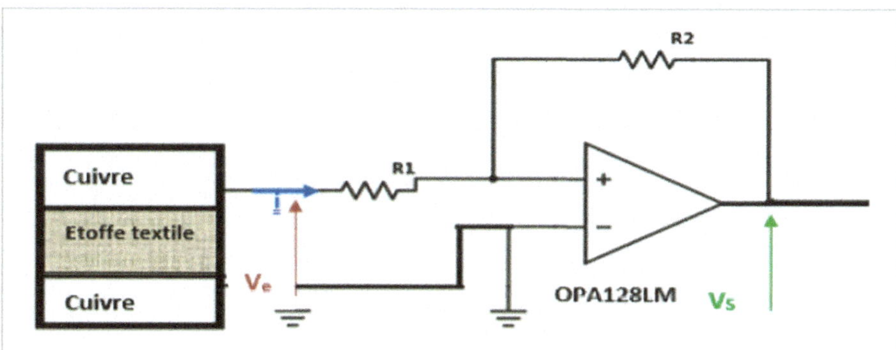

Fig. 1. SEM images of samples

3 Results and Discussion

3.1 Identification of the Obtained Nanocomposite

The SEM images of the obtained coated fabric with nanocomposite polymer/nanoclay show that the nanoclay is well intercalated in the organic matrix and forms a unique structure of a new hybrid material.

The chemical composition inside any organic phase showed the presence of clay elements (Ca2, K, Al, and Si). Since the composition of the organic resins can contain only C, H, and O, the presence of clay in these zones is confirmed to be inside the structure of the organic matrix, which again confirms the formation of nanocomposites. A similar operation was done for the mineral phase to confirm the presence of the resin components (C, H, and O). (Figure 2 sample number 5).

3.2 Dielectric Properties

Dielectric Strength is a measure of the electrical strength of a material as an insulator. Dielectric strength is defined as the maximum voltage required to produce a dielectric breakdown through the material and is expressed as Volts per unit thickness. A higher dielectric strength represents a better quality of insulator.

In this work we have to study the effect of nanocomposites on the dielectric properties of the fabric and to prove that the resulting resistivity can only be increased in comparing with the fabric resistivity.

So, we used capacitance measurements to determine the electric permittivity with a new measuring device as presenting in the top.

Sample coated with resin only, presents a slightly higher resistance than the reference, but a lower one than the rest of the samples with nanocomposite contents. For the same deposited quantity of nanocomposite, any increase of the clay percentage leads to an enhancement of the electrical resistances. This is probably due to the high electric resistance of the ceramics nanoparticles.

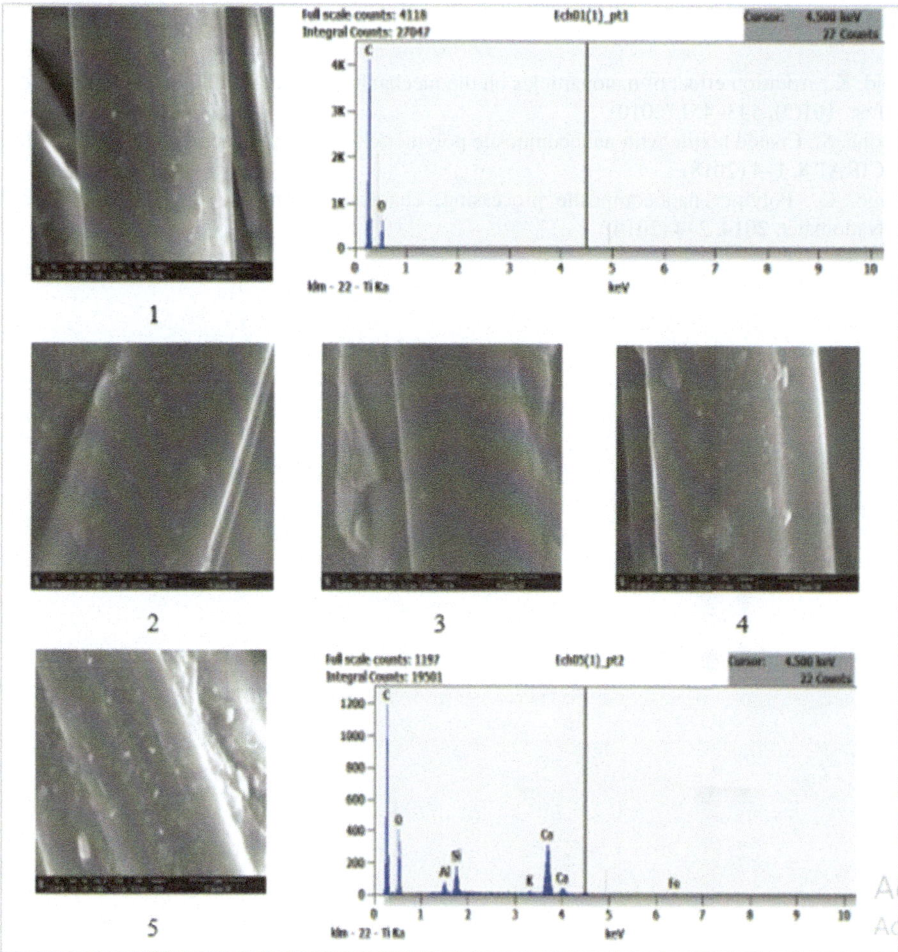

Fig. 2. SEM images of samples

4 Conclusion

Dielectric and electrical properties should be correlated with data obtained from such analytical observation. It is also necessary to create a forum for research and promote international cooperation to facilitate and accelerate our research and to put polymer nanocomposites practical as dielectrics and electrical insulation. The nano charges created a network of connectivity between the fibers for the passage of current. This explains the improvement observed in the electrical resistances of composites in the presence of clay nanoparticles.

References

Abid, K.: Addition effect of nanoparticles on the mechanical properties of coated fabric. J. Text. Inst. **101**(5), 443–451 (2010)

Ezzine, S.: Coated textile with nanocomposite polymer/clay: surface properties characterization. CIRAT **8**, 1–4 (2018)

Mago, G.: Polymer nanocomposite processing, characterization, and applications 2013. J. Nanomater. **2014**, 2–4 (2014)

Plasma and Chitosan Pre-treatments for Improvement of Color Strength and Fastness Properties in Printing of Cotton Fabric Using Direct Dyes

Aminoddin Haji(✉) and Rezvan Mahmoudi Hashemi

Department of Textile Engineering, Yazd University, Yazd, Iran
ahaji@yazd.ac.ir

Abstract. In this study, O_2/Ar plasma was used for surface activation and attachment of chitosan on cotton fabric. The effects of plasma treatment and chitosan attachment on the color strength, and fastness properties of the samples printed with a direct dye was investigated. The results showed that both plasma treatment and subsequent chitosan attachment enhanced the color strength of the printed cotton. Increasing the chitosan concentration and plasma treatment time increased the color strength of the printed samples. The effect of chitosan coating was more pronounced when the chitosan was applied after plasma treatment. All fastness properties were improved when the cotton sample was treated with plasma and coated with chitosan.

Keywords: Cotton · Plasma · Chitosan · Printing · Fastness

1 Introduction

Cotton is the most important natural fiber which is used in textile industry and can be dyed with a wide range of dye classes including direct, reactive, vat, azoic, and sulfur dyes (Andreaus et al. 2019). Direct dyes are a class of synthetic dyes with easy application on cotton and low price, but they have a major drawback of low to moderate wash fastness. When used for cotton printing, another drawback is that they need long steaming times and stain on the white backgrounds during the after-washing step (Kanik and Hauser 2004).

Due to the repulsion between the surface negative charge of cotton fibers and the sulphonic groups in the structure of direct dyes, they cannot be absorbed well during the printing process and the wet fastness of the end product is not good. Surface modification of the cotton with positively charged moieties that leads to ionic bond formation with the anionic dyes is one of the most studied approaches to solve this problem (Baaka et al. 2017).

The modification of cotton for improvement of its dyeing and printing abilities with different synthetic and natural dyes has received considerable attention in recent years. Plasma treatment (Gorjanc et al. 2018; Haji 2019), crosslinking with polycarboxylic

S. Msahli and F. Debbabi (Eds.): CIRATM 2020, SPM 17, pp. 129–134, 2022.
https://doi.org/10.1007/978-3-031-08842-1_21

acids (Haji et al. 2018), surface grafting (Haji 2013; Haji 2017; Haji et al. 2016; Patil et al. 2019), and cationisation (Arivithamani and Giri Dev 2018; Hauser 2018; Hauser and Kanik 2003; Kanik and Hauser 2002; Kanik and Hauser 2004; Marie et al. 2016; Nallathambi and Venkateshwarapuram Rengaswami 2017; Rym et al. 2016; Samanta et al. 2016) are examples of the methods investigated with the aim of improvement of cotton dyeing, printing and fastness properties.

Plasma treatment can modify the surface properties of textile fibers without imposing significant alteration to their bulk properties. Oxygen plasma can enhance the wettability and improve the dyeability of cotton fibers (Sun and Stylios 2004). However, grafting of a bio-polymer like chitosan which brings amine functionalities to the surface of cotton fibers, can further enhance its dyeability and printability with anionic dyes due to the ionic attraction between the dye anions and the cationized amine groups (Kampeerapappun et al. 2010; Ke et al. 2021; Mansour and Ben Ali 2021).

Chitosan can be simply applied on the surface of cotton or chemically bonded to the cellulose chains using crosslinking agents (El-tahlawy et al. 2005), oxidation pretreatment (Ke et al. 2021), plasma activation (Haji et al. 2016), etc.

In this study, effect of Argon/Oxygen plasma treatment and subsequent grafting of chitosan on the plasma-activated fibers on the printing of cotton with a direct dye is investigated.

2 Experimental

2.1 Materials

Pure cotton fabric (Plain weave, 100 g/m^2, yarn count Nm = 40) was obtained from Mazandaran Textile Co., Iran. Medium molecular weight chitosan (75–85% deacetylated), Triton X-100 (nonionic surfactant), urea, and acetic acid were purchased from Sigma Aldrich (USA). C.I. Direct orange 26 was obtained from Alvan Sabet Company, Hamedan, Iran. The chemical structure of the dye is shown in Fig. 1. Gum tragacanth was obtained from a local market and used as thickener.

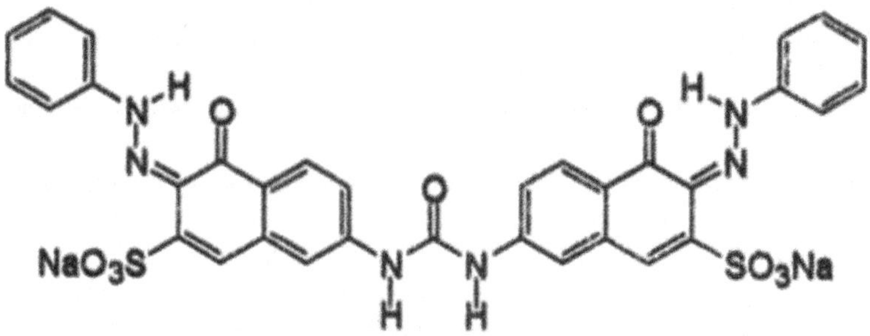

Fig. 1. Chemical structure of C.I. direct orange 26

2.2 Methods

Plasma treatment: Cotton samples were pretreated using radio frequency (13.56 MHz) low pressure plasma equipment (model: PlasmaDEJ VP06BF, BasaFan, Iran) with oxygen/argon mixture as the processing gas. The sample chamber was evacuated to 50 mTorr and maintained at this pressure during the treatment. Oxygen and argon with a flow rate of 50 sccm (Standard Cubic Centimeters per Minute) were used in the plasma treatment process. Plasma was generated at 100 W for 1–5 min. Then, atmospheric air was introduced into the chamber and the plasma treated sample was removed.

Chitosan treatment: Plasma treated samples were immediately impregnated in solutions of chitosan with different concentrations (0.125, 0.25, and 0.5% w/v) containing 1% v/v acetic acid for 15 min. Then the samples were padded with 100% wet pick up and dried at 85 °C for 5 min. The dried samples were thoroughly washed with a solution containing 1% w/v Triton X-100 at 50 °C for 15 min to remove non-reacted chitosan from the surface of the fabric samples.

Printing process: To prepare the stock thickener, 10 g of gum tragacanth powder was mixed with 50 mL of distilled water and stored overnight. The stock printing paste was prepared by mixing 23.5 g of the stock thickener, 1 g urea, and 0.5 g direct dye. 2.5 g of the above mentioned printing paste was used for printing of each cotton sample (5 cm * 15 cm). Manual printing was done using a simple flat screen (fineness 110 thread/cm). Printed samples were dried at 80 °C and steamed on a sample steamer at 100 °C. Finally, the samples were scoured using a solution containing 1% w/v Triton X-100 at 50 °C for 15 min followed by a cold wash to remove non-absorbed dye and thickener from the surface of the fabric samples.

Color strength measurements: The reflectance of the printed samples were measured on a Color-eye 7000A spectrophotometer (X-rite, USA) using illuminant D65 and 10° standard observer. Color strength (K/S) of each sample was calculated using Kubelka-Munk equation:

$$K/S = (1 - R)^2/2R \tag{1}$$

where R is the observed reflectance at wavelength of maximum absorbance.

Color fastness tests: Color fastness of the dye samples against washing, artificial light and rubbing were measured according to ISO 105-C01 (2006), ISO 105-B02 (2014), and ISO 105-X12 (2016) standards, respectively.

3 Results and Discussion

Figure 2 shows the effect of plasma treatment time on the color strength of the printed cotton samples (without chitosan coating). It can be seen that the plasma treated samples showed a higher color strength compared with the untreated sample. The color strength increased by increasing the plasma treatment time. Plasma treatment enhanced the wettability and increased the surface roughness of the cotton fibers which caused better adhesion of the printing paste and improved the penetration of the dye molecules into the cotton fibers (Haji et al. 2016).

Figure 3 shows the effect of application of chitosan onto cotton fabrics prior to the printing on the color strength of the printed samples. It can be seen that chitosan treatment increased the color strength. Increasing the chitosan concentration in the range of 0.125–0.5% w/v increased the color strength of both plasma-treated and untreated cotton fabrics. The amine groups of the chitosan made the surface of the cotton positively-charged which absorbed more anionic dye molecules and increased the color strength of the printed samples. The plasma-treated-chitosan-coated samples showed higher color strength values compared with the samples which were coated with the same concentration of chitosan, but without plasma pre-treatment. This finding can confirm higher fixation of chitosan on cotton fibers after the plasma activation.

Table 1 shows the fastness properties of selected samples after printing with the same recipe using C.I. direct orange 26. As expected the wash fastness of the direct dye on the untreated sample is low, the light fastness is moderate, and the rub fastness is good. Plasma treatment slightly improved the fastness to rubbing (wet) and washing. The plasma-treated-chitosan-coated sample showed better fastness properties against rubbing, washing, and light. These improvements may be due to the ionic interactions between the anionic dye molecules and the amine functionalities introduced to the surface of the cotton fibers after coating with chitosan.

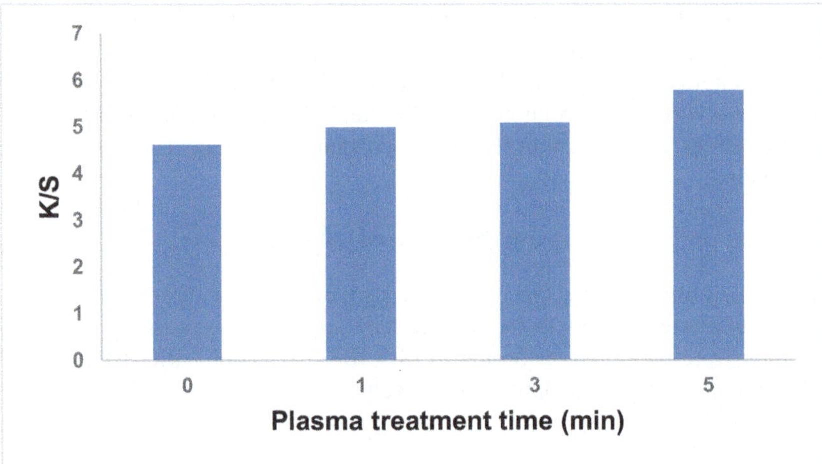

Fig. 2. Effect of plasma treatment time on color strength of the printed cotton fabrics

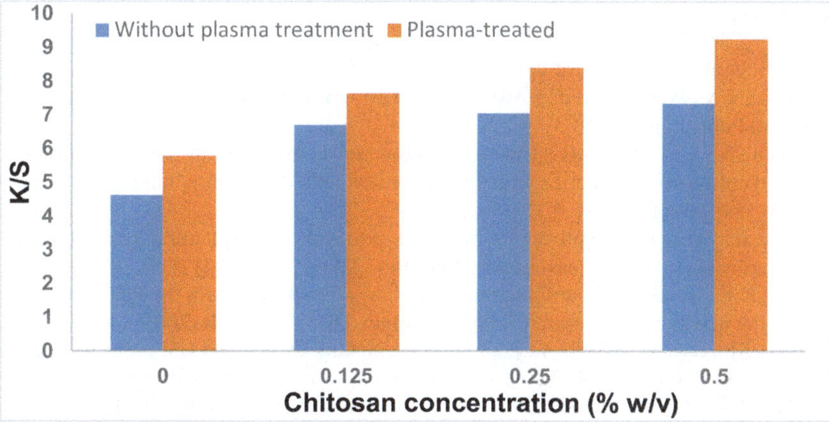

Fig. 3. Effect of chitosan concentration on the color strength of the untreated and plasma treated cotton fabric after printing with C.I. direct orange 26

Table 1. Fastness properties of different printed samples

	Untreated	5 min plasma-treated	5 min plasma-treated-chitosan-coated (0.5% w/v)
Rub fastness-dry-color change	5	5	5
Rub fastness-dry-staining	4	4	5
Rub fastness-wet-color change	4	4	4
Rub fastness-wet-staining	2	2–3	3
Light fastness	5	5	6
wash fastness-color change	2	2–3	3
Rub fastness-staining	2	2	3

4 Conclusion

Both plasma treatment and subsequent chitosan coating improved the color strength of the cotton fabrics printed with C.I. direct orange 26. Plasma treatment improved the fastness properties of the printed samples against rubbing (wet) and washing slightly. Attachment of chitosan onto the plasma-treated cotton fibers enhanced all the fastness properties of the printed fabrics. Plasma treatment and attachment of chitosan can be an environmentally friendly approach for preparation of cotton goods for printing with direct dyes to achieve higher color strength and better fastness properties.

References

Andreaus, J., et al.: Processing of cotton and man-made cellulosic fibers. In: Cavaco-Paulo, A., Nierstrasz, V.A., Wang, Q. (eds.) Advances in Textile Biotechnology, 2nd edn., pp. 185–238. Woodhead Publishing (2019)

Arivithamani, N., Giri Dev, V.R.: Characterization and comparison of salt-free reactive dyed cationized cotton hosiery fabrics with that of conventional dyed cotton fabrics. J. Clean. Prod. **183**, 579–589 (2018)

Baaka, N., et al.: Green dyeing process of modified cotton fibres using natural dyes extracted from Tamarix aphylla (L.) Karst. leaves. Nat. Prod. Res. **31**(1), 22–31 (2017)

El-tahlawy, K.F., et al.: The antimicrobial activity of cotton fabrics treated with different crosslinking agents and chitosan. Carbohyd. Polym. **60**(4), 421–430 (2005)

Gorjanc, M., Mozetič, M., Vesel, A., Zaplotnik, R.: Natural dyeing and UV protection of plasma treated cotton. Eur. Phys. J. D **72**(3), 1–6 (2018). https://doi.org/10.1140/epjd/e2017-80680-9

Haji, A.: Eco-friendly dyeing and antibacterial treatment of cotton. Cellul. Chem. Technol. **47**(3–4), 303–308 (2013)

Haji, A.: Improved natural dyeing of cotton by plasma treatment and chitosan coating; optimization by response surface methodology. Cellul. Chem. Technol. **51**(9–10), 975–982 (2017)

Haji, A.: Dyeing of cotton fabric with natural dyes improved by mordants and plasma treatment. Prog. Color Colorants Coat. **12**(3), 191–201 (2019)

Haji, A., Nasiriboroumand, M., Qavamnia, S.S.: Cotton dyeing and antibacterial finishing using agricultural waste by an eco-friendly process optimized by response surface methodology. Fibers Polym. **19**(11), 2359–2364 (2018). https://doi.org/10.1007/s12221-018-8657-2

Haji, A., Qavamnia, S.S., Bizhaem, F.K.: Salt free neutral dyeing of cotton with anionic dyes using plasma and chitosan treatments. Ind. Text. **67**(2), 109–113 (2016)

Hauser, P.J.: Cationized cotton: opportunities and challenges. AATCC Rev. **18**(3), 40–45 (2018)

Hauser, P.J., Kanik, M.: Printing of cationized cotton with acid dyes. AATCC Rev. **3**(3), 25–28 (2003)

Kampeerapappun, P., et al.: Effect of chitosan and mordants on dyeability of cotton fabrics with Ruellia tuberosa Linn. Chiang Mai J. Sci. **38**(1), 95–104 (2010)

Kanik, M., Hauser, P.J.: Printing of cationised cotton with reactive dyes. Color. Technol. **118**(6), 300–306 (2002)

Kanik, M., Hauser, P.J.: Printing cationized cotton with direct dyes. Text. Res. J. **74**(1), 43–50 (2004)

Ke, G., Zhu, K., Chowdhury, M.H.: Dyeing of cochineal natural dye on cotton fabrics treated with oxidant and chitosan. J. Nat. Fibers **18**(3), 317–329 (2021)

Mansour, R., Ben Ali, H.: Investigating the use of chitosan: toward improving the dyeability of cotton fabrics dyed with Roselle (Hibiscus sabdariffa L.). J. Nat. Fibers **18**(7), 1007–1016 (2021)

Marie, M., et al.: Optimization of dyeing of cationized cotton fibers with safflower extracts. Int. J. Sci. Res. **5**, 1184–1192 (2016)

Nallathambi, A., Rengaswami, G.D.V.: Industrial scale salt-free reactive dyeing of cationized cotton fabric with different reactive dye chemistry. Carbohydr. Polym. **174**(Supplement C), 137–145 (2017)

Patil, A.A., et al.: The use of poly(amido)amine dendrimer in modification of cotton for improving dyeing properties of acid dye. Int. J. Clothing Sci. Technol. **31**(2), 220–231 (2019)

Rym, M., et al.: Dyeing properties of cationized and non-cationized cotton fabrics dyed with Vitis vinifera L. leaves extract. J. Text. Inst. **107**(4), 525–530 (2016)

Samanta, A.K., et al.: Eco-friendly salt-free reactive dyeing of cotton (muslin) fabric after cationization with amino acid from soya. Text. Res. J. **86**(20), 2179–2192 (2016)

Sun, D., Stylios, G.K.: Effect of low temperature plasma treatment on the scouring and dyeing of natural fabrics. Text. Res. J. **74**(9), 751–756 (2004)

Using Response Surface Methodology to Optimize Experimental Parameters of Plasma Induced Graft-Polymerization of Acrylic Acid on PET Braided Structure

Nesrine Bhouri[1(✉)], Faten Debbabi[2], Esther Rohleder[3], Boris Malhtig[4], and Saber Ben Abdessalem[1]

[1] Textile Materials and Processes Research Unit, University of Monastir, Monastir, Tunisia
Bhouri.nesrine@yahoo.fr
[2] Textile Engineering Laboratory, University of Monastir, Monastir, Tunisia
[3] Research Institute of Textile and Clothing, Niederrhein University of Applied Sciences, Monchengladbach, Germany
[4] Faculty of Textile and Clothing Technology, Niederrhein University of Applied Sciences, Monchengladbach, Germany

Abstract. PET is a highly crystalline polymer that does not contain chemical reactive groups. Plasma treatment is widely adopted method to improve surface reactivity while preserving the bulk properties. In this paper, PET-braided structures are functionalized using plasma-induced graft polymerization of acrylic acid. The effect of plasma variables (Gas nature, Plasma power, Gas flow rate, and Exposure time) and acrylic acid graft conditions (acrylic acid concentration and impregnation time) were varied and the optimization of carboxylic acid concentration and the graft yield were established using response surface methodology. Optimum conditions were found under an oxygen atmosphere with a flow rate of 100 sccm and the plasma power set at 72 W for 5 min, after impregnation in solution of 100% AA for 10 min.

Keywords: PET braided structure · Plasma treatment · Plasma-induced graft polymerization · Acrylic acid

1 Introduction

PET is highly crystalline, markedly hydrophobic and does not contain chemically reactive groups, which makes it resistant to straight chemical modification (Del Hoyo-Gallego et al. 2016). However, chemical modification by graft polymerization is an attractive way in which desired monomer can be grafted onto the surface with a covalent bond, giving it good stability. It was used to improve dyeability, antistatic properties, moistures regain or impart antibacterial properties to the fibers (Abdolahifard et al. 2011). Surface modification is a widely adopted method due to its ability to improve the biocompatibility of a material surface while preserving the bulk properties undamaged (Keshvari et al. 2008). Surface modification of PET fibers and fabrics has been carried

S. Msahli and F. Debbabi (Eds.): CIRATM 2020, SPM 17, pp. 135–141, 2022.
https://doi.org/10.1007/978-3-031-08842-1_22

out using plasma, UV, ozone, and radiation-induced graft polymerization of various monomers (Gupta et al. 2001). The grafting of such hydrophilic monomer leads to a surface with suitable chemical functionality for biomolecule interaction at the interface. However, these techniques still have some shortcomings, such as the photosensitizers used in UV grafting have to be thoroughly removed from the polymer surface as a pre-condition for any biomedical application. furthermore, a long gamma irradiation time can damage the mechanical properties of the polymer (Keshvari et al. 2008). Plasma induced graft polymerization is an attractive way of modifying the surface chemistry and morphology of polymeric materials (Cireli et al. 2007). The plasma-induced graft polymerization of monomers onto PET has been investigated by several researchers and changes in the physical behavior and surface morphology have been reported (Gupta et al. 2010). In the current study, we attempted to optimize the experimental variables of the plasma-induced graft polymerization of acrylic acid (AA) onto PET braided structures using response surface methodology. The effects of plasma treatments (gas nature, plasma power, gas flow rate, exposure time) and acrylic acid graft conditions (AA concentration; impregnation time) were investigated.

2 Materials and Methods

PET braids are made with 16 non-texturized PET yarns using a HERZOG circular braiding machine. After braiding, they were washed by Soxhlet using solvents to remove all impurities. The braided threads were functionalized using cold plasma (Diener electronic, Germany) at low pressure. After plasma treatment, a modified grafting procedure was employed. After the grafting reaction, the sample was removed and soxhlet extracted with water to remove any homo-polymer adhering to its surface. In order to study the effect of manufacturing conditions, we carried out several tests following the Taguchi experimental plan with six factors (Gas nature, plasma power, gas flow rate, Exposure time, acrylic acid concentration, Impregnation time) and three levels (Table 1).

Table 1. Experimental plan of acrylic acid grafting

	−1	0	1
Gas nature: X1	Oxygen	Nitrogen	Argon
Plasma power: X2 (W)	30	60	100
Gas flow rate: X3 (sccm)	20	60	100
Exposure time: X4 (min)	1	3	5
Acrylic acid concentration: X5 (%)	20	50	100
Impregnation time: X6 (min)	1	5	10

Each experiment was replicated three times and the graft yield was calculated at the end of each reaction using the Eq. (1).

$$\text{Graft yield} = \frac{W_1 - W_0}{W_0} \times 100 \quad (1)$$

where: W_1 is the weight of grafted PET yarns (g); W_0 is the weight of non-grafted PET yarns (g).

The density of carboxylic functions (µmol of COOH per gram of support) is determined by a staining method using the cationic dye Oxidized Toluidine Blue (TBO) (Saxena et al. 2010). Treated and untreated PET samples were placed at 30 °C for 6 H in a beaker containing a TBO solution (pH = 10). Samples were then rinsed in a hydroxide sodium solution (pH = 10) and immersed in a 50% (v/v) acetic acid solution for 30 min in the dark. The concentration of carboxylic functions is determined by using the Lambert beer law after measuring the absorbance with a spectrophotometer at 634 nm.

3 Results and Discussion

To study the efficiency of the grafting of acrylic acid onto PET braids as a function of experimental parameters, the main effect curves of COOH concentration (Fig. 1) and graft yield (Fig. 2) were established. Results show that the COOH functions of grafted surfaces are more produced in an oxygen atmosphere. It can be seen also that the acrylic acid concentration does not affect the COOH concentration of treated PET braids and an increase of the gas flow rate, the exposure time and the impregnation time increase significantly the COOH functions. On the other hand, results indicate that grafting parameters (AA concentration and the impregnation time) affect more the graft yield of PET braids. In fact, when increasing the AA concentration and the impregnation time, a significant increase in the graft yield was seen. However, the grafting yield decreased at a limit concentration.

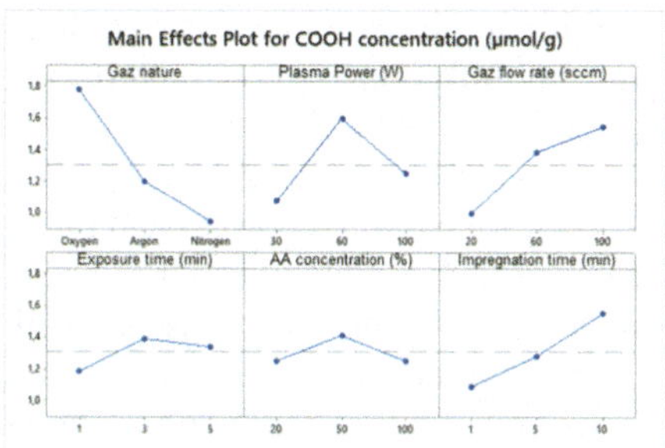

Fig. 1. Main effect plot of COOH concentration

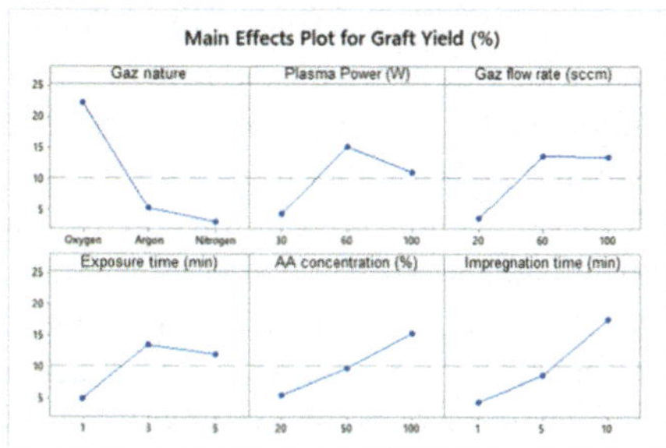

Fig. 2. Main effect plot for graft yield

The regression equations of the COOH concentration and graft yield on the function of experimental parameters were established with squared multiple correlation R2 respectively about 96.70% and 72.88%. So it can be deduced that the obtained models have good predictability. Indeed, generally, linear regression models with a coefficient of regression higher than 70% can be considered acceptable.

Although, the Minitab software provides a search for the best linear regression submodels including the minimum significant factors with the highest regression coefficient R^2 and R^2 (adj) and the lowest standard deviation of the S estimate. In this study, factors

Table 2. Regression analysis results for PET-AA grafting responses

[COOH] (μmol/g)											
	X1	C	X2	X3	X4	X5	X6	X2 * X2	X4 * X4	X5 * X5	X5 * X6
	Ox	−0,909	50,86 10^{-3}	15,47 10^{-3}	237,7 10^{-3}	7,08 10^{-3}	−2,6 10^{-3}	−0,372 10^{-3}	−33,1 10^{-3}	−0,099 10^{-3}	0,902 10^{-3}
	Ar	−1,333									
	N	−1,577									
P-value	0,000	0,000	0,015	0,000	0,027	0,896	0,000	0,000	0,031	0,012	0,011
S = 0,149; R^2 = 96,70%; R^2 (adj) = 92,52%											
Graft yield (%)											
	Ox	−10,85	0,951	0,1217	1,79	0,1255	1,506				
	Ar	−27,87									
	N	−30,05									
P-value	0,001	0	0,152	0,045	0,013	0,04	0,008	–	–	–	
S = 10,3384; R^2 = 72,88%%; R^2 (adj) = 60,83%											

included in the best sub-models for carboxylic groups concentration and the grafting yield as well as a novel R^2 were given in Table 2.

In order to identify optimal conditions of the surface functionalization process with plasma treatment and acrylic acid grafting process, contour plots are plotted as a function of all parameters included in the previously established models for each studied response. Contour plots allow a three-dimensional relationship to be represented in two dimensions, where points with the same response value are linked to give constant response contours and considering other variables constant at a central level.

Figure 3 shows that to maximize the concentration of carboxylic groups of grafted PET sutures, multiple solutions are available. In fact, for a plasma power (X2) ranging between 60 and 80 W, COOH concentration can reach 2.25 μmol/g when the exposure time (X4) is set at 4 min, the gas flow rate (X3) higher than 60 sccm, and the impregnation time in acrylic acid solution (X6) exceeded 5 min. Alternatively, the carboxylic group concentration can be maximized when the gas flow rate (X3) exceeded 65 sccm with exposure time (X4) higher than 1.5 min. from this figure, it can be also seen that acrylic acid concentration (X5) does not affect the [COOH] response optimization. These results coincide with our previously established model where the p-value of X5 is higher than 0.05.

As concerning the graft yield, Fig. 4 present that best results are obtained when the plasma power (X2) is set between 60 W and 85 W, the gas flow rate (X3) over 80 sccm, a plasma exposure time (X4) longer than 4.5 min, an acrylic acid concentration above 80% and finally an impregnation time (X6) higher than 6 min. At these settings, we provide an acrylic acid graft yield on PET braided structure exceeded 30%.

The optimum conditions, of the acrylic acid grafting process after surface modification with plasma treatment, were predicted by the response optimizer tool of Minitab17

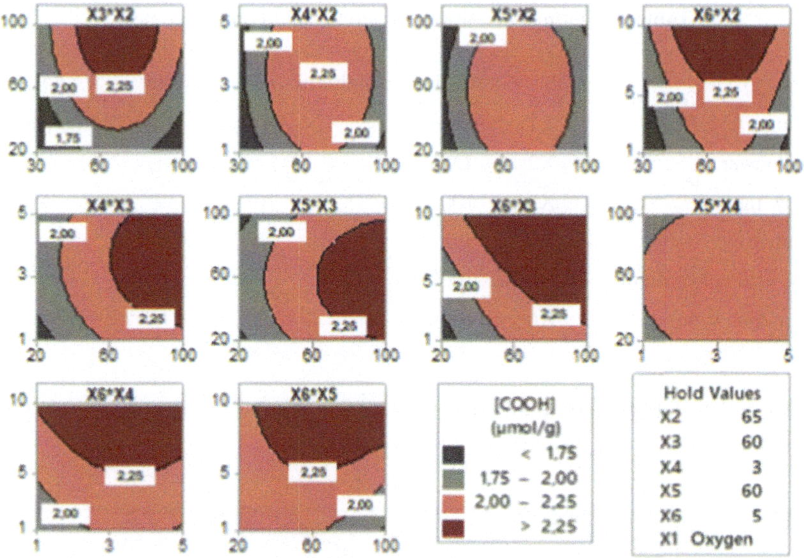

Fig. 3. Contour plots of the carboxylic group concentration

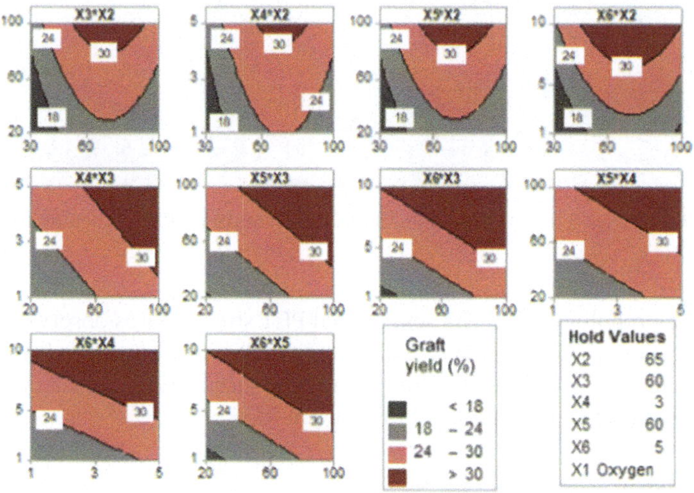

Fig. 4. Contour plots of the graft yield

software for a maximized response of the graft yield and the concentration of carboxylic groups. It allows us to estimate the values predicted by a desirability value varying from 0 to 1. Optimal conditions are shown in Table 3 with an overall desirability value to 0.87.

To confirm the established models, further samples were performed under the optimal conditions and carboxylic concentration groups and the grafting yield of the grafted samples were measured. Carboxylic acid concentration of the obtained samples are

Table 3. Optimal conditions of acrylic acid grafting process

Factors		Optimal conditions	Graft yield	[COOH]
Gas nature	X1	Oxygen	48.8%	2.60 μmol/g
Plasma power	X2	72 W		
Gas flow rate	X3	100 sccm		
Exposure time	X4	5 min		
Acrylic acid concentration	X5	100%		
Impregnation time	X6	10 min		

about 50.86% ± 2.62% and obtained graft yield is around 2.62 μmol/g ± 0.02 μmol/g. These results confirm the validity of both models.

Fig. 5. SEM images at a magnification of 3000X of (a) untreated PET structure; (b) AA-grafted PET structure

Surface morphology of acrylic acid grafted PET structure was observed using SEM at a magnification of X3000 and presented in Fig. 5. It is obvious that untreated braided surfaces are smooth and distinct. However, treated surface (Fig. 5-b) shows a uniform layer of acrylic acid on the surface of PET structure.

4 Conclusion

A new process for plasma-induced graft polymerization of acrylic acid on PET braided structure has been developed in order to improve their hydrophilicity. For this purpose, plasma experimental parameters were varied. Then graft polymerization of the acrylic acid was studied. The measurement of carboxylic group concentration and the graft yield were used to determine the physical characteristics of plasma-induced graft polymerization of acrylic acid on PET braids. The modeling and the optimization of these parameters using response surface design found that optimum properties were obtained in an oxygen atmosphere at 72 W plasma power, 100 sccm for 2 min, with impregnation in acrylic acid at a concentration of 100% for 10 min.

References

Abdolahifard, M., et al.: Surface modification of PET fabric by graft copolymerization with acrylic acid and its antibacterial properties. ISRN Org. Chem. **2011**, 265–415 (2011). 2090-5149

Cireli, A., et al.: Surface modification of polyester and polyamide fabrics by low frequency plasma polymerization of acrylic acid. J. Appl. Polym. Sci. **104**(4), 2318–2322 (2007). 00218995

Del Hoyo-Gallego, S., et al.: Construction of antibacterial poly(ethylene terephthalate) films via layer by layer assembly of chitosan and hyaluronic acid. Carbohydr. Polym. **143**, 35–43 (2016). 1879-1344

Gupta, B., et al.: Plasma-induced graft polymerization of acrylic acid onto PET films: characterization and human smooth muscle cell growth on grafted films. Biomaterials **23**, 863–871 (2001)

Gupta, B., et al.: Plasma induced graft polymerization of acrylic acid onto PET monofilament. Indian J. Fibre Text. Res. **35**, 9–14 (2010)

Keshvari, H., et al.: Collagen immobilization onto acrylic acid laser grafted silicone for using as artificial skin: in vitro. Iran. Polym. J. **17**(3), 171–182 (2008)

Saxena, S., et al.: Chitosan immobilization on polyacrylic acid grafted polypropylene monofilament. Carbohydr. Polym. **82**(4), 1315–1322 (2010). 01448617

Development of Fluorine-Free Highly Hydrophobic Fabric Finishing Agent by Silica Hydrosol: Optimizing Process Conditions

Mahdi Hasanzadeh[1](✉), Hossein Shahriyari Far[2], and Aminoddin Haji[1]

[1] Department of Textile Engineering, Yazd University, P.O. Box 89195-741, Yazd, Iran
m.hasanzadeh@yazd.ac.ir

[2] Department of Chemistry, Iran University of Science and Technology, Narmak, Tehran, Iran

Abstract. This work focused on the fabrication of superhydrophobic fabric through hydrolysis of tetraethoxysilane (TEOS). Affecting parameters on the morphology and hydrophobicity of fabric such as ammonia content (5–9 mL) and the treatment time (40–120 min) were investigated via experimental design approach. The morphology and the composition of superhydrophobic fabric were characterized by scanning electron microscopy (SEM) and energy dispersive X-ray (EDS). The results revealed the formation of spherical silica nanoparticles with average particle size of 270 nm throughout the fabric surface. Measurement of water contact angle (WCA) also showed the hydrophobicity of fabrics.

Keywords: Cotton fabric · Hydrophobic coating · Silica · Contact angle · Experimental design

1 Introduction

A surface with a water contact angle (at equilibrium) larger than 90° and a sliding angle lower than 10° is considered as a superhydrophobic surface. Nowadays superhydrophobic coatings have gained increasing attention due to their potential practical applications including self-cleaning (Zhang et al. 2013), bio-fouling (Xue et al. 2015), anti-icing (Hu et al. 2015), anticorrosion (Liu et al. 2016), and selective absorption (Xue et al. 2013). Cotton fabric has always been the principal clothing fabric due to its attractive characteristics such as softness, comfort, warmness and biodegradability. However, the large number of hydroxyl groups on cotton surface makes the fabrics water-absorbent and easily stained by liquids. Therefore, additional finishes are required to impart superhydrophobicity and self-cleaning properties to cotton fabrics. A large number of superhydrophobic surfaces have been prepared on various substrates by different methods (Roach et al. 2008), such as glass (Deng et al. 2011), metals (Liu et al. 2014), sponges (Wu et al. 2015), and textiles (Wu et al. 2013). Until now, it is gradually paid attention to the construction of hydrophobic fiber-based flexible substrates (Zhou et al. 2013). In general, surfaces with a certain extent roughness can be obtained by introducing inorganic nano-size particles such as SiO_2 (Wu et al. 2009), TiO_2 (Goncalves et al. 2009) and ZnO

S. Msahli and F. Debbabi (Eds.): CIRATM 2020, SPM 17, pp. 142–147, 2022.
https://doi.org/10.1007/978-3-031-08842-1_23

(Lakshmi and Basu 2009). Monodisperse silica nanospheres with remarkable colloidal stability are one of the most attractive materials. In order to realize superhydrophobicity, the desired surface roughness can be obtained by carefully controlling hydrolysis and condensation reactions during sol–gel processing. On the other hand, the surface chemistry may be commonly modified by introducing some hydrophobic compounds through the surface condensation reaction. In present, most of approaches for preparing SiO_2 nanoparticles were based on the Stöber method (Lakshmi and Basu 2009). However, the Stöber synthesis involves a large amount of organic solvent like alcohol as the reaction medium to prevent phase separation for producing a monodisperse shape. Roughened surfaces have been commonly obtained by introducing nano-size particles onto the pristine surface and the sol–gel technique has been reported as a promising tool for preparation of water repellent coatings, especially versatile for application on paper, textiles or wood.

There have been some reports on the preparation of silica nanoparticles in the alcohol-free sol–gel system. For example, Li et al. (Shang et al. 2010) prepared silica hydrosols via acid catalyzed hydrolysis and condensation of water glass and obtained superhydrophobic cotton fabrics by dip-coating the silica hydrosols and self-assembling HDTMS. But the mechanical properties of the treated cotton fabrics may be affected as a result of the introduction of hydrochloric acid. Menget al. (Meng et al. 2009) prepared highly monodisperse hybrid spherical silica nanoparticles with diameters ranging from 30 to 200 nm by a one-step emulsion polymerization in aqueous solution, opening a new broad avenue for the hybrid material.

Herein, we present a facile and non-fluorinated approach to fabricate superhydrophobic and breathable fabric by silica hydrosol. The formation of silica nanoparticles, morphology and hydrophobicity of treated samples were investigated by scanning electron microscopy (SEM), energy dispersive X-ray (EDS) and contact angle measurement. The influence of affecting parameters, including ammonia content and treatment time, on morphology of synthesized silica nanoparticles and the hydrophobicity of treated cotton fabric have been addressed.

2 Experiments

Tetraethylorthosilicate (TEOS), ammonium hydroxide (NH_4OH), and ethanol were obtained from Sigma Aldrich Co. All the chemicals were used as received without further purification. The plain-woven, desized and bleached cotton fabric used in this work as substrate.

The spherical silica nanoparticles were prepared by the modified Stober method. First, a mixture of TEOS (0.39 mol) and ethanol (2.35 mol) was prepared as a Solution A. Then, 2.94 mol distilled water, 2.35 mol ethanol, and NH_4OH were mixed together to form Solution B. The Solutions A and B were mixed at 40 °C for 40 min. The cotton fabrics were dip-coated with the as-prepared sol of silica nanoparticles at 30 °C for 5 min and then dried at 80 °C for 3 min.

The surface morphology of pristine and treated fabrics, as well as the composition of materials, were studied after gold coating of samples by scanning electron microscopy (Hitachi su3500) and energy-dispersive X-ray spectroscopy (EDS). The water contact angle (WCA) measurement was carried out at room temperature using home-made instrumentation including a microscope equipped with a CCD camera and PCTV vision software. A 5 μL water droplet was dropped to the five different locations of fabric surface and an average value was reported.

The hydrophobic properties of cotton fabrics were optimized with response surface methodology (RSM) by Design-Expert software (version 8.0.3) including analysis of variance (ANOVA). Based on preliminary experiments, NH_4OH content (X1) and treatment time (X2) were determined as critical factors with significance impact on hydrophobicity of cotton fabrics. A series of 9 silica sols were prepared by changing the ammonia content (5–9 mL) within 150 mL and treatment time (40–120 min). Table 1 shows the experimental design condition.

Table 1. Design of experiment (factors and levels)

Variables	Symbol	Coded values		
		−1	0	+1
Ammonia content (mL)	X1	120	125	130
Treatment time (min)	X2	5	7	9

3 Results and Discussions

3.1 Analysis of Variance

The experimental results concerning hydrophobicity of fabric in term of WCA (Y) using two significant parameters of ammonia content and treatment time are obtained. The results of the second-order response surface model in the form of analysis of variance (ANOVA) for WCA are obtained. In this work, statistical conclusions were obtained at 95% confidence level. Using 5% significance level, the factor is considered significant if the p-value is less than 0.05.

By linear regression analysis, the predicted response function (Y) was obtained and given as

$$Y = 124.79 + X_1 - 2.00X_2 + 0.75X_1X_2 + 0.72X_1X_1 + 4.72X_2X_2 \quad (1)$$

The obtained results showed that quadratic model was significant and have a good agreement with experimental data. The coefficient of determination (R^2) (0.72) also revealed an appropriate correlation between observed and predicted values. This can be also confirmed by comparing the actual values against the predicted responses by the models for hydrophobicity of cotton fabric (Fig. 1).

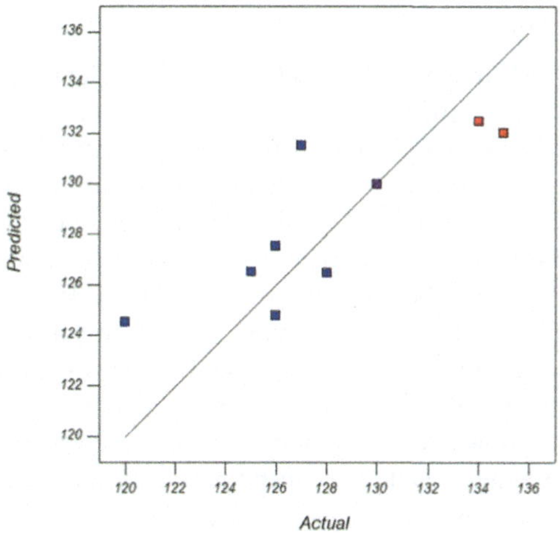

Fig. 1. Predicted water contact angle versus actual values.

3.2 Effects of Significant Parameters

Figure 2 shows the combined effect of varying ammonia content and treatment time on WCA of hydrophobic cotton fabric. It can be seen that at fixed treatment time, an increase in ammonia content causes the decrease followed by an increase in WCA. It is obvious that the maximum WCA could be achieved at low ammonia content and treatment time.

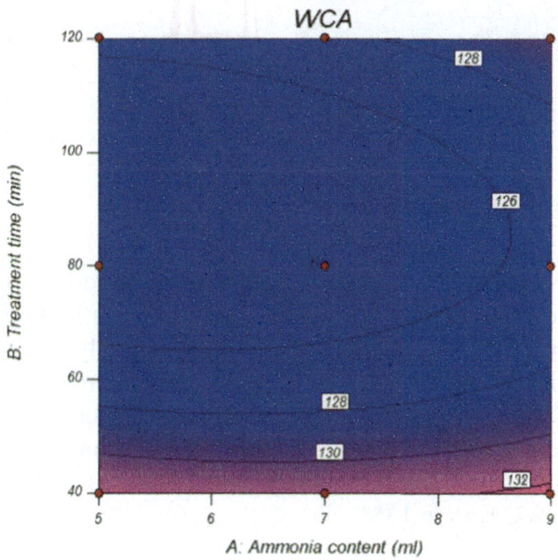

Fig. 2. Response surface contour plot for water contact angle of hydrophobic cotton fabric.

3.3 Morphological Study

Figure 3 shows the SEM images of hydrophobic cotton. The results exhibited a rough surface and the presence of silica nanoparticles on the surface of the fabric. The high-magnification (inset) SEM image showed that the uniform silica nanoparticles were deposited throughout the fabric surface with an average particle size of about 270 nm. The results of EDS spectra (Fig. 3b) confirm the presence of the Si element on the superhydrophobic fabric surface. The peaks of C and O elements were also detected in EDS spectra which are attributed to the polymer chain. It was anticipated that the addition of silica nanoparticles would increase the cotton surface roughness, leading to an additional increase in the water contact angle. The results obtained from WCA measurement show that the WCA increases from 121 to about 133°. According to SEM result, the surface of cotton fiber by its nature, shows the roughness and distribution of silica nanoparticles which made a contribution to an increase in the fiber surface roughness geometry.

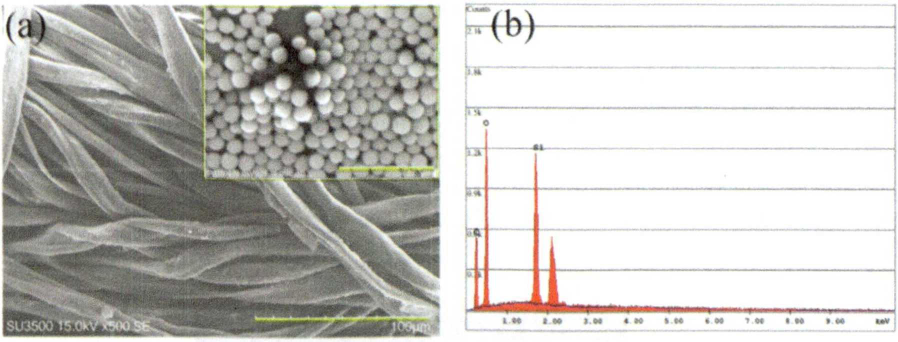

Fig. 3. (a) SEM images and (b) EDS spectrum of silica coated cotton fabric.

4 Conclusion

From the obtained results it can be concluded that the modification of cotton fabric by silica nanoparticle is a promising method to confer high hydrophobicity to cotton fabrics. In this way, silica-based compounds homogeneously dispersed on the fabric surface with the consequence of an increase of the surface roughness as well as with decrease the surface free energy. The effect of ammonia content and the treatment time, as the most important factor, were studied through experimental design approach. The obtained results showed that decreasing the treatment time as well as ammonia content leads to increase in WCA up to 135°.

References

Deng, X., Mammen, L., Zhao, Y., et al.: Transparent, thermally stable and mechanically robust superhydrophobic surfaces made from porous silica capsules. Adv. Mater. **23**, 2962–2965 (2011). https://doi.org/10.1002/adma.201100410

Goncalves, G., Marques, P.A.A.P., Pinto, R.J.B., et al.: Surface modification of cellulosic fibres for multi-purpose TiO2 based nanocomposites. Compos. Sci. Technol. **69**, 1051–1056 (2009). https://doi.org/10.1016/j.compscitech.2009.01.020

Hu, F., Xu, P., Wang, H., et al.: Superhydrophobic and anti-corrosion Cu microcones/Ni-W alloy coating fabricated by electrochemical approaches. RSC Adv. **5**, 103863–103868 (2015). https://doi.org/10.1039/c5ra20638c

Lakshmi, R.V., Basu, B.J.: Fabrication of superhydrophobic sol-gel composite films using hydrophobically modified colloidal zinc hydroxide. J. Colloid Interface Sci. **339**, 454–460 (2009). https://doi.org/10.1016/j.jcis.2009.07.064

Liu, N., Cao, Y., Lin, X., et al.: A facile solvent-manipulated mesh for reversible oil/water separation. ACS Appl. Mater. Interfaces **6**, 12821–12826 (2014). https://doi.org/10.1021/am502809h

Liu, W., Xu, Q., Han, J., et al.: A novel combination approach for the preparation of superhydrophobic surface on copper and the consequent corrosion resistance. Corros. Sci. **110**, 105–113 (2016). https://doi.org/10.1016/j.corsci.2016.04.015

Meng, Z., Xue, C., Zhang, Q., et al.: Preparation of highly monodisperse hybrid silica nanospheres using a one-step emulsion reaction in aqueous solution. Langmuir **25**, 7879–7883 (2009). https://doi.org/10.1021/la900458b

Roach, P., Shirtcliffe, N.J., Newton, M.I.: Progess in superhydrophobic surface development. Soft Matter **4**, 224 (2008). https://doi.org/10.1039/b712575p

Shang, S.M., Li, Z., Xing, Y., et al.: Preparation of durable hydrophobic cellulose fabric from water glass and mixed organosilanes. Appl. Surf. Sci. **257**, 1495–1499 (2010). https://doi.org/10.1016/j.apsusc.2010.08.081

Wu, J., Li, J., Deng, B., et al.: Self-healing of the superhydrophobicity by ironing for the abrasion durable superhydrophobic cotton fabrics. Sci. Rep. **3**, 1–6 (2013). https://doi.org/10.1038/srep02951

Wu, L., Li, L., Li, B., et al.: Magnetic, durable, and superhydrophobic polyurethane@Fe3O4@SiO2@fluoropolymer sponges for selective oil absorption and oil/water separation. ACS Appl. Mater. Interfaces **7**, 4936–4946 (2015). https://doi.org/10.1021/am5091353

Wu, W., Wang, X., Wang, D., et al.: Alumina nanowire forests via unconventional anodization and super-repellency plus low adhesion to diverse liquids. Chem. Commun. 1043–1045 (2009). https://doi.org/10.1039/b818633b

Xue, C.-H., Guo, X.-J., Ma, J.-Z., Jia, S.-T.: Fabrication of robust and antifouling superhydrophobic surfaces via surface-initiated atom transfer radical polymerization. ACS Appl. Mater. Interfaces **7**, 8251–8259 (2015). https://doi.org/10.1021/acsami.5b01426

Xue, C.H., Ji, P.T., Zhang, P., et al.: Fabrication of superhydrophobic and superoleophilic textiles for oil-water separation. Appl. Surf. Sci. **284**, 464–471 (2013). https://doi.org/10.1016/j.apsusc.2013.07.120

Zhang, X., Guo, Y., Zhang, Z., Zhang, P.: Self-cleaning superhydrophobic surface based on titanium dioxide nanowires combined with polydimethylsiloxane. Appl. Surf. Sci. **284**, 319–323 (2013). https://doi.org/10.1016/j.apsusc.2013.07.100

Zhou, H., Wang, H., Niu, H., et al.: Robust, self-healing superamphiphobic fabrics prepared by two-step coating of fluoro-containing polymer, fluoroalkyl silane, and modified silica nanoparticles. Adv. Funct. Mater. **23**, 1664–1670 (2013). https://doi.org/10.1002/adfm.201202030

Development of a Fluorescent Textile Marker

Fredj Saad(✉), Ayda Baffoun, and Mohamed Hamdaoui

National School of Engineers of Monastir, Textile Materials and Processes Research Unit (M.P.Tex), University of Monastir, Rue Ibn Aljazzar, 5019 Monastir, Tunisia
fredj93.saad@gmail.com

Abstract. Textiles are extremely important materials for everyday life, with a wide range of applications and properties. Due to the wide variations in quality, on the one hand, and the increasing customer awareness of quality and price, on the other hand, the availability of a simple tool to quickly test the correct identity of the purchased textile article would be a significant step forward in customer protection. The application of fluorescein (FL) in a textile filament polyamide for anti-counterfeiting applications is described. The fluorescent samples are characterized by Scanning Electron Microscopy SEM and optical spectroscopy in arrangement of reflectance and fluorescence. The treated samples show under UV light a yellowish green emission with a weak yellow coloration of polyamide fiber.

Keywords: Textile structure · Anti-counterfeiting · Conjugated organic polymers · Fluorescence

1 Introduction

The application of fluorescent molecules can be applied in a variety of different areas. However, in textile the application of fluorescent dyes results in a significant increase in the brightness of the colours, making the dyed materials more easily perceptible. Fluorescent materials have a unique advantage in the manufacture of sportswear and clothing for special services, such as firefighters and police, due to their increased perceptibility (Szuster et al. 2004). Recently, the application of fluorescent materials in medical diagnostics and biochemical investigations has been developing extremely dynamically (Satam et al. 2013). Among the many classes of highly fluorescent dyes, sets based on xanthene skeletons containing mainly rhodamine and fluorescein dyes have aroused considerable interest from chemists due to their excellent photophysical properties such as a high coefficient of absorption, excellent fluorescence quantum yield, good photostability and a relatively long emission wavelength. Rhodamine and fluorescein derivatives have been widely applied with success in many fields, including biology and medicine (Hong 2012). Fluorescein is widely used as a platform for various fluorescence probes and fluorescence markers due to its high fluorescence quantum efficiency in aqueous media and both its excitation wavelength and its emission wavelength are in the range of the visible region, which is beneficial for its detection. As fluorescein has been reported to be relatively non-toxic, among the most versatile chromophores, they have been used as fluorescence probes in an aqueous system of biological molecules.

S. Msahli and F. Debbabi (Eds.): CIRATM 2020, SPM 17, pp. 148–153, 2022.
https://doi.org/10.1007/978-3-031-08842-1_24

The actually paper is related to using the fluorescein dye in textile field as a new fluorescent material for anti-counterfeiting technology that will allow tagging and identifying legitimate items (Baatout 2019). Fluorescein with their interesting properties can be applied on polyamide in order to have a wire identification that can be used in textile articles such as composition labels, embroidered logos or even seams. Thus, its presence in textile support can only be detected after excitation at a very specific wavelength under UV light, which makes this material very effective in ensuring the traceability and detection of counterfeit items (Wang 2006; Lavis 2007).

The aim of this work was to examine the use of fluorescein as a fluorescent marker in the textile field. For this purpose, 100% polyamide filaments are treated with the organic molecule by the exhaustion technique. The prepared fluorescent sample is characterized by measuring the reflectance as a function of the absorption and emission wavelength of the marker in order to have an appreciation of the textile coloration in relation to their fluorescence intensity.

2 Materials and Methods

2.1 Materials

2.1.1 Fluorescein

Fluorescein dye is a synthetic organic compound available as a dark orange/red powder soluble in water and alcohol. Fluorescein dye has an absorption maximum at 490 nm and an emission maximum of 520 nm (in water), the color of its aqueous solution varies from green to orange.

Fluorescein and many of its derivatives exhibit multiple, pH-dependent ionic equilibria. Both the phenol and carboxylic acid functional groups of -fluorescein are almost totally ionized in aqueous solutions above pH 9 (Fig. 1). Acidification of the fluorescein dianion first protonates the phenol (pKa ~ 6.4) to yield the fluorescein monoanion, then the carboxylic acid (pKa ~ 2.1).

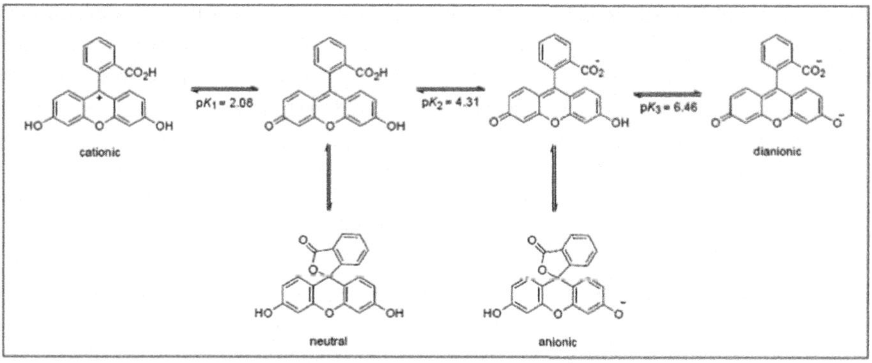

Fig. 1. pH dependence of fluorescein equilibria

With quantum yields of 0.36 and 0.93 respectively, only the monoanion and dianion of fluorescein are fluorescent, although excitation of neutral or cationic species produces emissions of the anion with effective quantum yields of 0.29 and 0.18, respectively. Another equilibrium involves the formation of a colorless, non-fluorescent lactone. Lactone is not formed in aqueous solution above pH 5 but may be the dominant form of neutral fluorescein in solvents such as acetone.

2.1.2 Polyamide

Polyamide 66 filaments of 110 Dtex titer are used. The nylon fibres contain both amine and carboxylic acid end-groups. Thus, by selecting the appropriate dyeing conditions, it can absorb anionic or cationic dyes. Nylon is usually dyed with anionic dyes via ionic bonds between the protonated amino terminal groups of the nylon (NH_3^+) and the residual anionic groups of the dye (Dye-X^-). As the number of terminal amino groups on the surface of nylon fibres is limited (0.011 amino ends/nm^2), high coverage and heavy shades in the dyeing process are only achieved to a limited extent when the molecular size of the dyes used is high. Furthermore, amino groups on the untreated fibers are capable of forming hydrogen bonds with the groups –OH and –COO of the dye molecules, therefore creating bridges between fiber and dye, consequently, increasing the affinity of dye to Nylon fiber (Fatma 2015).

2.2 Methods

A 20 ml dye bath, suitable for a 0.5 g nylon sample, containing fluorescent organic material with a concentration of 10^{-6} mol/L to 10^{-5} mol/L and acid donor for pH control has been prepared. The.dyeing was carried out for 60 min at 98 °C in a dyeing machine at the Ahiba Nuance laboratory (Ahiba, Datacolor). After dyeing, all samples were hot rinsed, then washed and finally air-dried.

2.2.1 Experimental Design

The full factorial design was used in this study. In this work, the concentration of the fluorescent tracer, the pH and the concentration of ammonium sulphate were adopted as controllable variables. Table 1 presents the factors and their different levels of experimentation. The experimental analysis of variance (ANOVA) was as follows using the Minitab 17 statistical software. This software was used to study the effects of the different parameters on the reflectance and the possible interactions between them.

Table 1. Control factors and their levels

Levels	Factor		
	Ammonium sulfate (g/L)	Fluorescein concentration (10^{-6} mol/L)	pH
1	0	1	5.5
2	2	5	6.5
3	4	10	7.5

2.2.2 Characterizations

The reflectance and fluorescence measurements were given by the Datacolor 400TM spectrophotometer in the visible part of the spectrum (400–700 nm). All measurements were performed under the conditions of a D65 illuminant and 10° observer.

SEM of the treated fabrics was studied using a scanning electron probe microanalyzer (type JSM-IT 100) Japan. The specimens in the form of the fabrics were mounted on the specimen stabs and coated with a thin film of gold by the sputtering method. The micrographs were taken at a magnification of 1500 using 15 kV accelerating voltage.

3 Results and Discussions

3.1 Main Effects Diagram for Reflectance and Fluorescence

The objective of this study is to determine the optimal conditions of the fluorescein application process on polyamide filaments using the full factorial methodology. This design output data is obtained by performing fluorescent fiber tests for the indication of reflectance using a spectrocolorimeter. It includes the measurement of the reflectance value and the fluorescence at the wavelength of 460 nm and 550 nm respectively. Factors that influence the reflectance and fluorescence of the treated samples are evaluated using factorial plots: mainly effects and interactions. The main effects of each (pH, fluorescein concentration and ammonium sulfate concentration) on the reflectance and fluorescence of treated polyamide are shown in Fig. 2.

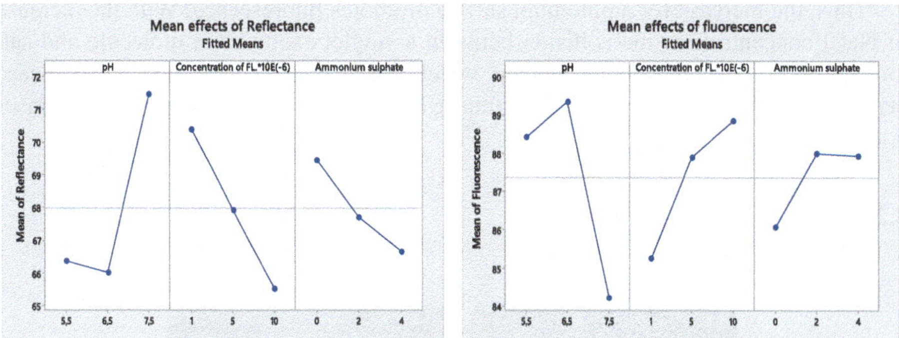

Fig. 2. Main effects diagram for reflectance on the left and fluorescence on the right

3.1.1 Effect of pH, Concentration of Fluorescein and Quantity of Ammonium Sulphate on the Reflectance

At pH 5.5, the reflectance is 67.13% while increasing the pH to 6.5 the reflectance decreases to a value of 66.39%. This reduction is explained by the important ionic interactions between the amine groups of polyamide and the carboxylic acid of fluorescein at pH 6.5. Indeed, our organic molecule is negatively charged at slightly acidic pH and the polyamide is positively charged which leads to one more ionic interaction than hydrogen interactions and van der Waals forces. At pH 7.5, the organic molecule presents the two following groups COO^- and OH and the polyamide become uncharged. Therefore, only hydrogen interactions and van der Waals forces will take place and not ionic interactions. This explains the increase in reflectance to 70.82%.

By increasing the amount of fluorescent tracer in the bath, the reflectance decreases and reaches a minimum reflectance value of 64.87% at a molar concentration of 10^{-5} mol^{-1}. This decrease can be explained by the increase in the number of carboxylic and alcoholic groupings of fluorescein in the bath and thus improve the accessibility of these groupings. This facilitates ionic interactions, hydrogen interactions and van der Waals forces with polyamide amine groups.

The role of ammonium sulfate is to maintain the pH of the bath at an acidic pH throughout the dye reaction. While increasing its concentration in the bath from zero to 4 g L^{-1} the reflectance decreases by a small margin from 68.94% to 67.34% respectively. This reduction can be explained by the exchange of the anion associated with ammonium ion in the fiber with a dye anion.

3.1.2 Effect of pH, Concentration of Fluorescein and Quantity of Ammonium Sulfate on the Fluorescence

For fluorescence, the acidic or even slightly acidic pH corresponds to a good emission under UV light where the negatively charged organic molecule reacts with the amino group of the nylon through ionic bonds. An increase in fluorescein concentration from 10^{-6} mol L^{-1} to 10^{-5} mol/L^{-1} leads to an elevation of the reflectance at the tracer emission wavelength from 85.31% to 89.87% respectively.

Thus, the increase for ammonium sulfate promotes fluorescence. With the increase of NaCl concentration, the collision between a singlet excited dye molecule and salt ions in solution becomes more frequent, which causes fluorescent molecules to release heat back to the environment and to transit back to the ground state in the form of non-radiation.

3.2 SEM of Polyamide Filaments

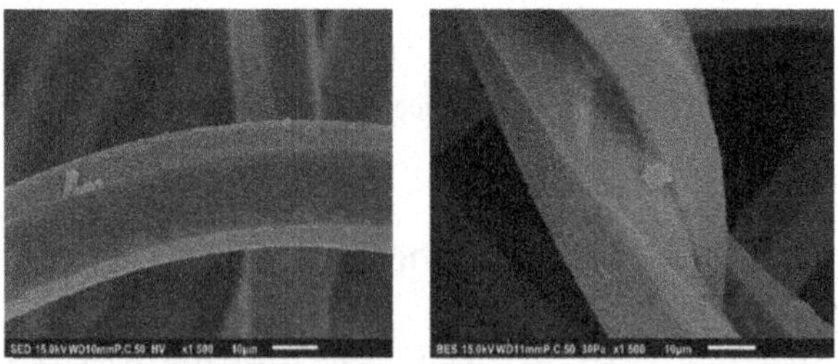

Fig. 3. The SEM micrographs polyamide filaments under magnification (×1500); in the left figure shows the SEM micrographs of untreated polyamide; in the right figure shows the SEM micrographs of treated polyamide.

The SEM analysis shows the presence of the fluorescent agent into the polyamide filaments (Fig. 3).

4 Conclusion

In this study, organic fluorescent molecule was successfully incorporated into polyamide 66 filaments for use as an anti-counterfeiting agent. SEM results showed that a large number of granular fluorescent particles were present on the surface of the filaments. In addition, the effect of pH, fluorescein concentration and the amount of ammonium sulfate on the reflectance and fluorescence of the treated samples has been clearly studied using Minitab 17.

References

Fatma, A.: Solvent free fluorescein dye and its application use Microwave. Int. J. Chem. Technol. Res. **7**, 164–169 (2015)

Hong, Z.: Advances in modifying fluorescein and rhodamine fluorophores as fluorescent chemosensors. The Royal Society of Chemistry, vol. 19 (2012)

Baatout, K.: Luminescent cotton fibers coated with fluorescein dye for anti-counterfeiting applications. Int. J. Mater. Chem. Phys. **234**, 304–310 (2019)

Wang, L.: Study on the fluorescence properties of fluorescein dye incorporated into SBA-15. Opt. Mater. **28**, 1232–1234 (2006)

Lavis, L.D.: Tuning the pKa of fluorescein to optimize binding assays. Anal. Chem. **79**, 6775–6782 (2007)

Satam, M.A., Raut, R.K., Sekar, N.: Fluorescent azo disperse dyes from 3-(1,3-benzothiazol-2-yl)naphthalen-2-ol and comparison with 2-naphthol analogs. Dyes Pigments **96**(1), 92–103 (2013)

Szuster, L., Kaźmierska, M., Król, I.: Fluorescent dyes destined for dyeing high-visibility polyester textile products. Fibres Textiles Eastern Europe **45**, 70–75 (2004)

Surface Characterization of Textile Reinforcements

Elaissi Arwa[1](✉), Alibi Hamza[2], and Ghith Adel[1]

[1] Textile Materials and Research Process (ENIM), University of Monastir, Monastir, Tunisia
elaissiarwa@gmail.com

[2] Laboratory of Thermal and Energetic Studies (LESTE), Tunisia National Engineering School of Monastir, University of Monastir, Monastir, Tunisia

Abstract. Natural fibers represent renewable materials which, nowadays, are experiencing a great revival. They are low-density materials yielding considerably lightweight composites with highly specific properties. However, the disadvantage of these fibers is the low adhesion with most polymers. In composites, the matrix as well as the reinforcing preserve their physical and chemical properties, offer a better combination of mechanical properties. In the present study, we describe the modification of reinforcement surface by alkali treatment and its influence on the adhesion between the components of the composite, which is evaluated using zeta potential and contact angle tests to estimate the specific surface-chemical changes.

Keywords: Reinforcements · Zeta potential · Adhesion · Modification

1 Introduction

During the last years, industrial waste is becoming a huge problem around the world and more particularly in developing countries due to the increase in production resulting in population growth, urbanization, industrialization, and economic development (Elaissi et al. 2020). Composites had attracted the attention of the majority of researchers as they represent a new class of materials of high interest. More importantly, different types of composites containing essentially natural fibers and natural polymers have been proposed. Currently, the use of natural fibers as reinforcement is becoming the main challenge. Indeed, natural fibers represent renewable materials which, nowadays, are experiencing a great revival. A composite material is an assembly of reinforcement with a matrix. The functionalization of reinforcement surfaces improves the performance of a material (Jölly et al. 2015).

In the present study, we describe the modification of reinforcement surface for composites and its influence on the adhesion between the components of the composite, which is evaluated using zeta potential and contact angle tests.

2 Materials and Methods

Cotton fiber waste is collected by the Tunisian combers of a spinning mill. Waste tow fibers are also collected from the local building construction industry.

S. Msahli and F. Debbabi (Eds.): CIRATM 2020, SPM 17, pp. 154–158, 2022.
https://doi.org/10.1007/978-3-031-08842-1_25

The cotton-based textile waste and cellulose tow must receive a pickling treatment to remove any impurities, and a bleaching treatment to obtain active chemical sites within the fibrous material.

Tow fibers (about 3 mm long) were stripped for 2 h in a flask containing a mixture of 1.5 ml/L NaOH, 3 g/L Na_2CO_3, and 4 g/L non-ionic detergent (ratio of liquor 1:40) to clean the fibers by removing starch and waxes from the surface. The fibers are then rinsed hot and then cold and finally, the fibers are dried in a laboratory oven (Tka et al. 2018).

To improve fiber-fiber and fiber-iron adhesion, the fibers were also subjected to a cationization process carried out with dimethyl diallyl ammonium chloride and a diallyl amine copolymer, under alkaline conditions (Bigand et al. 2011).

For the surface treatment, the fibers were dissolved in a cationized/water mixture at a temperature of 50 °C. The mixture was stirred for 30 min, followed by cooling. Then the fibers were immersed in an acetic acid solution which was followed by washing with distilled water. Two percentages of agents were considered: 3% and 6%. These percentages of agents were chosen to take into account the chemical reactivity and the functional groups present in the fibrous material. A comparative study was carried out to obtain the optimized percentage giving the best adhesion in the prepared fibrous materials (Jabli et al. 2017).

After the chemical treatments, the fibrous materials are mixed using a Shirley waste analyzer. Then, the fibers are opened and mixed by a card to obtain a homogeneous layer of fibers. The needling operation is maintained to consolidate the sail.

Nonwoven were prepared with three percentages of tow and cotton: 20% tow/80% cotton, 50% tow/50% cotton, and 80% tow/20% cotton.

Reinforcements, after chemical modification, were analyzed using many techniques including zeta potential and contact angle.

3 Electrokinetic Phenomena

We establish the phenomenon of electrophoresis, that is, the study of the movement of charged particles suspended in a liquid under the influence of an electric field. The negatively charged particles will move towards the anode and on the contrary, the positively charged particles will move towards the cathode. Once the balance of forces is reached, the speed of movement of the particles is measured (Guo et al. 2009). A Zetasizer 3000 from the company Malvern is used to determine the zeta potential of resin. While measuring the zeta potential of fibers we used a zeta-meter from the company ZetaCAD (Fig. 1).

4 Wettability Assessment

Wettability studies the spread of a drop of liquid on a solid surface.

This phenomenon depends on the relative amplitudes of molecular forces located in the liquid (cohesive) and between the liquid and the solid (adhesive).

The device responsible for this study is the DIGIDROP (Grundke et al. 1995).

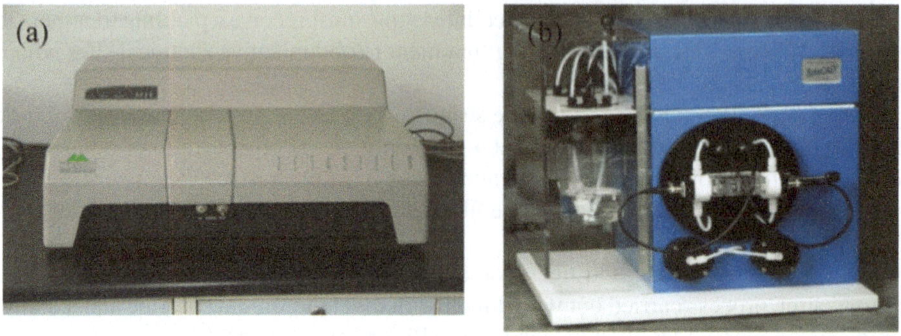

Fig. 1. (a) Zetasizer machine and (b) Zetameter machine

It will be given directly by measuring the tangent to the liquid-vapor interface and the solid surface using a goniometer by applying the following equation:

$$\theta = 2 * \arctan(2h/D) \quad (1)$$

with h and D are the height h and the diameter D of the drop.

5 Results and Discussions

The diameter of tow fibers is approximately 41.73 μm. Figure 2a and Fig. 2b show the microscopic view of flax fiber and tow fibers respectively. The figure proves that the fibers are similar in morphology and compositions. In these cases, no significant difference can be seen between the spectra of flax and tow fibers. There is only a slight decrease of these two peaks for tow fibers.

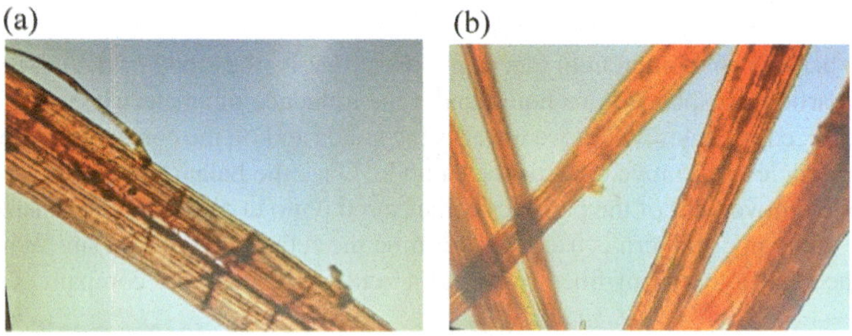

Fig. 2. (a) Microscopic image of flax fibers, (b) microscopic image of tow fibers

6 Zeta Potential

The zeta potential varies with pH and becomes more positive and negative in amplitude with acidic and basic pH, respectively, for aqueous dispersions where H+ and OH− are major ionic constituents (Guo et al. 2009). The sign of the zeta potential of the resin used is negative and it is approximately −60 mV. Therefore, for better adhesion between the reinforcement and the matrix, it is necessary to work with nonwovens having a positive zeta potential (Delgado et al. 2007). To obtain a stable suspension, modified samples (3% and 6% cationization treatment) adhere better to resin (Fig. 3).

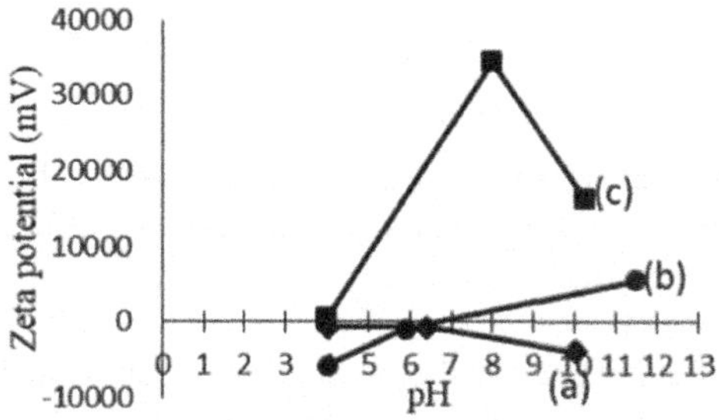

Fig. 3. Schematic representation of zeta potential vs. pH plots of: (a) untreated non-woven, (b) cationized non-woven 3% and (c) cationized non-woven 6%

7 Contact Angle

In our study, we are looking for a partial wetting that is less than 90° so that the support is hydrophilic with a certain adhesion between the reinforcement and the matrix.

Table 1. Contact angles measured with the resin

Sample	ɵR (°)	CV (%)
20/80 untreated	101,2	3,189
50/50 untreated	117,1	5,533
80/20 untreated	106,08	7,407
80/20 cationised 3%	88,59	9,680
50/50 cationised 3%	89,95	3,465
20/80 cationised 3%	86,869	9,129
20/80 cationised 6%	89,61	4,070
50/50 cationised 6%	88,3	8,342
80/20 cationised 6%	89,068	2,858

The contact angles between the resin and samples are lower for treated supports compared to untreated supports (Table 1). This indicates that the treatment makes the fibers more hydrophilic.

The purpose of the chemical modifications made to the substrates is to improve the wettability and to modify the cellulose to account for it with the resin. More precisely, it is the hydroxyl groups of the cellulose that will have to be modified, otherwise, the polarity of the latter will prevent the resin from chemically bonding with the cellulose. It is necessary to substitute or modify the hydroxyl groups of the cellulose which let the resin, which is hydrophobic and non-polar, adheres to it well (Behari et al. 2012).

8 Conclusion

The basic surface character of fibers would be enhanced by the presence of reaction sites. This increase was not only proved by zeta potential measurements, but also by the contact angle method. The creation of reactive sites improves the adhesion between the reinforcement and the matrix.

In this article, the effect of chemical modification of cellulosic fibers on the adhesion of composites and the results of the electrokinetic tests carried out on the different nonwovens and the binder. The described test methodology and results are a valuable tool for the qualitative prediction of adhesion. Electrokinetic analyzes have shown that the cationizing agent admits better results than untreated materials in terms of adhesion.

In a future study, it would be interesting to examine the mechanical properties to use these materials in exterior applications.

References

Behary, N., et al.: Adsorption of surfactin produced from Bacillus subtilis using nonwoven PET (polyethylene terephthalate) fibrous membranes functionalized with chitosan. Colloids Surfaces B Biointerfaces **90**, 137–143 (2012)

Bigand, V., et al.: Cationisation of galactomannan and xylan hemicelluloses. Carbohydr. Polym. **85**(1), 138–148 (2011)

Delgado, A.V., et al.: Measurement and interpretation of electrokinetic phenomena. J. Colloid Interface Sci. **309**(2), 194–224 (2007)

Elaissi, A., et al.: Development of abrasives from non-woven based on used textiles. J. Nat. Fibers (Aout 2020), **15**, 1–15 (2020)

Grundke, K., et al.: Physico-chemical properties of surface-modified polymers. J. Adhesion Sci. Technol. **9**(3), 327–350 (1995)

Guo, L., et al.: Zeta potential and surface physico-chemical properties of atmospheric air-plasma-treated polyester fabrics. Text. Res. J. **79**(15), 1371–1377 (2009)

Jabli, M., et al.: Dimethyl diallyl ammonium chloride and diallylamin co-polymer modified bio-film derived from palm dates for the adsorption of dyes. Sci. Rep. **7**(1), 1–12 (2017)

Jölly, I., et al.: Chemical functionalization of composite surfaces for improved structural bonded repairs. Compos. Part B Eng. **69**(7), 296–303 (2015)

Tka, N., et al.: Amines modified fibers obtained from natural Populus tremula and their rapid biosorption of Acid Blue 25. J. Mol. Liq. **250**, 423–432 (2018)

Advance in Clothing and Leather Industries

Analysis of the Effect of Stone Washing in the Residual Bagging Height

Abir Ben Fraj[1,2](✉), Mouna Gazzah[2], and Boubaker Jaouachi[1,2]

[1] National Engineering School of Moanstir, University of Monastir, Monastir, Tunisia
abir1991_benfraj@yahoo.fr

[2] Textile Engineering Laboratory, University of Monastir, Monastir, Tunisia

Abstract. The purpose of this study is to determine the effect of process parameters of stone washing on the residual bagging height. Six denim fabrics were processed with the quantity of pumice stone of 100% to 200%, treatment temperature from 30 °C to 60 °C and treatment time from 30 min to 60 min. The results of the main effect plot show that the minimum value of residual bagging height was obtained with 100% pumice stone at 30 °C for 30 min.

Keywords: Residual bagging height · Stone washing · Denim fabric · Main effects plot

1 Introduction

Denim became the most considered style today. It is the most popular clothing material in the past three decades. Finishing treatments give the denim fabric comfort and softness. For denim there are several finishing treatments, either conventional treatment such as desizing, pumice stones, enzymatic washing, bleaching or modern treatments such as ozone, laser… Denim manufacturers have used pumice stones for a soft feel and a worn look. The main drawbacks of this procedure are the difficulty of removing residual pumice and equipment damage. Despite these disadvantages, stone washing is still used because this method gives denim the additional effect of a faded or worn look as it abrades the surface of the denim removing some dye particles from the surfaces of the yarn.

Until now, overall studies still focus on the effects of treatment in the mechanical, physical and sensorial properties (Ansari 2017; Khan and Mondal 2013; Sarkar et al. 2014). However, there are no studies dealing with the effect of the finishing treatments in the bagging behaviors. The bagging is unaesthetic and undesirable phenomenon, which appears at the knee and elbow after applying repetitive loads and/or after several uses. Hence, the aim of this work is to investigate the effect of pumice stones wash on the residual bagging height property of denim fabrics. Thus, a Taguchi experimental design was applied in order to evaluate the main contribution of this kind of washing treatment on the behavior of bagged fabrics.

S. Msahli and F. Debbabi (Eds.): CIRATM 2020, SPM 17, pp. 161–165, 2022.
https://doi.org/10.1007/978-3-031-08842-1_26

2 Materials and Methods

2.1 Materials

Six denim fabrics having different fiber blend ratios were prepared into 50 × 100 cm^2 and sewed as tubes. Although these samples have different characteristics, they present the same weave (twill 3/1).

The enzyme ULTRA REDEP, a nonionic agent, is very active for the dispersion of indigo molecules in the treatment bath. It has a strong anti-redeposition power, which keeps the indigo in suspension and prevents its redeposition on the fibers. Pumice stones are light, highly porous and rocky substances that will float on the water. It is used to scrub the surface of the fabric to achieve a surface pattern effect within a color contrast and a softer touch.

2.2 Methods

2.2.1 Stone Washing

In order to know the effect of stone washing parameters on the bagging parameters, and based on the literature, we chose a different percentage of pumice (Qp) (100%, 150%, 200% o.w.f) at different temperature values (30, 45, 60 °C) and test durations (30, 45, 60 min). The process was carried out in a liquor containing 1g/l anti-redeposition agent (ULTRA REDEP) at pH = 7 with the fabric samples in the bath ratio of 1/10 in an industrial drum washing machine.

2.2.2 Test of Bagging

The bagging deformation test of fabric samples was investigated using dynamometer type Lloyd (Fig. 1). During a bagging test, a removable hemisphere in the upper jaw applied a load on the sample up to a certain distance with a defined speed 60 mm/min. The number of the applied cycles was fixed at 5 cycles. After this test, the residual bagging height was measured after a relaxation time equal to 30 min. This test was repeated for 5 times. This method was inspired from literature (Doustar et al. 2009; Sülar and Seki 2018; Şengöz 2004).

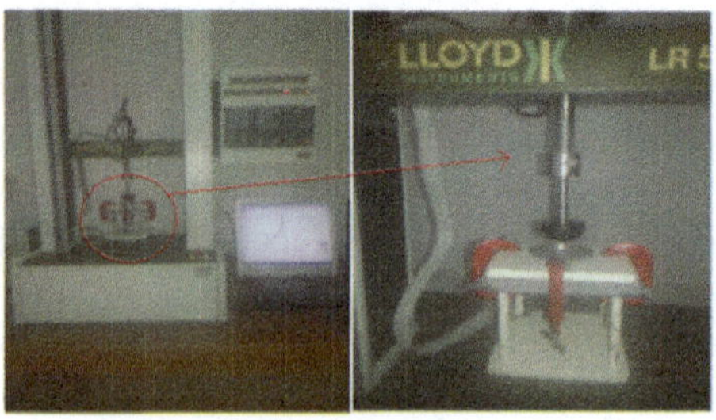

Fig. 1. Fabric bagging machine.

3 Results and Discussion

The objective of this study is to investigate the effect of the stone washing parameters such as temperature, time and quantity of pumice stone on the residual bagging height of samples. Thus, the residual bagging height for the treated and untreated samples were measured and analyzed. The obtained results relative to the 18 experiments of the used Taguchi design (Table 1) are analyzed using Minitab software 14.1.

Table 1. Results of residual bagging height

Samples	Qp (%)	Temperature (°C)	Time (min)	Residual bagging height (mm)	CV (%)
s11s	100,00	30,00	30,00	8,34	5,03
s12s	150,00	45,00	45,00	9,60	7,15
s13s	200,00	60,00	60,00	9,64	1,77
s21s	150,00	30,00	30,00	9,92	2,01
s22s	200,00	45,00	45,00	10,36	2,83
s23s	100,00	60,00	60,00	9,25	2,08
s31s	100,00	30,00	45,00	8,34	3,02
s32s	150,00	45,00	60,00	10,83	4,31
s33s	200,00	60,00	30,00	7,74	3,52
s41s	200,00	30,00	60,00	9,16	3,12
s42s	100,00	45,00	30,00	9,52	1,81
s43s	150,00	60,00	45,00	9,99	1,91
s51s	200,00	60,00	30,00	9,99	3,20
s52s	100,00	30,00	45,00	8,65	5,20
s53s	150,00	45,00	60,00	9,10	3,43
s61s	150,00	45,00	30,00	11,16	2,92
s62s	200,00	60,00	45,00	10,08	1,29
s63s	100,00	30,00	60,00	11,02	2,08

From the experimental observations, it was found that the process parameters had considerable effect in the residual bagging height.

Figure 2 show that the rise of the quantity of pumice stone and temperature causes the rise in the residual bagging height. It may be concluded that stone washing with 100% (o.w.f) presents the minimum value of residual bagging height, and then the increase in the quantity of pumice stone causes the increase in the residual bagging height up to 150%. This result is explained by the effect of the stone washing on the friction and abrasion denim fabric behaviors, the samples are solicited by abrasion with pumice stones which rub the surface of the fabric and degrade or remove the blue color of the fabric. The rubbing effect of the pumice stones on the fabric surface caused the degradation of fibers. However, this damage rises with the rise of the quantity of pumice stone. In fact, the action of the abrasion of the fabric surface will be more aggressive and induce more rubbing that will degrade and damage the fibers. These cracks in the fibers affect enormously the mechanical properties of bagged fabrics and as consequence the elastic recovery (Ben Fraj et al. 2021). While from 150% to 200% we will have a decrease in residual bagging height. Besides, the same evolution seems to happen for the temperature input parameter the residual bagging height increases from 30 °C to 45 °C and decreases up to 60 °C. However, the time of the process has no significant effect on the residual bagging height.

Also, it is noted from Table 2 that the quantity of the pumice stone is the most influencing factor on residual bagging height. This is followed by temperature and time.

Table 2. Response Table for residual bagging height

Factor	Level			Delta	Rank
	1	2	3		
Qp (%)	9,325	10,127	9,621	0,802	1
Temperature (°C)	9,387	10,121	9,565	0,734	2
Time (min)	9,561	9,705	9,807	0,246	3

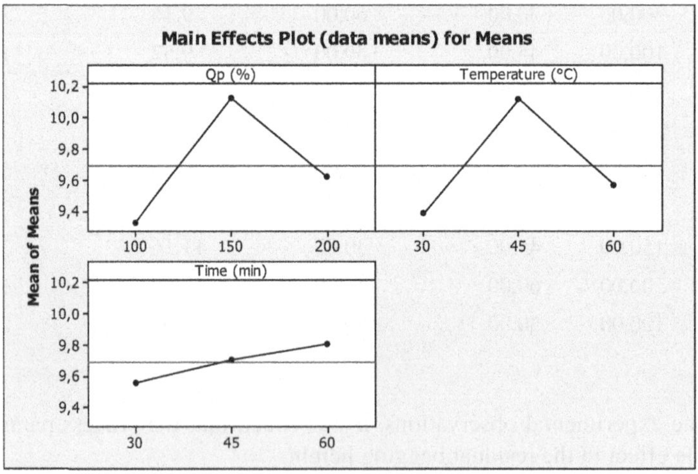

Fig. 2. Main effect evolutions for residual bagging height as function of the studied input parameters.

4 Conclusion

The aim of this study is to evaluate the effect of parameters of stone washing on the residual bagging height of denim garment. The finding obtained show that the quantities of pumice stone and temperature affect bagging behavior, but time is not an influencing factor. The ideal condition for this treatment to minimize the residual bagging height it will be studied in our future research work.

References

Ansari, I.Z.: Impact of stone wash and acid wash on the physical properties of denim. Int. J. Eng. Res. **6**(12), 499–501 (2017). ISSN 2319-6890

Ben Fraj, A., Jaouachi, B., Dominique, A.: Effect of stone washing parameters on denim behavior bagging. J. Nat. Fibers (2021). https://doi.org/10.1080/15440478.2021.1875358. ISSN 1544-046X

Doustar, K., Shaikhzadeh Najar, S., Maroufi, M.: The effect of fabric design and weft density on bagging behavior of cotton woven fabrics. J. Textile Inst. **101**(2), 135–142 (2009). ISSN 1754-2340

Khan, M.M.R., Mondal, M.I.H.: Physico-mechanical properties of finished denim garment by stone-bleach treatment. J. Chem. Eng. **28**(1), 36–40 (2013). ISSN 2221-7436

Sarkar, J., Khalil, E., Solaiman, Md.: Effect of enzyme washing combined with pumice stone on the physical, mechanical and color properties of denim garments. Int. J. Res. Advent Technol. **2**(9), 65–68 (2014). ISSN 2321-9637

Şengöz, N.G.: Bagging in textiles. Text. Prog. **36**(1), 1–64 (2004). ISSN 1754-2278

Sülar, V., Seki, Y.: A review on fabric bagging: the concept and measurement methods. J. Text. Inst. **109**(4), 466–484 (2018). ISSN 1754-2340

Moisture Management Behavior of a Cotton/Polyester Double-Sided Knit

Imene Ghezal[1,2(✉)], Ali Moussa[1,2], Imed Ben Marzoug[1,3], Ahmida El-Achari[4,5], Christine Campagne[4,5], and Faouzi Sakli[1,3]

[1] Textile Engineering Laboratory, University of Monastir, 5070 Ksar-Hellal, Tunisia
elghezalimene@hotmail.com
[2] National Engineering School of Monastir, University of Monastir, 5019 Monastir, Tunisia
[3] Higher Institute of Technological Studies of Ksar-Hellal, 5070 Ksar-Hellal, Tunisia
[4] Université Lille Nord de France, 59000 Lille, France
[5] ENSAIT, GEMTEX, 2 Allée Louise et Victor Champier, 59100 Roubaix, France

Abstract. The present research investigates the effect of a coating treatment on liquid moisture management performance of a double sided-knitted fabric. The moisture management tester (MMT) from SDL Atlas, USA was used. Studied characteristics are absorption rate (AR ($\%.s^{-1}$)), maximum wetted radius (MWR (mm)), spreading speed (SS ($mm.s^{-1}$)), one way transport index (R (%)), and overall moisture management capacity (OMMC). The tested fabric is composed of a cotton inner layer and a polyester outer layer. The polyester face was coated in order to obtain a waterproof breathable fabric. Results show that the uncoated samples have an excellent ability to manage sensitive perspiration with an OMMC equal to 0.893. After the coating treatment fabric wettability and moisture management capacity decreased significantly.

Keywords: Double-face fabric · Coated fabric · Thermophysiological comfort · Moisture management tester

1 Introduction

Thermophysiological comfort is related to moisture transmission through fabrics (Hu et al. 2005; Boughattas et al. 2019). It is the result of a thermal equilibrium between the human body and the surrounded environment (de Vasconcelos et al. 2017). Fabric moisture transmission is one of the most important characteristics when coming to the evaluation of a garment ability to provide thermophisiological comfort. For this reason, it is important to select conveniently the type of used fibers and fabric structure (Kumar et al. 2015). Natural and synthetic fibers are both used to produce apparels that provide thermophysiological comfort to the wearer (Kumar et al. 2015). Cotton fiber is characterized by its high capacity to absorb moisture (de Vasconcelos et al. 2017; Cruz et al. 2017). The most used synthetic fiber is polyester. It is known for its low moisture absorption which explains its quick drying and good ability to evacuate moisture by capillary action (de Vasconcelos et al. 2017; Kumar et al. 2015; Ghezal et al. 2020).

S. Msahli and F. Debbabi (Eds.): CIRATM 2020, SPM 17, pp. 166–171, 2022.
https://doi.org/10.1007/978-3-031-08842-1_27

Waterproof breathable fabrics were developed in order to provide both comfort and protection to the wearer (Mukhopadhyay and Midha 2008; Ghezal et al. 2019). However, some techniques used to improve a fabric waterproofness and resistance to win decrease its comfort property.

This research seeks to evaluate the moisture management capacity of double-sided knitted structure before and after undergoing a coating treatment. The outer-face of the fabric was made with a polyester multifilament and the inner-face with a cotton yarn. In this study, the moisture management tester (MMT) method was used. Seven indexes were determined and studied in order to evaluate liquid moisture management of the used fabric before and after the coating treatment.

2 Materials and Methods

2.1 Textile Material and Coating Process

In this research, the tested fabric is a double-sided jersey knit with a cotton inner-face and a polyester outer-face. An Interlock circular knitting machine type Albi Ram GT was used to produce the fabric. The textile thickness and mass per unit area are respectively equal to 256 $g.m^{-2} \pm 0.815\%$ and 1.063 mm $\pm$ 1.437%. A coat (2.71% fluorocarbon resin/412 $g.m^{-2}$) was applied to the PES outer face of the fabric by the aim of a coating screen. Samples were dried for 20 min at 90 °C and cured for 13.5 min at 150 °C..

2.2 Moisture Management Testing

The liquid moisture transport properties of the tested knitted structure were evaluated before and after the coating treatment. Tests were done according to the AATCC Test Method 195 (AATCC Test Method 2011) and by using the moisture management tester (MMT) from SDL Atlas, USA (Fig. 1).

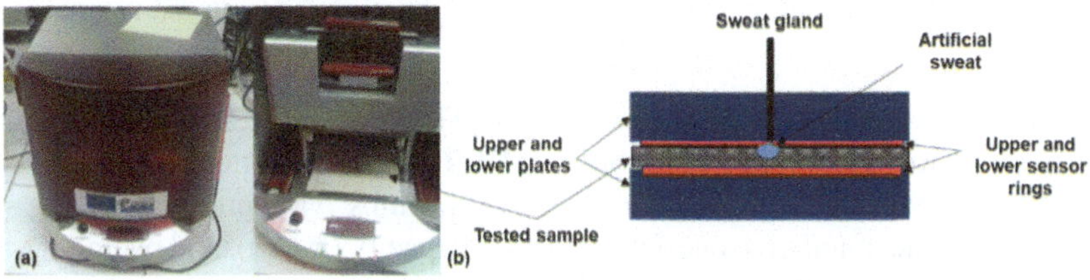

Fig. 1. Moisture Management Tester M 290 (MMT) (a) Plates and sensors sketch (b) (Hu et al. 2005; Jhanji et al. 2014)

This instrument is principally formed of sensors that measure the electrical conductivity of non-wetted and wetted samples. Water content as well as liquid transport properties of the fabric were then determined.

3 Results and Discussions

Samples were tested before and after the coating treatment. The evolution of water content (%) in time is represented by Fig. 2. Time zero corresponds to the instant in which the synthetic sweating is dropped on the cotton face. The inner-face facilitates liquid transfer from the cotton to the polyester side. The synthetic sweating diffuses then rapidly in the polyester surface by capillarity. The liquid diffusion speed is higher in the polyester side than in the cotton one. This is the result of the hydrophobic nature of polyester fiber that facilitates liquid diffusion in the fabric outer-face by wicking.

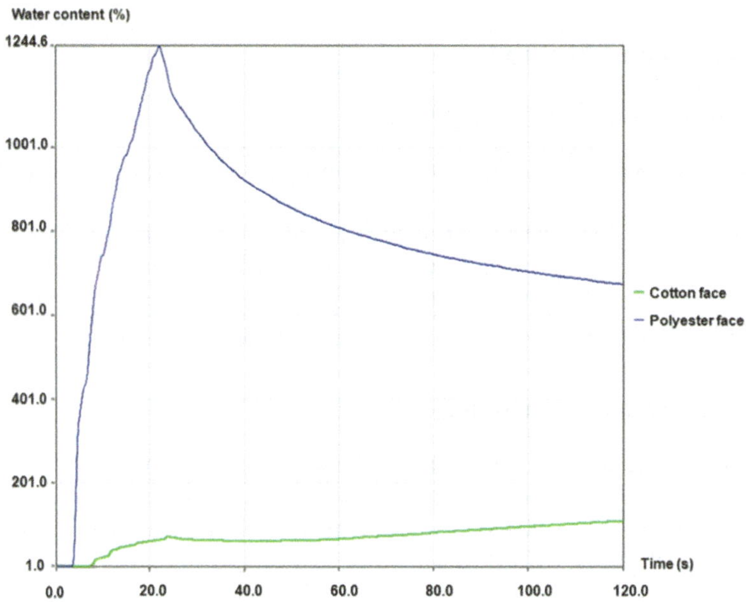

Fig. 2. Evolution of water content vs. time for the uncoated knitted structure.

Seven indices were then measured with the MMT. These indexes were absorption rate (AR_b ($\%.s^{-1}$)), top maximum wetted radius (MWR_t ($\%.s^{-1}$)), bottom maximum wetted radius (MWR_b ($\%.s^{-1}$)), top spreading speed (SS_t ($mm.s^{-1}$)), bottom spreading speed (SS_b ($mm.s^{-1}$)), accumulative one-way transport capacity (R (%)), and overall moisture management capacity (OMMC). Obtained results for uncoated and coated samples are presented in Table 1. The highest indexes were obtained with the uncoated samples. Before the coating treatment, the double-sided knit showed a good wettability with a moderate absorption rate ($AR_b = 63{,}911\%.\ s^{-1}$). A large spreading of the synthetic sweating was also noticed on both sample surfaces. Liquid moisture diffuses in fabrics by wicking and wetting mechanisms (Bhatia and Malhotra 2016). From Fig. 3 and Table 1 it is noticeable that moisture diffusion in the polyester side is better than in the cotton one ($MWR_t = 10$ mm; $MWR_b = 25$ mm). The hydrophobic nature of polyester fibers enhances wicking mechanism. However, cotton hydrophilic structure retains water molecules. This restricts the liquid diffusion in the fabric inner- side (Vasconcelos et al. 2017).

Table 1. MMT obtained indices for the double-sided knit before and after the coating treatment.

Indices	Uncoated fabric	Coated fabric
AR_b (%.s^{-1})	63.911 ± 2,27%	-
MWR_t (mm)	10	10
MWR_b (mm)	25	5
SS_t (mm/s)	1,001 ± 2,747%	0,410 ± 8,75%
SS_b (mm/s)	6,593 ± 7,72%	1,589 ± 7,675%
R (%)	749,365 ± 7,081%	-
OMMC	0.893 ± 1.47%	0.234 ± 4.284%

High obtained indexes values are the result of the structure and composition of the tested knit. Generally, double sided fabrics are known for their high moisture management attributes (Troynikov and Wardiningsih 2011).

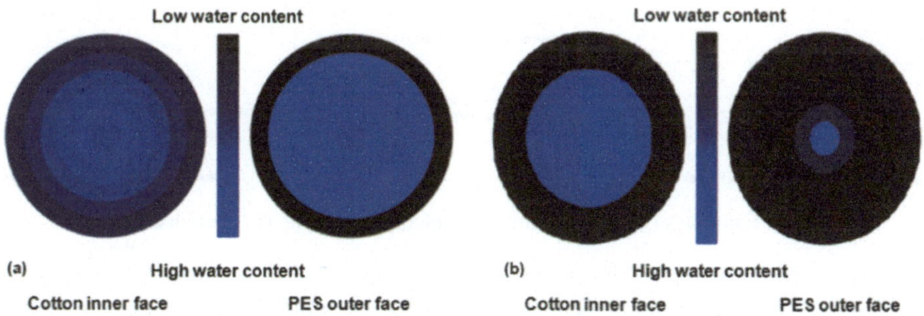

Fig. 3. Water location vs. time: (a) uncoated fabric, (b) coated fabric.

After the coating treatment, the maximum top wetted radius was equal to 10 mm. The same value was obtained before the coating treatment. However, the maximum bottom wetted radius, top and bottom spreading speeds, and the overall moisture management capacity decreased significantly. So, it can be deduced that the coating treatment restricted the wettability of the double sided fabric surface. It also decreased its moisture management capacity.

Fabric porosity influences significantly its moisture absorbency (Bedek et al. 2011). The coat paste fulfilled inter-yarns and inter-fibers spaces in the fabric surface. As a result, fabric porosity decreased significantly which affected the fabric moisture management capacity.

The grading table presented by the AATCC Test Method 195 (AATCC Test Method 2011) enables the conversion of obtained indexes to grades and the generation of a grading summary table also called a finger print (Fig. 4). The uncoated fabric overall moisture management capacity (Fig. 4(a)) was equal to 0.893. Based on the AATCC Test Method 195 (AATCC Test Method 2011), this value can be converted into a grade equal to 5. So, it may be concluded that the tested knitted structure has an excellent moisture management capacity.

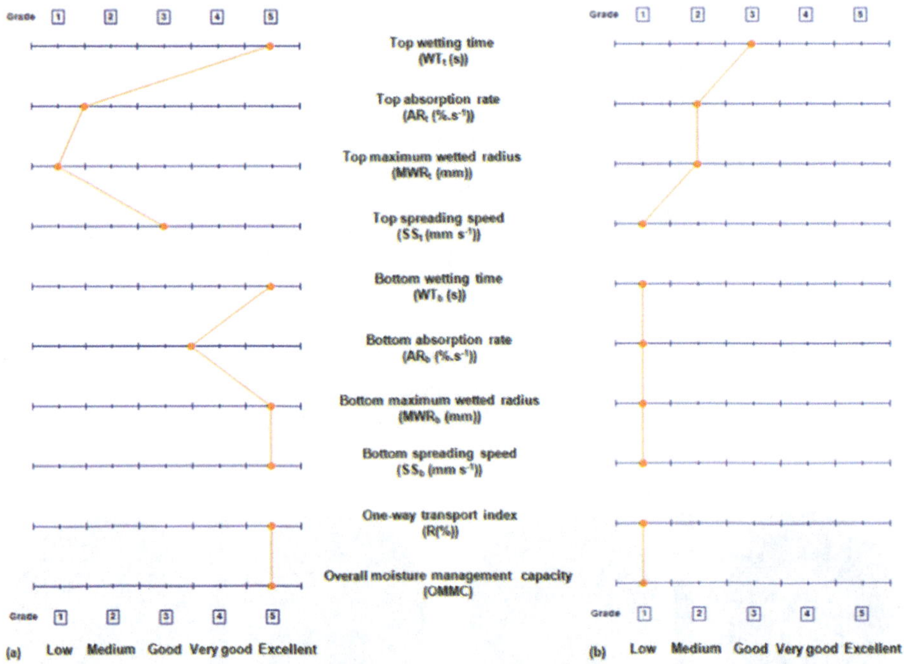

Fig. 4. Examples of a finger prints: (a) uncoated sample, (b) coated sample.

Based on the finger print presented by Fig. 4(b), the applied coat enhanced the fabric hydrophobicity. However, fabric moisture management capacity decreased significantly.

Fabric structure, yarn geometry, and fiber type affect the apparel ability to evacuate moisture (Kumar et al. 2015). Sensitive perspiration can be evacuated from fabrics by wetting and wicking mechanisms (Bhatia and Malhotra 2016). The coating treatment fulfilled inter-fibers and inter-yarns spaces diminishing fabric porosity. It also modified its surface chemical structure. This explains the decrease in moisture management capacity after the coating treatment.

4 Conclusion

The liquid moisture management properties of a double-sided knitted structure designed to be waterproof and breathable were evaluated before and after undergoing a screen coating process. The applied coat was a mixture of a fluorocarbon resin and an acrylic paste. Results showed that the cotton/polyester double-sided fabric has an excellent moisture management capacity which makes it appropriate to be used as an under-layer in outdoor garments. After the coating treatment, fabric hydrophobicity was ameliorated. However, the coated double-face fabric presented low bottom maximum wetted radius, top and bottom spreading speeds, and overall moisture management capacity. This is the result of fabric porosity and surface chemistry modification. To conclude, the coating treatment enhanced fabric hydrophobicity, nonetheless, it affected negatively its moisture management properties.

References

AATCC Test Method 195: Liquid moisture management properties of textile fabrics (2011)

Bedek, G., et al.: Evaluation of thermal and moisture management properties on knitted fabrics and comparison with a physiological model in warm conditions. Appl. Ergon. **42**, 792–800 (2011). ISSN: 0003-6870

Bhatia, D., Malhotra, U.: Thermophysiological wear comfort of clothing: an overview. J. Text. Sci. Eng. **6**(2), 1–8 (2016). ISSN: 2165-8064

Boughattas et al.: Moisture management properties of double face DENIM fabrics. Int. J. Appl. Res. Text. Special Issue CIRAT 8, 38–43 (2019). ISSN: 2356-5586

Cruz, J., et al.: Study of moisture absorption characteristics of cotton terry towel fabrics. Procedia Eng. **200**, 389–398 (2017). ISSN: 1877-7058

de Vasconcelos, F.B., et al.: Moisture management evaluation in double face knitted fabrics with different kind of constructions and fibers. J. Fashion Technol. Text. Eng. (3), 1–5 (2017). ISSN: 2329-9568

Ghezal, I., et al.: Evaluating the mechanical properties of waterproof breathable fabric produced by a coating process. Cloth. Text. Res. J. **37**(4), 235–248 (2019). ISSN: 1940-2473

Ghezal, I., et al.: Development and surface state characterization of a spacer waterproof breathable fabric. Fibers Polym. **21**(4), 910–920 (2020). ISSN: 1875-0052

Hu, J., et al.: Moisture management tester: a method to characterize fabric liquid moisture management properties. Text. Res. J. **75**(1), 57–62 (2005). ISSN: 1746-7748

Jhanji, Y., et al.: Moisture management properties of plated knit structures with varying fiber types. J. Text. Inst. **106**(6), 663–673 (2014). ISSN: 405000

Kumar, P., et al.: Moisture management behavior of modified polyester wool fabrics. Fashion Text. **2**(5), 1–17 (2015). ISSN: 2198-0802

Mukhopadhyay, A., Midha, V.K.: A review on designing the waterproof breathable fabrics Part I: fundamental principles and designing aspect of breathable fabrics. J. Ind. Text. **37**(3), 225–262 (2008). ISSN: 1530-8057

Troynikov, O., Wardiningsih, W.: Moisture management properties of wool/polyester and wool/bamboo knitted fabrics for the sportswear base layer. Text. Res. J. **81**(6), 621–631 (2011). ISSN: 1746-7748

Improvement in Comfort Properties of Gloves for Fighter Jet Pilots

Usman Ahmed(✉), Yasir Nawab, Ali Afzal(✉), Muhammad Umair, and Tanveer Hussain

School of Engineering and Technology, National Textile University, Faisalabad, Pakistan
usman.knit@gmail.com, aliafzalch89@gmail.com

Abstract. Fighter jet costumes have no cooling systems, especially as it becomes very difficult to sit in the cockpit while flying on hot deserts and in humid environment. The sweating appears underarm, feet and on hands. After sweating on hand-palm, the fighter jet pilot losses grip during operation. To prevent this slippage the pre-existing glove of pilot combat uniform was re-engineered. It was observed that existing gloves were good in fire resistant, but they have poor comfort properties. In this study, four samples (Lenzing (100%), Lenzing/Protex (70/30%), Protex/Lenzing (70/30%), Protex (100%)) of the knitting interlock structure were prepared on a flat knitting machine having 13 gauge. It was observed that samples with the blend composition (Protex/Lenzing (70/30)) showed the best results for thermo-physiological comfort properties. It has comparatively better air permeability and very good flame resistance with mean after flame time 4 s. The manufactured gloves showed air permeability, moisture management and flame resistance values better than existing commercially available gloves. The manufactured gloves were light weight, comfortable and provide proper grip to the wearer.

Keywords: Knitting · Fire resistant · Flame retardant · Comfort properties · Gloves

1 Introduction

Gloves are used to cover the hands. They protect a person's hand from the external environment. Gloves are available in many shapes. Their shapes are dependent on the end use like kitchen glove is totally different from a biker glove. Different knitted structures and chemical finishes are used to enhance the functional properties of gloves. Biker gloves are shown in the (Fig. 1) (Leslie et al. 1996).

1.1 Pre-existing Gloves

Commercially available samples of fighter jet pilot were taken from the market. They were examined in the textile laboratory to find out the properties of material used. It was found that, they are cut and sewn glove.

S. Msahli and F. Debbabi (Eds.): CIRATM 2020, SPM 17, pp. 172–181, 2022.
https://doi.org/10.1007/978-3-031-08842-1_28

Fig. 1. Gloves

The glove has the leather (as shown in Fig. 2) on the front side of the glove to improve the grip. And the back side of the glove has plain interlock structure (as shown in Fig. 2). It was found after testing that pre-existing glove has poor comfort properties. It has no ability to absorb the moisture and has no passage for air to pass from the glove.

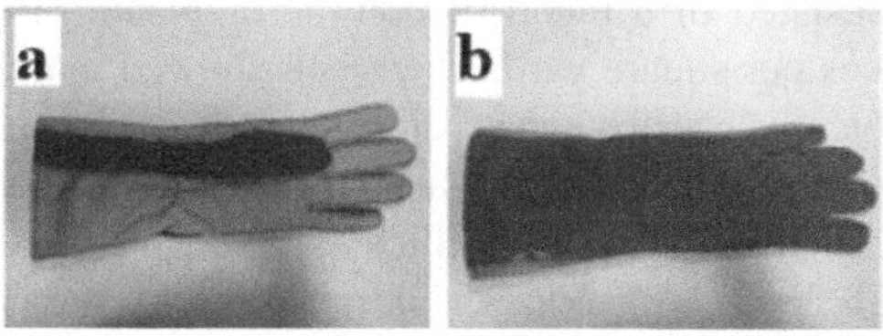

Fig. 2. Preexisting glove (a) front (palm) side, (b) back side

1.2 Literature Review

By using fiber (Kevlar, high temperature resistant polyester or oxidized polyacrylonitrile) a glove was prepared that offered the resistance to heat and abrasion. A knitted structure was selected to increase the flexibility and grip. The result shown that the prepared fabric for gloves has high resistance to heat and abrasion (frederick A. Vero, Easthampton 1998). Thermal properties of different knitted structures were tested in this research. Plain rib, interlock and single jersey structure were selected. The result showed that interlock and rib have a high thermal resistance than single jersey structure. But single jersey structure has a greater water vapor permeability (Oğlakcioğlu and Marmarali 2007).

For industrial usage, a pair of gloves having properties of antistatic, oil resistance and moisture permeability was developed. To increase the flexibility an elastane yarn was used. The coating film of hydrophilic polyurethane was used to protect the glove from oil strains. It has the thickness varies from 0.05 to 0.45 mm (Kleinerman 1988). Aromatic polyamide fibres were used to develop the heat resistant gloves. Both sides of gloves were made from the same material. Two fabric layers (one woven and one knitted) were used to form the cut and sewn glove. The woven fabric was paced on outer side with coating of elastic flame resistance film, that was used to safe hand from external flame. The inner side knitted structure was used to provide the comfort to the wearer (Sidman and Aroms 1984).

A multilayer glove was formed having an ability to resist high pressure, heat flame and cuts. The glove had an ability to stop the liquid molecules and allow the gas molecules to pass. A durable and flexible material was used. The barrier insertion of permeable polytetrafluoroethylene (PTFE) was designed and fitted inner side of outer layer. This barrier insertion has ability to protect the wearer's skin (Chen 2006). The glove and arm guard were developed by using steel wire and fiber strand. To reduce the slippage a nitrile rubber coating was used. The end product was highly cut-resistant, nonabsorbent, highly slip-resistant, flexible and light in weight (Hoffmeyer et al. 2016).

A research was conducted on a problem that impermeable gloves and other clothing did not allow moisture to evaporate from the human body. If the moisture accumulates than it will cause different problems like, pain, allergy and high temperature of human body. The solution was recommended by researcher by placing two-layer lining in the impermeable clothing. This lining absorbed all moisture. The lining was consisted of vapor and liquid permeable material such as microporous Teflon (trademark) film, microporous polyurethane flick and a hydrophilic polymer coating (Praveen et al. 2018).

A thermally protective, flame retardant fabric was developed. Which includes a substrate that was consisted of a blend of non-thermoplastic fibers having a weight from 2.0 to 15.0 ounces per square yard. The result showed that the fabric should has a contact thermal protective performance value of at least 4.5 and a contact thermal protective performance efficiency greater than 1.1. End applications of the fabric can be protective garments, articles of furniture, vehicle components and building components etc. (Monfalcone et al. 2009). A fabric was developed with the help of oxidized poly acrylonitrile, polyvinyl halide, poly benzimidazole, p-aramid, m-aramid, fire resistant polyester and fire-resistant rayon. The oxidized poly acrylonitrile is comprised of approx. 25 to 85% of yarn and the other p aramids comprised less than 35% by weight as yarn. The developed yarn and fabric has a greater property to resist the flame (Sarzotti 2018).

A yarn and a fabric were prepared from a blend of fibers including non-FR cellulosic fibers, modacrylic fibers and aramid fiber blend together. The blend ratio was selected as 45% by weight of cellulosic fibers and modacrylic ratio was 0.8%. The author claimed that the developed fabric has a high flame resistant properties (Mio 2007). Knit fabrics and military apparel such as T-shirts were made from blended yarns including nylon and cotton. Such fabrics comprise a weight ratio of cotton to nylon which ranges from 55:45 to about 85:15, and these fabrics having a weight 3 to about 8 oz/yd. The formed knitted fabrics have good thermal protective properties. The prepared fabric did not melt or start dripping during flame test (Publication 2021). A flame-retardant fiber blend was developed comprising 40% to 60% by weight of a mod acrylic; 5% to 25% by weight of a natural cellulosic material; and 20% to 40% by weight of a FR viscose. The fiber blend was mainly suitable for the manufacturing of a general fabric constructed to be resistant to flame, electric discharge and molten metal hazards (Tutterow and Dunn 2008).

1.3 Research Gap

Fighter pilot gloves have excellent fire-retardant properties, but their comfort properties are poor. Our goal is to develop the glove that have optimized comfort and fire-retardant properties.

1.4 Research Objectives

- To analyze the preexisting gloves.
- To develop a glove that provides good comfort and fire-resistant properties.

1.5 Design of Experiment

The different ratios and fabric structure, that will provide considerable result at the end of research is stated in the below table (Table 1).

Table 1. Design of experiment

Factors (fiber)	Level (blend ratios)			
	1	2	3	4
Lenzing	100%		70%	30%
Protex		100%	30%	70%

2 Materials and Methods

Protex is comprised of mod acrylic. Mod acrylic is long chain synthetic copolymer composed of less than 85% to 35% by weight of acrylonitrile units. It is also known as mod acrylic. The softening temperature is allowed to stretch and molded into special shapes. Protex fibers are manufactured from dry spun or wet spun. Protex fibers are inherently flame resistant. Protex burn when it is in the flame, but it does not melt. Mod acrylic yarn has very good fire retardant and comfort properties. It is widely used in high performance protective clothing. It has outstandingly resistant to chemical, acid/ alkaline environment. It is non allergenic and safe from the attack of moths and mildews. The moisture regain of protex is 3.5% LOI (limiting oxygen index) is 28 to 32. Strength of yarn is 2.6 cN/den (dry) with elongation 28% (dry) (Zhu et al. 2006) (shown in Fig. 3(a)).

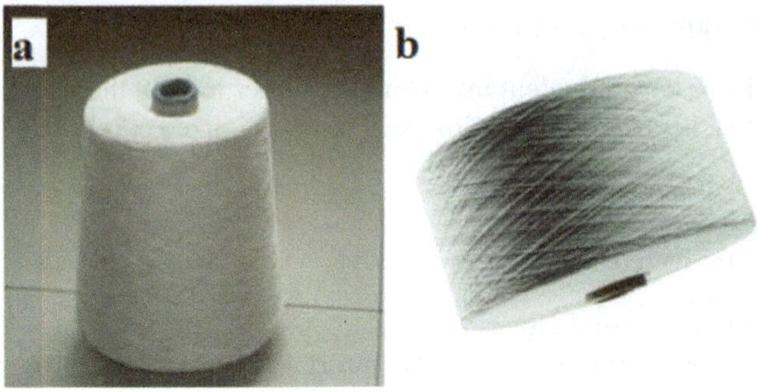

Fig. 3. (a) Protex yarn, (b) Lenzing yarn

Lenzing is produce from wood based viscose fibers. Lenzing has end application in clothing, home textile and technical textile. Lenzing has the LOI 26%. It has excellent fire-retardant characteristics. It drops strength above 149 °C and decomposes at 177 to 204 °C temperature and it doesn't melt or drip (Varga et al. 2011) (shown in Fig. 3(b)).

2.1 Machine Used for Knitting

2.1.1 Seamless Flat Knitting Machine

Fig. 4. Shima seiki glove knitting machine

It is shima seiki glove knitting machine, having ability to manufacturing a striper glove. It consist of flat bed for knitting a glove. It is fully electronic controlled machine. The amount for courses for each part is needed to insert for the formation of required size glove (Fig. 4).

2.1.2 Fully Fashioned Jacquard Knitting Machine

Fig. 5. Shima Shieki flat knitting machine

It is fully fashioned seamless flat knitting machine. It consists of 3 beds, two beds are at v-position, and one is placed above these two beds. The additional bed is for loop pressers (Fig. 5).

3 Results and Discussions

3.1 Air Permeability

The factors affecting air penetrability is porosity, fabric thickness, yarn linear density, yarn twist, and size of the pore. As shown in above bar chart (Fig. 6) the samples that have higher permeability are the fabric that contains lenzing fiber. The air permeability of developed samples is better than from reference sample. The air permeability of lenzing 100% is higher because it has chemically treated viscose with fiber length as compared to protex. Air permeability was found to be inversely proportional to surface density. By increasing the fiber length, the air permeability increased. The air permeability is inversely proportional to the mechanical properties of knitted fabric. When compactness and yarn in contact increases the interloping gaps decreases, due to this air permeability increases.

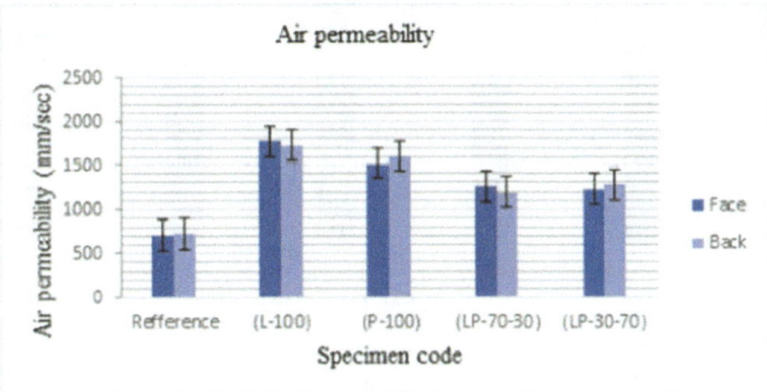

Fig. 6. Graphical representation of air permeability

3.2 Moisture Management

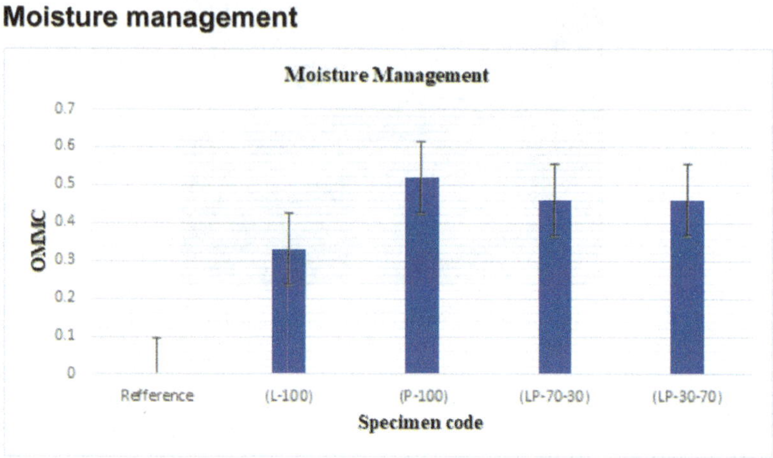

Fig. 7. Graphical representation of moisture management

The above Fig. 7 showed the values of OMMC. It has been seen from the bar chart that the values of reference sample is poor as compared to the develop samples. FR viscose has lower OMMC values due to its hydrophobic nature and protex has more value because of hydrophilic nature.

3.3 Thermal Resistance

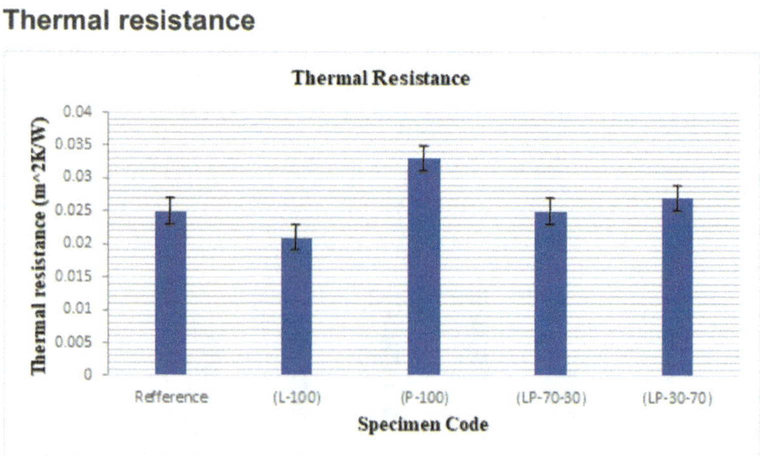

Fig. 8. Graphical representation of thermal resistance

The amount of resistance of a fabric to the flow of heat is known as thermal resistance. If the tightness factor is less, than more will be the thermal resistance. The interlock structure has a more thermal resistance value than single jersey fabric, that the reason it is preferred more for winter garments. The above Fig. 8 showed, the protex fabric has higher thermal resistance value, because it has short length fiber (they are also bulky in nature, so porosity become less, and it resists more heat).

3.4 Water Vapor Permeability

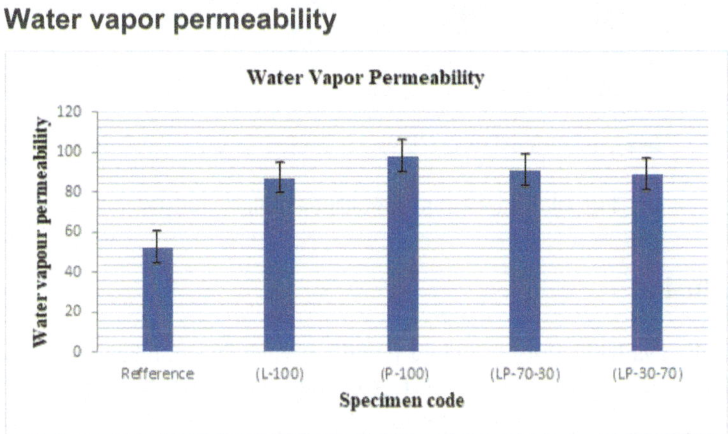

Fig. 9. Graphical representation of water vapor permeability

Water vapor permeability is the property of a material which permits the passage of water vapor through its structure. Protex 100% showed the highest value among all samples due to it hydrophilic nature. It can transfer moisture more quickly than lenzing (Fig. 9).

3.5 Vertical Flame Test

Before elaborating the figure below, here are some definitions that are necessary to understand. After flame time is the period for which sample continues to flame after eliminating the ignition source. Afterglow time is the period in which sample glow or burns after removal of ignition source and the termination of flame. Char length is the distance from fabric edge which is exposed to flame to the furthermost point of damage. The Fig. 10 showed that the lenzing has the lowest value of flame resistance because it is not inherently flame retardant. Lenzing is made up with natural raw material and during its manufacturing it is treated with flame resistant chemicals.

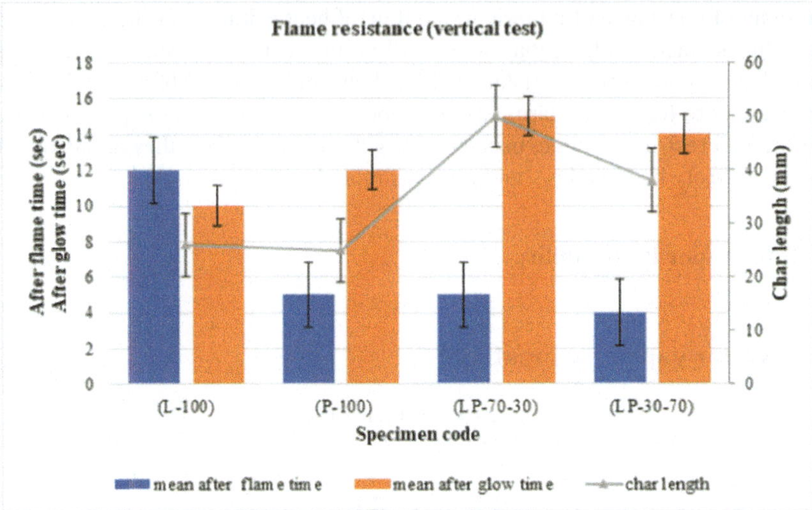

Fig. 10. Graphical representation of flame resistance

4 Conclusion

Commercially available glove of fighter jet pilot was analyzed. It has an excellent flame resistance property but poor in comfort properties. The lenzing and protex fibers and their blends were used to manufacturing a yarn. A fully fashioned knitting machine was used to develop a cut sewn knitted glove. Seamless glove knitting machine was used for the development of complete glove. It was seen that the samples of blend ratio fibers have overall improved properties than pure and reference sample. Like, they have comparatively better air permeability and very good flame resistance with few seconds of after flame time. The comfort properties were much better than existing glove.

References

Chen, F.B.: Multilayered gloves having enhanced barrier protection (Patent No. US 2006/0026737 A1) (2006). https://patents.google.com/patent/US20060026737?oq=US+2006%2F0026737+A1

Praveen, D., Jayasuriya, N., Satharasinghe, A.S.: Absorbable pad (Patent No. US 2018/0014983 A1) (2018). https://patents.google.com/patent/US20180014983A1/en?oq=US+2018%2F0014983+A1

Vero, F.A.: United States Patent (19) Patent Number: 6,021,523 "Heat And Abrasion Resistant Glove" (Patent No. 6,021,523) (1998). https://doi.org/US005485919A

Kleinerman, G.J., Woodcock, J.M.: glove (Patent No. 4888829) (1988). https://patents.google.com/patent/US4888829A/en?oq=4888829

Hoffmeyer, I.A., et al.: Protective garment (Patent No. US 8347422 B2) (2016). https://patents.google.com/patent/US8347422B2/en?oq=US+8%2C347%2C422+B2

Sidman, K.R., Aroms, I.J.: United States Patent (19) "Heat Resistant Protective Hand Covering" (Patent No. 4,433,439) (1984). https://doi.org/US005485919A

Leslie, L.F., Woods, J.A., Thacker, J.G., Morgan, R.F., McGregor, W., Edlich, R.F.: Needle puncture resistance of surgical gloves, finger guards, and glove liners. J. Biomed. Mater. Res. **33**(1), 41–46 (1996). https://doi.org/10.1002/(SICI)1097-4636(199621)33:1%3c41::AID-JBM7%3e3.0.CO;2-M

Oğlakcioğlu, N., Marmarali, A.: Thermal comfort properties of some knitted structures. Fibres Text. Eastern Europe **15**(5–6), 94–96 (2007). https://doi.org/10.1080/10803548.2010.11076842

Publication, A.: Antistatic gloves and process for making same (Patent No. US 2020/0299532 A1) (2021). https://patents.google.com/patent/US20200299532A1/en?oq=US+2020%2F0299532+A1

Zhu, R., Guckert, D., Lovasic, S.L.: Modacrylic/aramid fiber blends for arc and flame protection (Patent No. US 7065950 B2) (2006). https://patents.google.com/patent/US7065950B2/en?oq=7%2C065%2C950

Sarzotti, D.M.: Flame retardant yarns and fabrics comprising them (Patent No. EP 2619358 B1) (2018). https://patents.google.com/patent/EP2619358B1/en?oq=EP+2619358+B1

Tutterow, D.C., Dunn, C.S.: Flame resistant fabrics and garments made from same (Patent No. Wo 2008/027454 A1) (2008). https://patents.google.com/patent/WO2008027454A1/en?oq=Wo+2008%2F027454+A1

Varga, K., Noisternig, M.F., Griesser, U.J., Aljaz, L., Koch, T.: Thermal and sorption study of flame-resistant fibers. Lenzinger Berichte **89**(May), 50–59 (2011)

Monfalcone, V.A., Roberson, C.D., Fraser, Jr., L.L.: Thermally protective flame retardant fabric (Patent No. US 8501639 B2). In System and Method for Programming a Weighing Scale Using a Key Signal To Enter a Programming Mode (US 8501639 B2) (2009). https://patents.google.com/patent/US8501639B2/en?oq=US+8%2C501%2C639+B2

Mio, W., Mihoichi, M.: Flame retardent knit fabric (Patent No. US 2007/0190877 A1). In Us 2007/0197807 a1 (US 2007/0190877 A1) (2007). https://patents.google.com/patent/US20070190877A1/en?oq=US+2007%2F0190877+A1

Identify the Most Influential Operating Parameters Affecting Bending Stiffness of Knitted Fabric Using Fuzzy Decision Trees

Rania Baghdadi[1(✉)], Hamza Alibi[1], Faten Fayala[1], and Xianyi Zeng[2]

[1] National School of Engineering of Monastir, Laboratory of Thermal and Energetic Systems Studies, LESTE, University of Monastir, 5000 Monastir, Tunisia
baghdadi.rania@gmail.com

[2] National School of Arts and Textiles Industries (ENSAIT), GEMTEX Research Laboratory, University North Lille of France, 59000 Lille, France

Abstract. Stiffness is one of the most important utility properties of textile materials and plays a significant role in well-being due to its influence on physiological comfort. There are a lot of structural properties of textile materials also operating parameters (knitting + finishing) influencing stiffness. As part of our research, we proposed to help industry adjust the most relevant operating parameters prior to actual manufacturing to achieve the desired stiffness and satisfy consumers by the conception of a predictive artificial neural network's models to predict the bending stiffness of knitted fabrics.

Keywords: Finishing · Fuzzy decision tree · Knit fabrics · Bending stiffness · Textile · Well-being

1 Introduction

Stiffness is one of the most important utility properties of textile materials and plays a significant role in well-being (Kan and Zhou 2019). There are a lot of structural properties of textile materials also operating parameters (knitting + finishing) influencing stiffness and there are also statistically significant interactions between the main factors influencing the stiffness of knitted fabrics made from pure yarn cotton and viscose fibers and plated knitted with elasthane fibers.

As part of our research, help industry adjust the most relevant operating parameters prior to actual manufacturing to achieve the desired air permeability and satisfy consumers. Within the framework of the work presented, we used a selection procedure using the Fuzzy Decision Tree, to solve the problem of insufficient data and reduce the complexity of predictive models. The use of this tool allowed us to simultaneously consider qualitative, quantitative variables which are better treated as fuzzy variables. This was a real advantage since we have both types of variables.

S. Msahli and F. Debbabi (Eds.): CIRATM 2020, SPM 17, pp. 182–187, 2022.
https://doi.org/10.1007/978-3-031-08842-1_29

2 Material and Methods

2.1 Fuzzy Decision Tree

Complementary to more traditional statistical methods such as linear regression, discriminant analysis, decision trees are a tool for decision support and data mining, thanks to the representation of a set of choices in the graphical form of a tree. Fuzzy decision trees used in different domain (Tayefi et al. 2017; Yang et al. 2018). The various possible decisions are at the extreme's branches (the "leaves" of the tree) and are reached according to the decisions taken at each stage.

As part of this research, we use fuzzy decision trees. The choice naturally fell on the use of the free FisPro software during our research. This software makes it possible to create fuzzy inference systems, and to use them for reasoning purposes, in particular for simulating a physical or biological system.

2.2 Materials

An appropriate set of plain knitted fabrics was collected at a first stage. They were different variants of plain knitted fabrics of different compactness resulting from different linear densities of ground and Lycra yarns as well as from the different loop length, gauge of industrial circular knitting machines, etc. All fabrics were processed according to different finishing steps according to Table 1, so that 422 samples were obtained.

Table 1. Different finishing stages of cotton and viscose stretch plain knitted fabrics

Finishing stages	Cotton				Viscose			
	Half white without finishing	Light color	Dark color	Half white	Half white without finishing	Light color	Dark color	Half white
Scouring (at 95 °C during 25 minute)	X	X	X	X	X	X	X	X
Bleaching (at 95 °C during 25 minutes with NaOH+H2O2)	X	X	-	X	-	-	-	-
Bleaching (at 85 °C during 25 minutes with H2O2+Na2CO3)	-	-	-	-	X	X	-	X
Light dyeing (with reactive dye)	-	X	-	-	-	X	-	-
Dark dyeing (with reactive dye)	-	-	X	-	-	-	X	-
Softening (with cationic softener)	-	X	X	X	-	X	X	X
Softening (with silicone softener)	-	X	X	X	-	X	X	X
Anti-pilling	-	X	X	X	-	-	-	-
Squeezing	X	X	X	X	X	X	X	X
Heat setting (without tension) in a stenter	X	X	X	X	X	X	X	X
Compacting	X	X	X	X	X	X	X	X

2.3 Test Methods

The functional parameter, bending stiffness of these samples, was tested according to the ASTM D4032 standard on the Circular Bend Stiffness Tester (Fig. 1). This method gives an average value of forces related to instantaneous rigidity in all directions. The imposition of the maximum values of the forces necessary to push a rectangle of dimensions fixed by the standard thanks to a truncated cone equipped with a presser.

In this study, the tests carried out were concerning the determination of these parameters according to the test method mentioned in Table 2. Table 3 presents the statistical values of structural parameters of set fabrics. The values of qualitative structural parameters are presented in Table 4.

Fig. 1. Stiffness tester

Table 2. Structural parameters.

Structural parameters (a)	*Test method*
Yarn count (Nm)	EN ISO 8388
Lycra proportion (%)	ISO1833-12
Lycra yarn count (dtex)	EN ISO 8388
Weight per unit area (g/m²)	NF G07-150
Thickness (m)	NF G07-153
Loop length (cm)	EN 14970

Let's deal with the properties of "Bending stiffness' of a set of prototypes. We have 14 design factors and one criteria of wellbeing B1 (Bending stiffness) whose modalities are 'Low', 'Medium' and 'High', defined by the expert from fuzzy membership functions (Fig. 2).

Table 3. The statistical values of some structural (a) and functional (b) parameters

Structural parameters (a)	*Mean value*	*Standard deviation*	*Maximum value*	*Minimum value*
Yarn count (Nm)	50.26	7.83	80	20
Gauge	27.43	1.64	28	20
Lycra proportion (%)	1.6	2.41	10	0
Lycra yarn count (dtex)	10.6	15.04	44	0
Weight per unit area (g/m^2)	150.11	76.17	487.92	58.25
Thickness (m)	0.66	0.14	1	0.46
Jersey loop length (cm)	0.31	0.03	0.45	0.24
Cationic softener (%)	1.12	1.14	3	0
Silicone softener (%)	1.23	1.17	3	0
Anti-pilling (%)	0.50	0.40	1	0
Functional parameter (b)	*Mean value*	*Standard deviation*	*Maximum value*	*Minimum value*
Stiffness (N)	0.92	0.63	4.26	0.16

Table 4. Qualitative structural parameters

Structural parameters	*Linguistic value*
Yarn composition	'Cotton' or 'viscose'
Dyeing	'Undyed' or 'Light Dye' or 'Dark Dye'
Bleached cotton	'Unbleached' or 'Bleached'
Bleached viscose	'Unbleached' or 'Bleached'

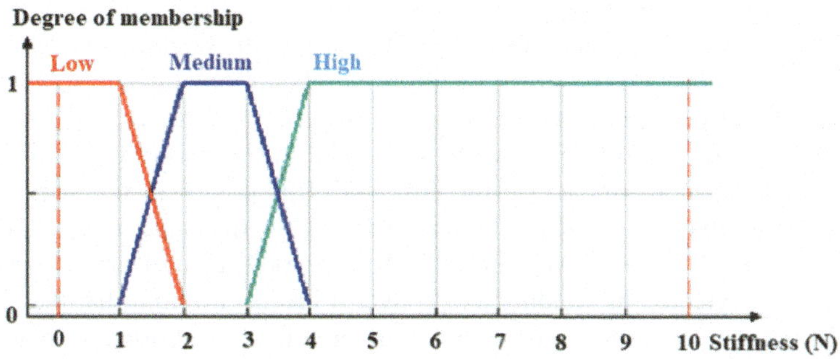

Fig. 2. Experimental device for stiffness property determination.

3 Results and Discussion

The use of the FisPro software allowed to obtain a decision tree relating to the bending stiffness in Fig. 3. Given the large size of the fuzzy decision tree obtained, we proposed to expose the final result in tabular form which highlights the different classes obtained from the tree with the design parameters (Table 5).

Table 5. Classification by decision tree method - bending stiffness

Classes	*Design Parameters*
1	Thickness (F14)
2	Gauge (F3), Weight per unit area (F13)
3	Percent Lycra (F5), Composition (F1)
4	LFA (F4), Cationic softening (F11), Silicone softening (F12), Cotton bleaching (F7)
5	Anti-pilling (F10), Viscose bleach (F8), Metric Number (F2), Lycra dtex (F6)
6	Dyeing (F9)

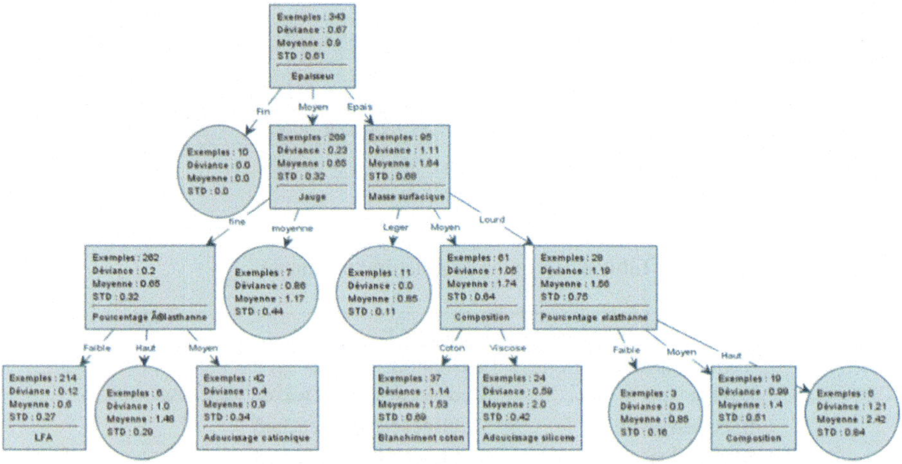

Fig. 3. First part of the decision tree relating to the 'stiffness'

From the previous figure and table, we can state that the thickness is the parameter that has the most impact on the stiffness, followed secondly by the gauge and the mass per unit area, then by the percentage of elastane and composition, etc.

Thus, it turns out that the parameters associated to the construction of the fabric are those that have more impact on the stiffness. This comes in the same direction as the research previously conducted concerning the stiffness parameter. Indeed, the studies directed by ABO et al. (Huang and Moraga 2004) came to show that rigidity was significantly affected by binding and gauge. It has been found that the gauge has a positive effect on the bending stiffness in both directions. Additional researches have studied the effects of thread length on the stiffness of single-knit fabrics. Diverse types of knits were used (jersey, Lacoste, ...). The absorbed length of yarn, the yarn count influences the dimensional stability as well as the length of bending. It has been found that the bending length declines progressively with the increase in the length of yarn absorbed for the same yarn count.

4 Conclusion

In this study, we are interested in the detection of the influence of the different input variables (design parameters) on the output variable (parameter bending stiffness). The goal is to rank these input parameters in order of importance to eliminate the least important ones and then simplify the build of predictive models. To solve the problem of insufficient data, a selection procedure has been proposed using the Fuzzy Decision Tree. The use of this tool allowed us to simultaneously take into account qualitative and quantitative variables (discrete or continuous).

Thus, it turned out that the design parameter that most influences bending stiffness is the thickness, followed by the gauge, the weight per unit area, the percent Lycra, the composition, and so on. These parameters are followed by the parameters of the finishing process.

References

Kan, C., Zhou, C.: High Performance Technical Textiles: Marine Textiles and Composites, Roshan Paul (2019). 978-1-119-32501-7

Tayefi, M., et al.: The application of a decision tree to establish the parameters associated with hypertension. Comput. Methods Programs Biomed. **139**, 83–91 (2017). https://doi.org/10.1016/j.cmpb.2016.10.020

Yang, W., et al.: Sex determination of skull based on fuzzy decision tree. In: Proceedings 4th Workshop on Advanced Research and Technology in Industry Applications, Advances in Engineering Research (2018). 978-94-6252-597-9

Huang, C., Moraga, C.: A diffusion-neural-network for learning from small samples. Int. J. Approx. Reason. **35**(2), 137–161 (2004). ISSN 0888-613X

Using Reverse Neural Networks (ANNi) to Predict the Structural Variables from a Fixed Value of the Crease Recovery Angle

Hamza Alibi[1(✉)], Rania Baghdadi[1], Faten Fayala[1], Abdelmajid Jemni[1], and Xianyi Zeng[2]

[1] National School of Engineering of Monastir, Laboratory of Thermal and Energetic Systems Studies, LESTE, University of Monastir, 5000 Monastir, Tunisia
alibi_hamza@yahoo.fr

[2] National School of Arts and Textiles Industries (ENSAIT), GEMTEX Research Laboratory, University North Lille of France, 59000 Lille, France

Abstract. Due to its influence on physiological comfort, the appearance of fabric has a significant role in well-being. For apparel application, crease recovery is considered as a significant property of textiles. In this paper, the objective is to help industrials to predict the input variables (structural) from a fixed value of the crease recovery angle thanks to the use of ANNi.

Keywords: Artificial neural network · Crease recovery · Reverse neural networks · Stretch finished knitted fabrics · Well-being

1 Introduction

Appearance of fabric has a significant role in well-being due to its influence on physiological comfort. In fact, it acts on the psychological and social well-being of consumers. Appearance has three main criteria namely style, color and texture, the choice of which is based on the socio-cultural situation. It is changed according to fashion, the age of the consumer, etc.

The well-being of the consumer depends on the sensations he feels when wearing it. Indeed, the human body is subject to environmental constraints (temperature, humidity) which condition his well-being. In addition, the appreciation of a textile article depends on the behavior of the latter following frequent use by the consumer. Thus, a textile product will be much more appreciated by the user, if it retains the characteristics before use (size, appearance, feel, etc.) and does not show any deterioration (dimensional variation, pilling, etc.) during use or when washing.

For apparel application, crease recovery is considered as one of the most significant properties of textiles. In a previous work (Baghdadi et al. 2016), an artificial neural network model based on the virtual leave one out approach was developed to predict crease recovery angle of finished stretch knit materials. In this article, the objective is to allow the industrialist to predict the input variables (structural) from a fixed value of the output (here crease recovery angle) thanks to the use of reverse neural networks (ANNi).

S. Msahli and F. Debbabi (Eds.): CIRATM 2020, SPM 17, pp. 188–192, 2022.
https://doi.org/10.1007/978-3-031-08842-1_30

2 Material and Methods

2.1 Reverse Neural Networks

Predict structural parameters in order to obtain optimal operating conditions or parameters, a mathematical description of such a process are required. Often, calculation of optimal parameters is difficult and requires special software, especially when process complexity is considered. This is the so-called inverse problem identification which must be solved to answer the following question: what are the controlled inputs that have resulted in this given output. Several applications based on inverse neural network models referred as were developed by several authors to optimize the performance of polygeneration systems parameters (Hernández et al. 2013), to optimize the operating conditions for compressor performance (Cortés et al. 2009), and so on (Hernández 2009; El Hamzaoui et al. 2011; Laidi et al. 2013; Márquez-Nolasco et al. 2017; Conde-Gutiérrez et al. 2018).

2.2 Materials

An appropriate set of plain knitted fabrics was collected at a first stage. They were different variants of plain knitted fabrics of different compactness resulting from different linear densities of ground and Lycra yarns as well as from the different loop length, gauge of industrial circular knitting machines, etc. All fabrics were processed according to different finishing steps according to Table 1, so that 422 samples were obtained.

Table 1. Different finishing stages of cotton and viscose stretch plain knitted fabrics

Finishing stages	Cotton				Viscose			
	Half white without finishing	Light color	Dark color	Half white	Half white without finishing	Light color	Dark color	Half white
Scouring (at 95 °C during 25 minute)	X	X	X	X	X	X	X	X
Bleaching (at 95 °C during 25 minutes with NaOH+H2O2)	X	X	-	X	-	-	-	-
Bleaching (at 85 °C during 25 minutes with H2O2+Na2CO3)	-	-	-	-	X	X	-	X
Light dyeing (with reactive dye)	-	X	-	-	-	X	-	-
Dark dyeing (with reactive dye)	-	-	X	-	-	-	X	-
Softening (with cationic softener)	-	X	X	X	-	X	X	X
Softening (with silicone softener)	-	X	X	X	-	X	X	X
Anti-pilling	-	X	X	X	-	-	-	-
Squeezing	X	X	X	X	X	X	X	X
Heat setting (without tension) in a stenter	X	X	X	X	X	X	X	X
Compacting	X	X	X	X	X	X	X	X

2.3 Test Methods

The functional parameter, crease recovery of these samples, was tested according to the ISO 2313 standard on the crease recovery Tester (Fig. 1). The principle of the test is to put a rectangular test tube under a load for 5 min. Then, let the specimen stand up freely, after having removed the load, and measure the recovery angle.

In this study, the tests carried out were concerning the determination of certain parameters according to the test method mentioned in Table 2. Table 3 presents the statistical values of quantitative structural parameters of set fabrics. Values of qualitative structural parameters are shown in Table 4.

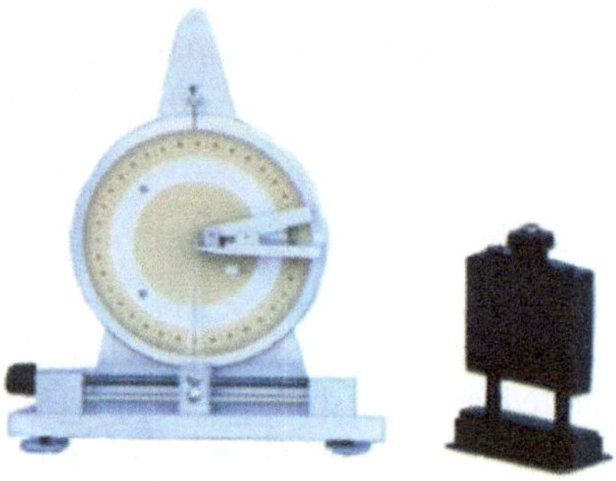

Fig. 1. Crease recovery tester

Table 2. Structural parameters.

Structural parameters (a)	*Test method*
Yarn count (Nm)	EN ISO 8388
Lycra proportion (%)	ISO1833-12
Lycra yarn count (dtex)	EN ISO 8388
Weight per unit area (g/m²)	NF G07-150
Thickness (m)	NF G07-153
Loop length (cm)	EN 14970

Table 3. The statistical values of some structural (a) and functional (b) parameters

Structural parameters (a)	*Mean value*	*Standard deviation*	*Maximum value*	*Minimum value*
Yarn count (Nm)	50.26	7.83	80	20
Gauge	27.43	1.64	28	20
Lycra proportion (%)	1.6	2.41	10	0
Lycra yarn count (dtex)	10.6	15.04	44	0
Weight per unit area (g/m²)	150.11	76.17	487.92	58.25
Thickness (m)	0.66	0.14	1	0.46
Jersey loop length (cm)	0.31	0.03	0.45	0.24
Cationic softener (%)	1.12	1.14	3	0
Silicone softener (%)	1.23	1.17	3	0
Anti-pilling (%)	0.50	0.40	1	0
Functional parameter (b)	*Mean value*	*Standard deviation*	*Maximum value*	*Minimum value*
Column crease recovery angle back/back (°)	143.19	18.72	192.00	93.67

Table 4. Qualitative structural parameters

Structural parameters	*Linguistic value*
Yarn composition	'Cotton' or 'viscose'
Dyeing	'Undyed' or 'Light Dye' or 'Dark Dye'
Bleached cotton	'Unbleached' or 'Bleached'
Bleached viscose	'Unbleached' or 'Bleached'

3 Results and Discussion

We were able to predict the input parameters that contributed to obtain a crease recovery angle in the column direction back/back given. After setting a given output value as well as some input parameters, an artificial reverse neural network using the software Neuro one V12 was applied to find the other structural parameters that contributed to this level of crease recovery angle column direction back/back. For an output value of 143.89°, the optimized structural parameters are described in Table 5.

A validating processes was applied using the test database. The experimental versus predicted values of test dataset is very well ($R^2 > 0.9$).

The levels of error (3%) are satisfactory and smaller than errors that normally arise due to experimental variation and instrumentation accuracy.

This small error in conjunction with a computing time of less than 2 s indicates that this strategy can be used with a high level of confidence for the knit wear industry.

Table 5. The results of the inverse neural network model for the crease recovery angle column direction back/back

Optimum input parameters	*Values*
Composition	Cotton
Nm	50
LFA (cm)	0.277
Proportion in Lycra (%)	0
Lycra (dtex)	0
Cotton bleaching	1
Dyeing	Undyed
Anti-pilling (%)	0
Silicone softening (%)	0
Mass per unit area (g/m^2)	139.887
Thickness (m)	0.526

4 Conclusion

In this study, the aim was to predict the structural variables from a fixed value of the crease recovery angle thanks to the use of reverse neural networks (ANNi). The optimization of these input parameters can facilitate the work of the development department in the textile industry also the manufacture of well-determined knitwear jerseys.

References

Baghdadi, R., et al.: A neural network model to predict crease recovery angle of finished stretch plain knitted fabrics. CIRAT-7, Hammamet, Tunisia, 11, 2016 (2016)

Hernández, J.A., et al.: Inverse neural network for optimal performance in polygeneration systems. Appl. Therm. Eng. **50**(2), 1399–1406, 1359–4311 (2013)

Cortés, O., et al.: Optimization of operating conditions for compressor performance by means of neural network inverse. Appl. Energy **86**(11), 2487–2493, 0306–2619 (2009)

Hernández, J.A.: Optimum operating conditions for heat and mass transfer in foodstuffs drying by means of neural network inverse. Food Control **20**(4), 435–438, 0956–7135 (2009)

El Hamzaoui, Y., et al.: Optimal performance of COD removal during aqueous treatment of alazine and gesaprim commercial herbicides by direct and inverse neural network. Desalination **277**(1–3), 325–337, 0011–9164 (2011)

Laidi, M., et al.: Optimal solar COP prediction of a solar-assisted adsorption refrigeration system working with activated carbon/methanol as working pairs using direct and inverse artificial neural network. Int. J. Refreg. **36**(1), 247–257, 0140–7007 (2013)

Márquez-Nolasco, A., et al.: Optimization and estimation of the thermal energy of an absorber with graphite disks by using direct and inverse neural network. J. Energy Resour. Technol. **140**(2), 1146–1159, 020906 (2017)

Conde-Gutiérrez, R.A., et al.: Optimal multivariable conditions in the operation of an absorption heat transformer with energy recycling solved by the genetic algorithm in artificial neural network inverse. Appl. Soft Comput. **2018**(72), 218–234, 1568–4946 (2018)

Determination of Interface Pressure Profiles Using Indirect Approach

Barhoumi Houda[1(✉)], Marzougui Saber[2], and Ben Abdessalem Saber[1]

[1] Textile Materials and Processes Research Unit, University of Monastir, Monastir, Tunisia
houda.barhoumi@enim.u-monastir.tn

[2] Laboratory of Textile Engineering, University of Monastir, Monastir, Tunisia

Abstract. The compression garments are characterized by the interface pressure they apply to a human body area. In this investigation, an indirect approach for measuring the interface pressure from a set of compression bands was developed. The Young modulus and the desired deformation values were determined. The calculated pressure values were compared against FlexiForce® sensor readings measured on 10 volunteers. After verifying the validity of this method, the interface pressure profiles along the leg were plotted.

Keywords: Interface pressure · Indirect method · Mechanical properties · Compression garment

1 Introduction

The concept of compression therapy stockings is based on a simple mechanical principle: By compressing the limb with graduated compression, strong at the ankle and decreasing as it goes up the leg, the compression stocking helps venous return, decreases venous pressure, prevents venous stasis and efficiently relieves aching and heavy legs (Partsch et al. 2008). Recently, there are a number of devices that measure the interface pressure from a compression fabric. These include using a direct testing approach where a sensor is placed on a patient's leg or a wooden cylinder model to observe pressure readings (Flaud et al. 2010). There are a few devices that allow a constant reliable observation with minimal costs, and while, these devices have proved useful in industry use, they are known to have weaknesses, such as limits in shape sizes.

This study presents an indirect testing approach that requires the use of a tensile tester commonly used in textile testing laboratories to evaluate the efficiency of compression therapy products.

2 Materials and Methods

Our indirect testing method was based on the use of a Lloyd S5 Testing Machine. In order to mimic the shape of the wanted leg area, customized grips (Fig. 1) were designed and fitted to the Lloyd S5 machine. Four jersey knitted compression bands (G1, G2, G3 and G4) were tested in this investigation. Their characteristics are presented in Table 1.

S. Msahli and F. Debbabi (Eds.): CIRATM 2020, SPM 17, pp. 193–197, 2022.
https://doi.org/10.1007/978-3-031-08842-1_31

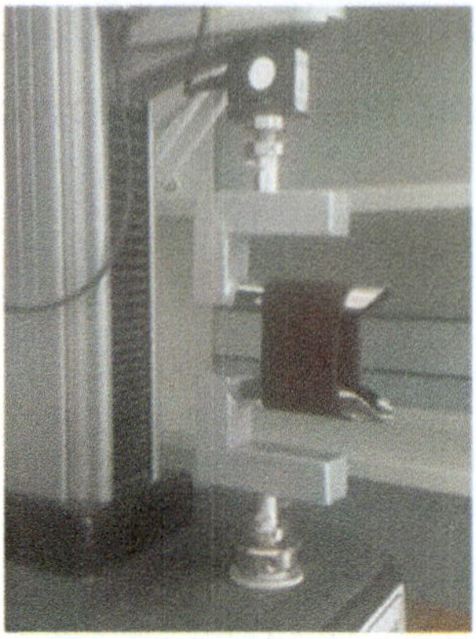

Fig. 1. Customized grips for Lloyd S5 testing machine.

Table 1. Characteristics of tested samples

Characteristics	Unit	Sample code			
		G1	G2	G3	G4
Yong modulus	MPa	0.696	0.786	0.633	0.411
Polyamide 6-6 count	dTex	77	77	77	77
Elastane percentage	%	10	15	15	8
Fabric weight	g/m^2	222	227	232	199
Fabric thickness	mm	0.742	0.95	0.87	0.8
Fabric circumference	cm	20.6	20.6	20.6	20.6

The tension – deformation curves of the four knitted bands are presented in the Fig. 2.

A new model (Eq. 1) based on Laplace's law has been developed in our previous work (Barhoumi et al. 2020). This model was used to calculate the theoretical interface pressure values (mmHg).

$$P = \frac{2\pi\varepsilon Ee}{C} \tag{1}$$

where P is the interface pressure in mmHg, E is the Young modulus in MPa, e is the sample's thickness in mm, C is the circumference of the body leg in mm and ε is the knitted band deformation at the wanted circumference.

In order to evaluate the effectiveness of the proposed indirect approach, it was necessary to compare the calculated data to direct pressure readings from a human limb. For this reason, ten healthy female subjects agreed to take part in this study. Their size and their average age are respectively 40 and 27 years old. A force-pressure sensor, "Flexi-Force A201" sensor, was used to measure the experimental interface pressure (Barhoumi et al. 2018).

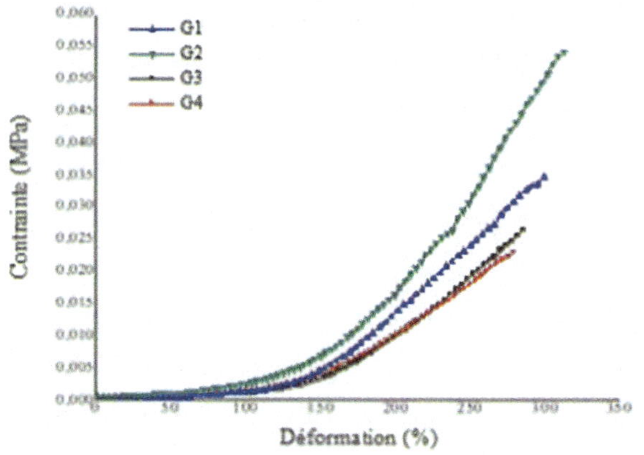

Fig. 2. Tension-Deformation curves of the different samples.

3 Results and Discussions

3.1 Assessment of the Indirect Approach

The standard deviation (SD) was used to quantify the amount of dispersion of experimental or direct interface pressure values. As shown in Table 2, for all the tested samples, the SD values are inferior to 1. This means that the distribution of the values around the mean direct interface pressure value is acceptable.

To validate the indirect approach, the error percentage between direct and indirect interface pressure values was calculated. We notice that it never exceeds 5% for all the tested samples Thus, there is no discrepancy between direct and indirect interface pressure values as the percentage error value can be evaluated as tolerable.

Table 2. Comparison between direct and indirect interface pressure values

C = 27.5 cm	Indirect pressure (mmHg)	Direct pressure (mmHg)		Error (%)
		Average value (mmHg)	SD	
G1	22.03	21.25	0.45	3.67
G2	22.9	22.1	0.31	3.62
G3	21.23	21.54	0.74	1.44
G4	17.95	18.8	0.62	4.52

3.2 Interface Pressure Profiles

The tight-fitting garment, such as the tested elastic bands, apply an amount of pressure to a specific area of the body through its own elasticity and stretch recovery performance. Hence, it is important to know the distribution profile of the interface pressure applied along the leg.

After validating the indirect approach, the interface pressure values are calculated and the average pressure profiles at four positions along the leg were plotted as shown in Fig. 3.

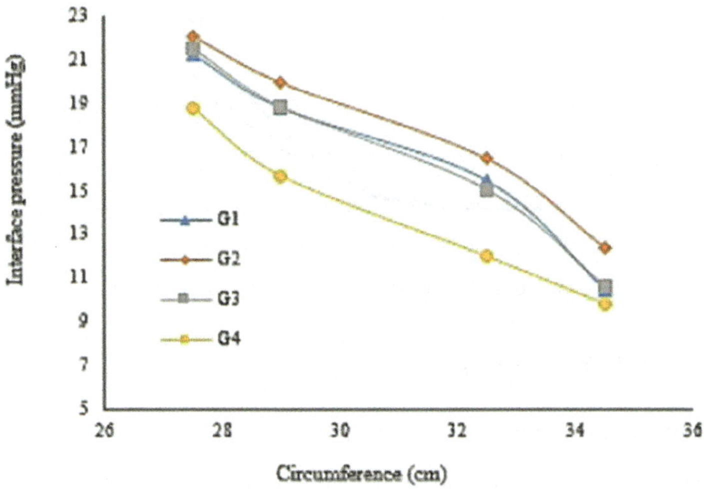

Fig. 3. Interface pressure profiles of the different samples using the indirect method

It is noticeable from the interface pressure profiles that the compression bands generate a variable pressure gradient along the lower limb. The body shape has a significant effect on the interface pressure distribution. Indeed, the human leg has an irregular geometry. Therefore, the tension applied and the pressure generated by the compression bands cannot be uniformly distributed. According to Eq. (1), the greater interface pressure on the lower limb will occur in the region with the lower circumference. As shown on Fig. 3, the performance of pressure levels in the ankle (circumference = 27.5 cm) is higher than those at the other positions.

For compression stockings to be effective, desirable pressure is of great importance. Inadequate interface pressure will be ineffective in venous disease reduction, and excessive pressure may compress arterioles without improving the venous flow (Partsch 2005). In literature, some study indicated that many compression stockings did not apply a sufficient amount of pressure on the wanted area, as claimed by the manufacturers (Best et al. 2000; Veraart et al. 1997). Therefore, implementing this testing approach could assist manufacturers of compression hosieries in gathering significant data as tensile testing machines are widely available across many physical testing laboratories.

4 Conclusion

In this study, we presented an indirect approach using custom grips that simulate the circumference of the body part. Based on the modified Laplace's law, the theoretical interface pressure values were calculated and compared with the experimental ones.

It was found that the proposed indirect approach presents a good agreement with the experimental results. This funding is of great interest to the manufacturers of compression garment because it is simple, not expensive, valid and adequate.

References

Barhoumi, H., Marzougui, S., Ben Abdessalem, S.: Clothing pressure modeling using the modified Laplace's law. Cloth. Text. Res. J. **38**(2), 134–147 (2020)

Barhoumi, H., Marzougui, S., Ben Abdessalem, S.: Influence of manufacturing parameters of knitted compression fabric on interface pressure. Indian J. Fibre Text. Res. **43**(4), 426–433 (2018). ISSN 0975-1025

Best, A.J., et al.: Graded compression stockings in elective orthopaedic surgery. An assessment of the in vivo performance of commercially available stockings in patients having hip and knee arthroplasty. J. Bone Joint Surg. **82**(1), 116–118 (2000). ISSN 0301-620X

Flaud, P., Bassez, S., Counord, J.-L.: Comparative in vitro study of three interface pressure sensors used to evaluate medical compression hosiery. Dermatol. Surg. **36**(12), 1930–1940 (2010). ISSN 1076-0512

Partsch, H.: The static stiffness index: a simple method to assess the elastic property of compression material in vivo. Dermatol. Surg. **31**(6), 625–630 (2005). ISSN 1076-0512

Partsch, H., et al.: Classification of compression bandages: practical aspects. Dermatol. Surg. **34**(5), 600–609 (2008). ISSN 1524-4725

Veraart, J.C., Pronk, G., Neumann, H.A.: Pressure differences of elastic compression stockings at the ankle region. Dermatol. Surg. **23**(10), 935–939 (1997). ISSN 1076-0512

The Effect of Bleach Washing Treatment on the Comfort Properties of Dyed Cotton Garments

Sarra Said[1(✉)], Sabri Halaoua[1], Imed Feki[1], Mohamed Hamdaoui[1], and Walid Sahraoui[2]

[1] Textile Materials and Processes Research Unit (MPTEX), University of Monastir, Monasir National School of Engineers, Monastir, Tunisia
sarrasaid@yahoo.fr
[2] TANIT Textile Services Company, Monastir, Tunisia

Abstract. Consumers who purchase textiles generally perceive the tactile properties of the fabric, indicating that they are more interested in comfort properties than other properties. In this investigation, the changes in comfort properties of dyed cotton garments were evaluated. Results showed that tensile properties, bending and compression behaviors, air permeability and surface roughness of dyed cotton fabrics were differently affected by the bleach washing treatment.

Keywords: Bleach treatment · Cotton garments · Comfort properties · KES · Tensile strength

1 Introduction

Comfort is one of the most desirable properties of textiles. It is described as the state of psychological, physiological and physical harmony between human beings and the environment (Bhatia and Malhotra 2016). This property is very important in helping users to choose an appropriate textile substrate. Analysts have found that customers prefer clothing that is not only beautiful but also comfortable to wear (Behera et al. 1997). Tactile properties, namely "sensorial comfort", or "handle" or "skin sensorial wear comfort" (Bensaid et al. 2006) are mainly due to the ratio of stresses generated by the fabric during use and the way it is distributed on the skin. The Kawabata Organizational Evaluation System (KES) was studied in the 1970's to objectively evaluate the sensorial comfort (Kawabata 1973).

Nowadays, the four KES modules have studied the low stress mechanical and surface properties by measuring the tensile, shear, bending and compression behaviors and surface characteristics of the fabric (Atalie 2018). Many researchers have studied the effect of finishing processes on the properties of textile substrates. However, there are few studies on their effects on dyed cotton garments. In this study, we investigated the effect of an ecofriendly bleaching treatment on the comfort of dyed cotton garments.

S. Msahli and F. Debbabi (Eds.): CIRATM 2020, SPM 17, pp. 198–203, 2022.
https://doi.org/10.1007/978-3-031-08842-1_32

2 Materials and Methods

2.1 Materials

Red sulfur dyed denim fabric was used in this investigation and its basic characteristics are shown in Table 1.

Table 1. Characteristics of untreated denim

Composition	100% cotton
Weave structure	Z-Twill 3, Warp effect
Surface density (g/m^2)	378 g/m^2
Warp density (yarn/cm)	29
Weft density (yarn/cm)	26

2.2 Methods

First of all, the red sulfur dyed denim fabrics were desized for 15 min at 45 °C to remove adhesives which already exists in the warp during weaving. Then the samples have been treated with an ecological Bleach washing using the hydrogen peroxide.

Tests were realized following cotton fabrics preconditioning at standard testing atmosphere (20 $\pm$ 2 °C and 65 $\pm$ 2% RH).

Air permeability (ISO 9237), Tensile strength (ISO 13934-2), Kawabata's Evaluation System for Fabrics (KES-FB2 (Bending), KES-FB3 (Compression), KES-FB4 (Surface)) have been used to characterize the effect of bleach washing on denim fabric. The tests were performed in the warp direction of the fabric and were repeated three times to reduce experimental errors.

3 Results and Discussion

3.1 Air Permeability

Air permeability is a physical parameter that presents the ability of the material to let air pass through.

Table 2. Air permeability of different samples

Samples	Untreated fabric	Bleached fabric
Average of "AP"	37.77	34.91
Standard deviation	0.85	0.84

The results in Table 2 showed a slight decrease in air permeability after bleaching treatment. High temperature and humidity of wash can release the tension of yarns, producing the shrinkage of denim garment in both the weft and warp directions (Haq 2015). After bleaching treatment, worn threads and hair threads are generated, which reduces the space between the threads resulting in a decrease in the air permeability of the bleached fabric (Du et al. 2019).

3.2 Tensile Strength

The tensile strength of the bleached cotton fabric is lower than that of the untreated as shown in Fig. 1.

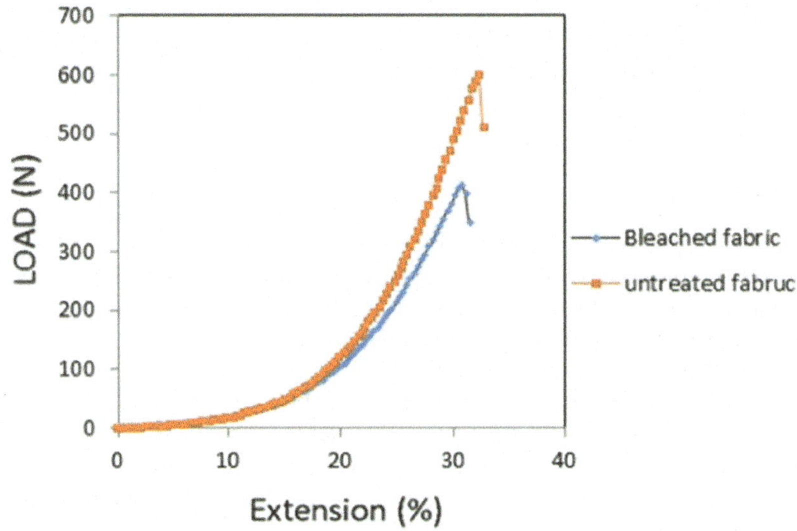

Fig. 1. Mechanical properties of the bleached and untreated garments

The decreasing of the tensile strength after bleach treatment is due to the oxidizing effect of hydrogen peroxide which produces HO_2^- ions with strong oxidizing action in alkaline condition (Du et al. 2019) and causes degrading effect on the fiber's surface. In fact, the finishing treatments of cotton garment are known to decompose cellulose at its end or in the middle of the cellulose chains, thus reducing the fabric's mechanical strength (Khan et al. 2012).

3.3 KES-FB2: Bending

Bending rigidity (B) indicates the flexibility of the fabric. A high value of bending rigidity (B) reflects a high resistance to bending movement. The bending hysteresis (2HB) reflects the ability of the fabric to recover from bending.

Bending rigidity (B) and hysteresis of bending property (2HB) have decreased after bleach washing treatment as compared to untreated fabric as shown in Table 3. Indeed, the

Table 3. Effect of bleach treatment on the bending properties of cotton fabrics

Samples	Untreated fabric	Bleached fabric
Average of B	0.6	0.625
Standard deviation of B	0.02	0.01
Average of 2HB	0.422	0.411
Standard deviation of 2HB	0.025	0.02

bleach treatment performed on our cotton fabrics causes a degradation of the cellulose, which leads to an increase of the amorphous regions and causes a decrease in the bending stiffness.

3.4 KES-FB3: Compression

We determined the EMC compressibility given by the relation below (Eq. 1) to evaluate the effect of compression on the different bleached samples.

$$\mathrm{EMC} = \frac{T_o - T_M}{T_o} \tag{1}$$

where:

- T_M: The final thickness measured under a pressure of 50 gf/cm^2.
- T_0: the initial thickness of the specimen measured under a pressure of 0.5 gf/cm^2;

Table 4. Compressibility (EMC) of different samples

Samples	Untreated fabric	Bleached fabric
Average of EMC	0.6	0.625
Standard deviation of EMC	0.02	0.01

The compressibility rate is higher in bleach washing with hydrogen peroxide as compared to untreated fabric as shown in Table 4 and Fig. 2.

The presence of pilosity resulting from the partial extraction of fibers by bleach treatment had increased the compressibility of the fabric as compared to the untreated garment.

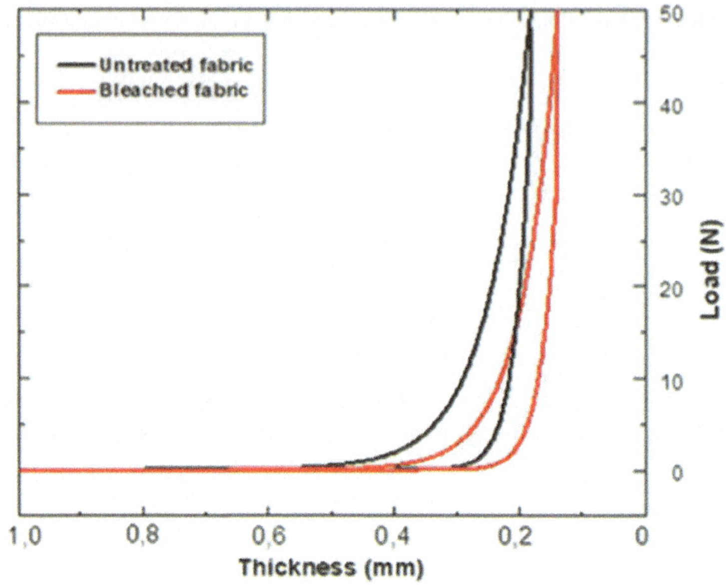

Fig. 2. Kawabata compression test curves of bleached and untreated fabrics.

3.5 KES-FB4: Surface

MIU reflects the resistant of the fabric. A high MIU value indicates high friction. SMD is an indicator of roughness and uniformity. The higher is the SMD value, the greater is the roughness and unevenness.

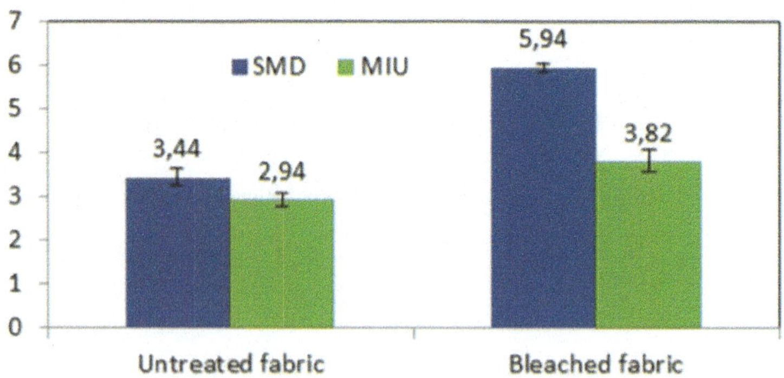

Fig. 3. SMD and MIU values of bleached and untreated fabrics

As shown in Fig. 3 MIU value of bleached fabric has increased as compared to untreated garment. Indeed, bleached fabrics are less smooth, rougher and more irregular than untreated one. Fabrics treated with hydrogen peroxide were found to have the highest SMD value. The highest value refers to the geometrically rough surface of the bleached fabric.

4 Conclusion

The purpose of the present study was to evaluate the effect of the Bleach eco-friendly finishing treatment on the comfort properties of cotton dyed garment. It was also noted that bleached cotton samples are more rigid and rougher than untreated tissue (high value of SMD and MIU). The air permeability, tensile strength and bending rigidity decrease after the bleach treatment. Eco-friendly finishing treatments provide an effective way to create desirable fashionable effects. Though, a control tests are required after finishing treatments to evaluate the overall properties of dyed fabrics in order to respond to consumer demands.

References

Atalie, D., Tesema, A.F., Rotich, G.K.: Effect of weft yarn twist level on thermal comfort of 100 per cent cotton woven fabrics. Res. J. Text. Apparel **22**(3), 180–194 (2018)

Behera, B., et al.: Comfort properties of fabrics woven from ring-, rotor-, and friction-spun yarns. J. Text. Inst. **88**(3), 255–264 (1997)

Bensaid, S., et al.: The effect of pattern construction on the tactile feeling evaluated through sensory analysis. J. Text. Inst. **97**(2), 137–145 (2006)

Bhatia, D., Malhotra, U.: Thermophysiological wear comfort of clothing: an overview. J. Text. Sci. Eng. **6**, 250 (2016)

Du, W., et al.: Comparative study on the effects of laser bleaching and conventional bleaching on the physical properties of indigo kapok/cotton denim fabrics. Appl. Sci. **9**(21), 4662 (2019)

Haq, U.N., et al.: Investigation of the bulk, surface and transfer properties of chlorine bleached denim apparel at different condition. Eur. Sci. J. **11**(12) (2015)

Kawabata, S.: Characterization method of the physical propperty of fabrics and the measuring system for hand-feeling evaluation. Sen'i Kikai Gakkaishi (J. Text. Mach. Soc. Japan) **26**(10), P721–P728 (1973)

Khan, M.M.R., et al.: Sustainable washing for denim garments by enzymatic treatment. J. Chem. Eng. **27**, 27–31 (2012)

New Graphical Predictions of Some UV Radiation and Water Shielded Attributes of Polymeric Supports with Direct Implication in Comfort Performance

Narcisa Vrinceanu[1](✉), Noureddine Ouerfelli[2], and Diana Coman[1]

[1] Faculty of Engineering, Department of Industrial Machines and Equipment, "Lucian Blaga" University of Sibiu, 10 Victoriei Boulevard, 550024 Sibiu, Romania
vrinceanu.narcisai@ulbsibiu.ro

[2] Laboratoire de Biophysique et Technologies Médicales Institut Supérieur des Technologies Médicales de Tunis, University of Tunis El Manar, 9, Av. Dr. Zouhaier Essafi, 1006 Tunis, Tunisia

Abstract. It is well-known that the textile industry focuses onto the design of responsive textile supports adapted to environmental conditions. The literature reported the surface tailoring of polymeric supports with nano-oxides, indicating a better absorbing of UV radiation, time stability against UV and water permeability, without modification of the mechanical properties, enhancing the performance and comfort. The BET model (Brunauer, P. H. Emmet, and E. Teller) - a theoretical model for gas adsorption allowing a multilayer adsorption was used as starting point. The main result and characterizing aspect of the study consist of making some mathematical predictions in terms of water and UV adsorption by a multilayer polymeric support. Up to our knowledge, this scientific argument has not been still approached. The theoretical/mathematical model for the porosity and water and UV permeability of polymeric supports, depends on the structural parameters such as pore radius and volume, volume-specific surface area. The validity of the model was confirmed by experimental results using 100% polyester supports coated with a formulation of 3% of nano-oxide suspension.

Thus, a framework to predict the water permeability of porous media given/created by the polymeric surface engineered/coated with nano-oxides, was developed.

Furthermore, the extensive pore structure parameters as variable are determined by this input variables are extracted by pore network model using a commercial software, and the permeability as the output. Based on these results, the pore scale simulations can be performed to determine the moisture flow attributes, such as the relative and absolute permeability, with direct implication in optimizing the textile comfort. Moreover, the achievements show that formulation of 3% of nano-oxide suspension as coating onto the polymeric support enhanced the water permeability properties compared with the other compositions, a fact which would promise improved attributes in terms of comfort.

Keywords: Comfort · Polyester · Nano-oxides · Mathematical model · Water angle

S. Msahli and F. Debbabi (Eds.): CIRATM 2020, SPM 17, pp. 204–209, 2022.
https://doi.org/10.1007/978-3-031-08842-1_33

1 Introduction

Permeability properties are of utmost importance for textiles used in clothing where they contribute to the comfort of the wearer. The comfort of summer clothing depends especially on its ability to dissipate excess water vapors on the protection against dangerous influence of UV-light. The water vapor permeability is a polymeric support attribute allowing the management of the moisture to vapor state, enhancing the comfort performance of textile material. Thus, water permeability is a vital quality in end-use applications as sport garments, underwear products, t-shirts, socks, and others (Bagherzadeh et al. 2012; Box and Behnken 1960). The water permeability of textile supports is strongly influenced by the porosity of the polymeric supports. Water vapor permeability is defined as the volume of water in liters, which is passed in one minute through 100 cm^2 (10 cm × 10 cm) of the fabric at a pressure difference of 10-mm head of water (TS 391 EN ISO 9237 1999).

In order to change the porosity of the surface, nanoparticles have been used to obtain a new engineered support, in other terms, a new porosity. The novel functionalized surface will increase the protection and comfort of the customer. C.-H. Xue et al. reported the surface tailoring of polymeric supports with nano-oxides, indicating a better absorbing of UV radiation and dissipating the surface charge, time stability against UV and water permeability, without modification of the mechanical properties, offering enhanced performance and comfort (Xue et al. 2011; Broasca et al. 2013).

In 1938, S. Brunauer, P. H. Emmet, and E. Teller proposed a theoretical model for gas adsorption (BET model) that allows a multilayer adsorption (Gibson et al. 2020). Starting from this point, the main result and characterizing aspect of the study consists of making some mathematical predictions in terms of water and UV adsorption by a nano-oxides coated polymeric support. Up to our knowledge, our scientific argument has not been still approached.

2 Methodology

The research in porous media focuses onto the comprehension of two contrary environments: micro-space consisting of empty spaces, and the macro-scaled pores.

Actually, water and UV permeability of polymeric supports, as macroscopic transport attributes of the porous medium are of the key interest for textile comfort and rely on the microscopic structure of interconnected pore space.

The type of medium influences the shape and complexity of pores. In other terms, the pores can be elongated and interwoven showing a high porosity and anisotropy like in fibrous media (Koponen et al. 1998; Shou et al. 2011).

Within this background/context, the determination of water permeability of polymeric supports is highly complex and difficult. The study approaches a theoretical model for the porosity and predicted the water and UV permeability of polymeric supports.

A mathematical model was created to predict the total porosity and the water permeability of a polymeric structure depending on the structural parameters such as pore radius and volume, volume-specific surface area.

For this purpose, a theoretical model of porous surfaces on BET (Brunauer, Emmett, and Teller) method was used, and the validity of the model was confirmed by experimental results using 100% polyester supports coated with a formulation of 3% of nano-oxide suspension.

The experimental data were used as input for some mathematical models predicting the optimum level of modification of polymeric surface with different formula of nano-oxides, yielding stable properties, without overloading the polymeric surface, and enhancing the textile comfort.

Basically, the total surface area includes all external and internal surface areas accessible to moisture/water vapors (Gibson et al. 2020). Because the amount of water passing through both the pores (meso, micro, and nanopores) was calculated, theoretical values of water permeability fitted to the experimental values.

Thus, a framework to predict the water permeability of porous media given/created by the polymeric surface engineered/coated with nano-oxides, was developed.

Furthermore, the extensive pore structure parameters as variable are determined by this input variables are extracted by pore network model using a commercial software, and the permeability as the output.

3 Results and Discussion

3.1 The Determination of Size Distribution of Nano Powders

This measurement was performed in order to identify the nano-oxides particles distribution onto the surface of the polymeric support. For the reliability only three data have been used. The device for quantifying the size distribution is an analyzer using laser radiation and a measurement optical system, by detecting their distribution within the range of 150 nm–1500 nm (Fig. 1).

It is notable that the dimensions of the nanoparticles are between 300–700 nm, result which is in good agreement with the literature (Anžlovar et al. 2008).

On the other hand, a previous study highlighted that the impregnation with nano-oxides leads to higher hydrophobicity, due to the augmentation of θ (contact angle) of coated samples (Broasca et al. 2013). It is considered that this increase cannot be correlated with the porosity/rough surface of the fabric, which is comparable for uncoated and coated samples.

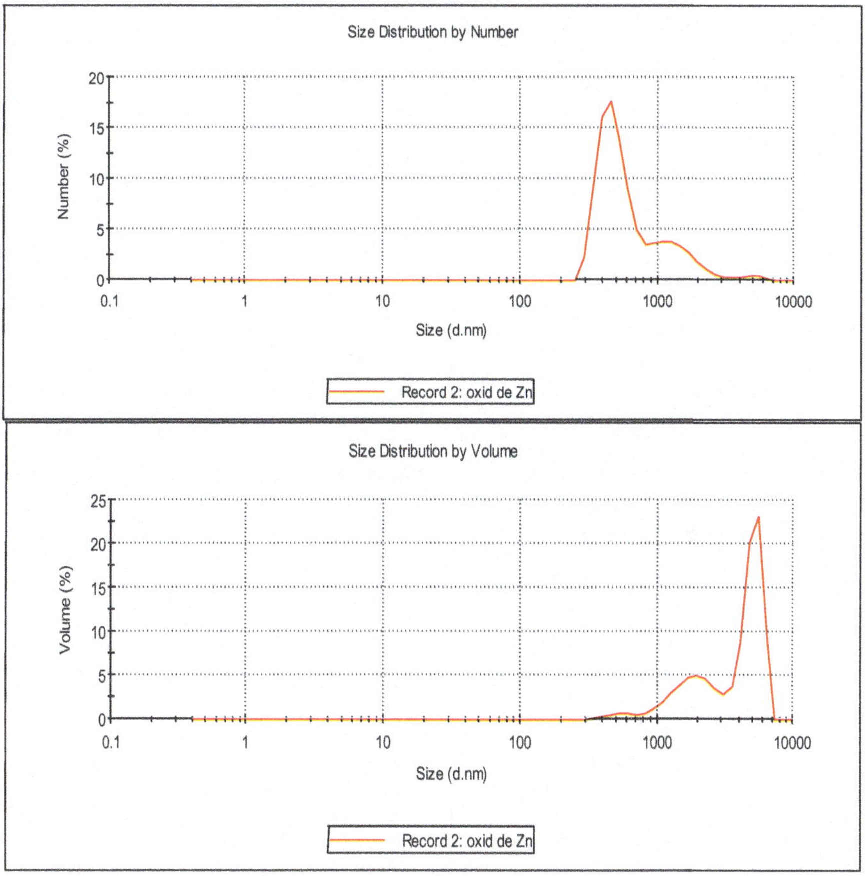

Fig. 1. The distribution of nano-oxides particles in terms of treatment conditions

That is why this study proposed a mathematical prediction aiming at a satisfactory homogeneity of the coating. In this respect, the figures below are plotting the dependence of contact angle versus the concentration of nano-oxide concentration of emulsion, sustaining the idea of levelness of coating (Fig. 3).

It is noticed that the contact angle varies in the same manner before and after UV irradiation, according to a non-linear dependence. In the same time, there is a maximum achieved at a concentration of 3% of nano-oxides emulsion. Again, a non-linear dependence is notable. By comparison to Fig. 2, in this case a minimum point is achieved at a concentration of 3% of nano-oxides emulsion (Figs. 4 and 5).

Most probably a linear or parabolic causal correlation between contact angle and concentration of nano-oxides emulsion of coating is noticed. It is noteworthy to mention that the minimum achieved is recorded without nano-oxides.

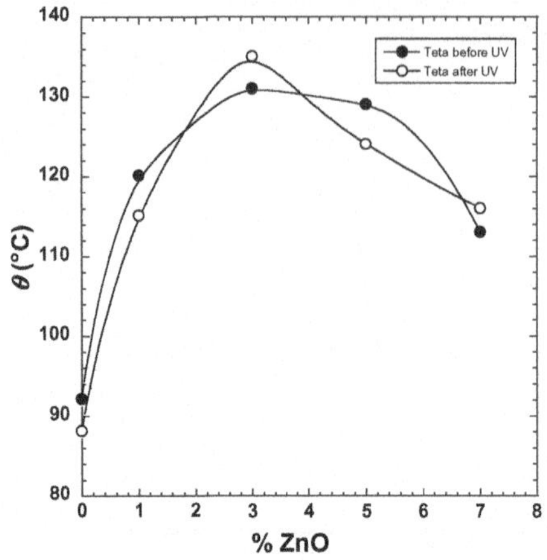

Fig. 2. The correlation between contact angle and concentration of nano-oxides emulsion of coating (expressed as θ)

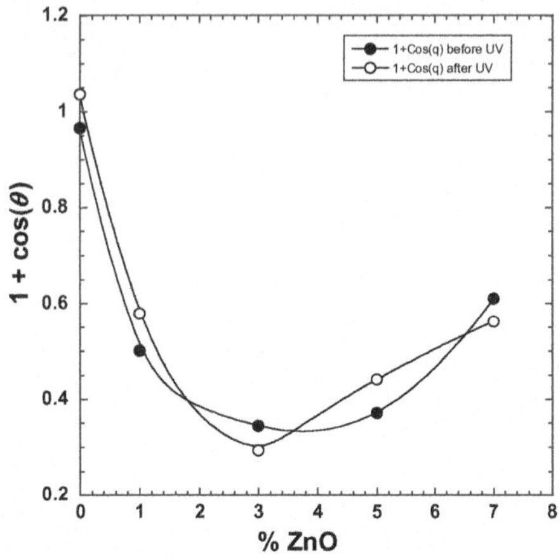

Fig. 3. The correlation between contact angle and concentration of nano-oxides emulsion of coating (expressed as cos (θ))

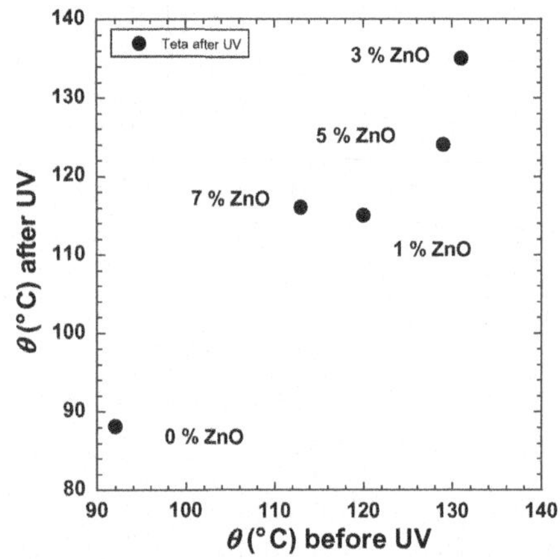

Fig. 4. The correlation between contact angle and concentration of nano-oxides emulsion of coating (expressed as θ angle)

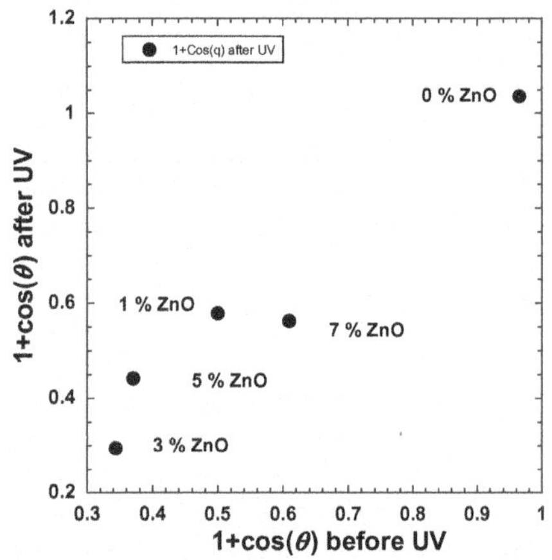

Fig. 5. The correlation between contact angle and concentration of nano-oxides emulsion of coating (expressed as 1 + cos (θ))

4 Conclusive Remarks

Thus, the modelling of the results show that the porosity obtained by coating of polymeric surfaces with nano-oxides suspensions is inversely proportional to water permeability, indicating that the optimized model could be robust and efficient in predicting water permeability. Starting from the fact that the increase of contact angle (hydrophobicity) cannot be correlated with the porosity/rough surface of the fabric, this study proposed a mathematical prediction aiming at a satisfactory homogeneity of the coating. The experimental data were used as input for some mathematical models predicting the optimum level of modification of polymeric surface with different formula of nano-oxides, yielding stable properties, without overloading the polymeric surface, and enhancing the textile comfort.

References

Anžlovar, A., et al.: Nanocomposites with nano-to-sub-micrometer size zinc oxide as an effective UV absorber. National Institute of Chemistry, Ljubljana, Slovenija (2008)

Bagherzadeh, R., Gorji, M., Latifi, M., et al.: Evolution of moisture management behavior of high-wicking 3D warp knitted spacer fabrics. Fibers Polym. **13**, 529–534 (2012). https://doi.org/10.1007/s12221-012-0529-6

Box, G., Behnken, D.W.: Some new three level design for the study of quantitative variables. Technometrics **2**, 455–475 (1960)

Broasca, G., et al.: Characterization of ZnO coated polyester fabrics for UV protection. Appl. Surf. Sci. **279**, 272–278 (2013)

Gibson, N., et al.: Volume-specific surface area by gas adsorption analysis with the BET method. Elsevier (2020). ISBN: 9780128141823

Koponen, A., et al.: Permeability of three-dimensional random fiber webs. Phys. Rev. Lett. **80**(4), 716–719 (1998)

Shou, D., et al.: Hydraulic permeability of fibrous porous media. Int. J. Heat Mass Transfer **54**, 4009–4018 (2011)

Xue, C.-H., et al.: UV-durable superhydrophobic textiles with UV-shielding properties by coating fibers with ZnO/SiO_2 core/shell particles. Nanotechnology **22**, 415–603 (2011)

TS 391 EN ISO 9237 (1999)

Durability and Comfort Assessment of Casual Male Socks

Antoneta Tomljenović[1](✉), Zenun Skenderi[1], Ivan Kraljević[2], and Juro Živičnjak[1]

[1] Faculty of Textile Technology, University of Zagreb, Prilaz baruna Filipovića 28a, 10000 Zagreb, Croatia
antoneta.tomljenovic@ttf.unizg.hr

[2] Jadran Hosiery, Vinka Žganeca 2, 10000 Zagreb, Croatia

Abstract. On male socks are placed high demands of usage durability, comfort and design. Therefore, it is very important to select yarns for their production. In this paper, the properties of casual male socks, made of high content of cotton in full plating by yarns of different composition (polyamide 6.6, Lycra/cotton, Lycra/polyamide 6.6) were compared, by applying the proposed methodology for evaluation quality of socks according to the standardized test methods. With particular reference to usage properties, it was carried out the assessment of dimensional stability of socks after one and five repeated washing and drying cycles; propensity to surface pilling and abrasion resistance using the Martindale abrasion tester comparing two methods; as well as testing of thermo-physiological comfort of socks before and after five repeated washing and drying cycles by measuring of thermal resistance on the thermal foot.

Keywords: Socks · Usage durability · Thermal resistance · Textile testing

1 Introduction

A sock is an item of clothing worn on the feet and often covering the ankle and some part of the calf. Male socks usually come in calf and over-calf lengths. For men, four major types of socks exist - dress, casual, sport and work. Casual socks are worn with sport wear and in less formal situations. They are made from bulky yarns, usually wool or acrylic, or from cotton for softness and comfort. They may be blended with polyamide or Lycra for improved fit, durability and shrink resistance. Casual socks, regardless of their construction, normally are calf length and typically dark-colored. Socks are knitted, giving them stretch and the ability to conform to the foot and leg. Generally, plain knit stitch is used in the foot area and a rib stitch is used in the leg area, usually at the top of the socks. Both the toe and heel areas should be smooth, otherwise irritation could occur during wear. Toe and heel reinforcements are also important features. Polyamide yarn is often knitted into these areas to prolong wear.

The number of European standards related to testing of knitted fabrics and socks are low (Tomljenović et al. 2016). With increasing demand for garment comfort, there are many studies related to the thermal comfort properties of fabrics. However, there

S. Msahli and F. Debbabi (Eds.): CIRATM 2020, SPM 17, pp. 210–215, 2022.
https://doi.org/10.1007/978-3-031-08842-1_34

are significantly less studies examining the thermal comfort of socks, which are next-to-skin-type garments, using e.g. thermal foot model (Skenderi et al. 2017), sock-shaped hot plate (Cimilli et al. 2010) or direct measurement at the feet of human subjects (West et al. 2021). Therefore, it is necessary to expand the research in that field. Socks are one of the textile product having the least life among clothing since they can be produced with lower cost than other textile products and so much consumed and used according to fashion and needs during human life (Hashan et al. 2017). As, on male socks are placed high demands of usage durability, comfort and design, it is very important to harmonize and select yarns for their production, in addition to other production requirements. Therefore, the properties of three groups of casual male socks, made of high content of cotton in full plating by yarns of different composition (textured polyamide 6.6, Lycra wrapped with cotton, and Lycra wrapped with textured polyamide 6.6) were compared in this paper, by applying the proposed methodology for evaluation quality of socks according to the standardized test methods. The influence of socks fiber content, changed with the applied plating yarns, on their usage durability and thermal comfort were analyzed.

2 Materials and Methods

The investigation was carried out on three groups of calf length fine casual, black coloured male socks of the same size (EU 42), made of high content of cotton in full plating by synthetics yarns of different composition. Textured polyamide 6.6 yarn (designation: PA 6.6 44dtex f13 × 2, linear density: ca. 10tex) was knitted into toe and heel areas to prolong wear of all socks samples. For reinforcement and support of socks in the foot and leg area following planting yarns were used - textured polyamide 6.6 yarn (designation: PA 6.6 44dtex f13 × 2, linear density: ca. 10tex), Lycra wrapped with cotton (designation: Lycra/cotton, linear density: 10tex) and Lycra wrapped with textured polyamide 6.6 (designation: Lycra/PA 6.6, 22dtex/PA 6.6 78dtex f23/1). The soft good quality single ring spun cotton yarn (linear density: 29.4 tex) add softness, strength and comfort of all socks. Lycra wrapped with textured polyamide 6.6 yarn of higher linear density (20 tex) and designation (130 dtex/PA 6.6 78 dtex f23/1) was knitted into the ribbing at the top of the all socks to prevent their falling down. Plain knit stitch was used in the foot and leg area and rib stitch was used at the top of the socks.

The socks were made in Jadran Hosiery, Croatia, using Lonati sock knitting machine E14 of cylinder diameter 95 mm (3 ¾″) with 168 needles. Toes seams were placed high over the toe, toe and heel areas were smooth, and square heel shape in sock that is equally important to comfort were made. The socks were ironed at a temperature of 120 °C using a Cortese machine. Characteristics of male socks produced are shown in Table 1, including values of fiber content, weight of one sock, values of moisture regain and mass per unit area of knits.

After the conditioning (at temperature: 20 ± 2 °C and air relative humidity: 65 ± 4%) and sampling of socks (Fig. 1a) of the almost same weight, the following were determined:

Table 1. Characteristics of three groups of calf length casual male socks

Sock sample	Fiber content (%)			Platting yarn		Weight of sock (g)	Moisture regain (%)	Mass per unit area (g/m²)
	Cotton	PA 6.6	Lycra	Foot and leg	Toes and heel			
1	78	21	1	PA 6.6	PA 6.6	19.9	6.30	189.9
2	91	6	3	Lycra/cotton	PA 6.6	19.8	6.96	199.9
3	78	19	3	Lycra/PA 6.6	PA 6.6	19.9	6.63	237.6

- plain knit mass per unit area according to the ISO 3801 expressed in g/m^2.
- plain knit moisture regain according to the ASTM D 2654-89a. Knits samples are conditioned in the standard atmosphere of air temperature of 20 ± 2 °C and relative humidity of 65 ± 4% for 24 h, after which they are weighted, dried in an oven at 105 °C and reweighed. The difference between the mass of conditioned and the mass of oven-dried samples is calculated as moisture regain and expressed in percentage.
- dimensional stability in the length and width direction, measured in the foot and leg area of socks after one and five repeated washing and drying cycles according to the procedure 4N (normal agitation during heating, washing and rinsing at 40 °C) of EN ISO 6330 with non-phosphate ECE reference detergent (without optical brightener) and open-air drying (procedure A, line dry). The percentage change in the length and width of the socks was calculated, and the state of whether the dimension has decreased (shrinkage) was expressed my means of minus.
- plain knit propensity to surface fuzzing and pilling according to the Martindale method (EN ISO 12945-2). The socks were rubbed with standardized wool abradant and visually assessed by comparing with photographs (degrees of pilling: 1 - 5), after 125, 500, 1000, 2000, 5000 and 7000 rubbing cycles.
- plain knit abrasion resistance by determination of specimen breakdown using the Martindale abrasion tester according to the EN 13770, method 1, using two different specimen holders - standard according to the EN ISO 12947-2 and modified according to the EN 13770 (Fig. 1b). When using this method, the specimens sampled from heel and sole area of socks, moves according to the Lissajous curve, and standard woven wool fabric is abraded over the entire surface.
- thermal resistance of socks on the *thermal foot* (Fig. 1c) before and after five repeated washing and drying cycles. The thermal foot is divided into 13 segments that are individually heated. In each segment, heaters and temperature sensors are installed. Temperature of the measuring unit is 35 °C ± 0,5 °C. Thermal resistance for each segment and/or total thermal resistance of sock could be determined by special algorithm. The apparatus constant (*Rcto*) must first be determined for defined environment conditions (air temperature: 20 ± 2°C, relative humidity: 65 ± 4% and air speed: 1 m/s). The specimen to be tested is then placed on an electrically heated foot and total thermal resistance *Rctt* of apparatus and sock was measured. From the difference

between *Rctt* and *Rct0*, thermal resistance of the tested sock *Rct* should be calculated from Formula (1):

$$Rct = Rctt - Rct0 \tag{1}$$

where: *Rct* is the thermal resistance of sock ($m^2 \cdot °C/W$); *Rct0* is the apparatus constant, ($m^2 \cdot °C/W$) and *Rctt* is total thermal resistance of apparatus and sock ($m^2 \cdot °C/W$).

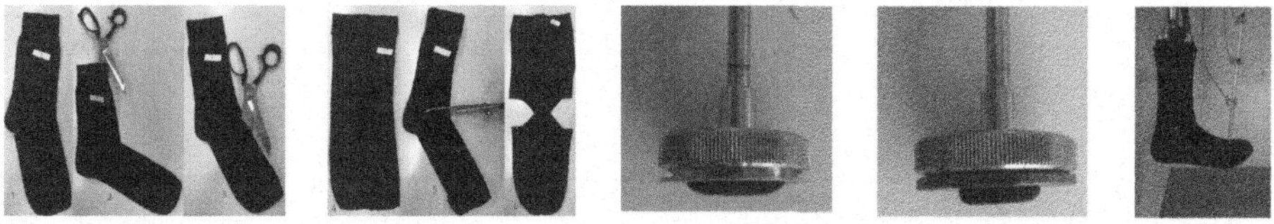

(a) (b1) (b2) (c)

Fig. 1. Socks testing: a) sampling, b) samples prepared for abrasion resistance testing: 1- according the EN ISO 12947-2; 2- according to the EN 13770, c) thermal foot

3 Results and Discussion

The results obtained by the investigation are presented in Tables 1, 2, 3, 4 and 5. Lycra plating threads change the structure of plated knit socks as the construction of such socks were tighter and heavier (Table 1). All tested socks samples showed shrinkage after washing and drying (Table 2). Increasing the number of washing and drying cycle increases the shrinkage of all sock samples. Sock sample 2 that contain the highest amount of cotton shrink at most, and shows the highest value of water vapor absorption (Table 1).

Table 2. Changes in dimensions of socks in the length and width directions

Sock sample	Dimensional stability (%), 1x washed		Dimensional stability (%), 5x washed	
	Length	Width	Length	Width
1	−5	−7	−6	−8
2	−9	−10	−12	−13
3	−2	−9	−6	−11

The results presented in Table 3 indicate a high propensity to surface pilling of all socks samples tested. In the socks of the first and third groups that contain higher amount of synthetic plating threads it is worth mentioning a somewhat lower propensity to surface pilling, but also higher abrasion resistance of knit from the sole area of the socks (Table 4) indicating their better aesthetic characteristics and usage durability. Sock

Table 3. Visually assessed propensity to surface pilling of socks

Number of pilling rubs	Sock sample 1	Sock sample 2	Sock sample 3
125	4	4	4
500	3/4	3/4	3/4
1000	3/4	3/4	3/4
2000	3	3	3
5000	3	2	2/3
7000	2/3	1/2	2

sample 3, plated with Lycra wrapped with polyamide 6.6, shows the highest abrasion resistance in the sole area. Only by sock sample 2, plated with Lycra wrapped with cotton, endpoint was determined with the appearance of a hole (Table 4). By applying of modified holders adapted to the requirements of the EN 13770, method 1, a lower number of rubs to reach endpoint is required.

Table 4. Abrasion resistance of heel and sole area of socks - determination of specimen breakdown

Sock sample	EN ISO 12947-2		EN 13770, method 1	
	Description of the endpoint/ number of rubs to reach endpoint			
	Heel	Sole	Heel	Sole
1	Thinning/35000	Thinning/35000	Thinning/6500	Thinning/6500
2	Thinning/35000	Hole/30000	Thinning/6500	Hole/6000
3	Thinning/35000	Thinning/45000	Thinning/6500	Thinning/7000

Description: hole – develops when one thread is broken causing a hole to appear; thinning – the cotton spun staple yarn wears away leaving a base of the synthetic filament yarn.

Arithmetic mean values of the thermal resistance measured on three socks within the sock groups are presented in Table 5. The technique described enables the thermal resistance *Rct* of a sock to be determined by subtracting the thermal resistance of the boundary air layer above the surface of the test apparatus from that of a test specimen plus boundary air layer, both measured under the same conditions. Non-washed socks samples 1 and 3, plated with textured multifilament polyamide 6.6 and Lycra wrapped with textured multifilament polyamide 6.6 in the socks foot and leg area, are more elastic and better fit with the foot and leg. Therefore, the thermal resistance of mentioned tested socks is higher. After five repeated washing and drying cycles, total thermal resistance increases in almost all tested sock samples. Due to determined shrinkage of socks (Table 2) and the increase in open porosity of their structure on the thermal foot, decreases the thermal resistance of the tested socks (Table 5). It was also determined that, a higher thermal resistance was shown for socks plated with Lycra thread (samples 2 and 3) after five repeated washing and drying cycles.

Table 5. Thermal resistance of socks before and after five repeated washing and drying cycles

Sock sample	*Rctt* ($m^2 \cdot °C/W$)		*Rct* ($m^2 \cdot °C/W$)		(*Rct*/ *Rctt*) x 100 (%)	
	Non-washed	5x washed	Non-washed	5x washed	Non-washed	5x washed
1	0.127005	0.134184	0.015870	0.007562	12.676742	5.650739
2	0.131486	0.132872	0.010784	0.009747	8.220660	7.378982
3	0.135114	0.133988	0.012585	0.009688	9.314884	7.239121

Description: *Rctt* - total thermal resistance of apparatus and sock; *Rct* - thermal resistance of the tested sock.

4 Conclusion

On the basis of the results obtained, the applicability of the proposed methodology was confirmed and concluded that when choosing the yarns for production of socks is necessary to consider their comfort and expected durability in the conditions of use.

Acknowledgment and Funding. This paper is funding by the Croatian science foundation within the Project IP-2016-06-5278 (Comfort and antimicrobial properties of textiles and footwear, principal investigator: Prof. Zenun Skenderi, PhD).

References

ASTM D2654-89a:1989 Test Methods for Moisture in Textiles

Cimilli, S., et al.: A comparative study of some comfort-related properties of socks of different fiber types. Textile Res. J. **80**(10), 948-957 (2010). ISSN 1746-7748

EN 13770:2002: Textiles - Determination of the abrasion resistance of knitted footwear garments

EN ISO 12947-2:2016: Textiles - Determination of the abrasion resistance of fabrics by the Martindale method - Part 2: Determination of specimen breakdown

EN ISO 12945-2:2020: Textiles - Determination of fabric propensity to surface fuzzing and to pilling - Part 2: Modified Martindale method

EN ISO 6330:2012: Textiles - Domestic washing and drying procedures for textile testing

ISO 3801:1977: Textiles - Woven fabrics - Determination of mass per unit length and mass per unit area

Hashan, Md.M., et al.: Functional properties improvement of socks items using different types of Yarn. Int. J. Text. Sci. **6**(2), 34–42 (2017). ISSN 2325-0100

Skenderi, Z., et al.: Thermophysiological wear comfort of footwear. J. Leather Footwear **66**(3), 12–21 (2017). ISSN 0450-8726

Tomljenović, A., et al.: Durability assessment of functionally printed knitted fabrics for T-Shirts. Fibers Text. Eastern Europe **24**(4), 129–138 (2016). ISSN 1230-3666

West, A.M., et al.: Are running socks beneficial for comfort? The role of the sock and sock fiber type on shoe microclimate and subjective evaluations. Text. Res. J. **91**(15–16), 1698–1712 (2021). ISSN 1746-7748

Relation Between Calcium Carbonate Properties and Mechanical Properties of the Expanded Layer of PVC Synthetic Leather

Mouna Stambouli[1](✉), Walid Chaouech[1], Sondes Gargoubi[1,2], Riadh Zouari[1], Aweb Baccar[3], and Slah Msahli[1]

[1] Textile Engineering Laboratory of ISET Ksar Hellal, University of Monastir, Monastir, Tunisia
Mounastambouli@gmail.com
[2] Chimitex Plus, Sousse, Tunisia
[3] PLASTISS: Industry of Synthetic Leather Situated in Sayada, Monastir, Tunisia

Abstract. The main objective of this study was to study the relation between carbonate calcium and mechanical properties of PVC synthetic leather. Filler is an additive that is used for plastic material essentially to reduce cost.

Although, filler are primarily used to make PVC synthetic leather cheaper, they also influence a variety of parameters during production and technical properties of the final product.

Polyvinylchloride calcium carbonate composites (PVC/$CaCO_3$) with different particle size and types were prepared.

The mechanical tests showed that the addition of the $CaCO_3$ fillers could have an effect on the mechanical properties, the tear resistance, the break strength and the elongation at break of the PVC synthetic leather increased with decreasing $CaCO_3$ particle size, which was attributed to increased interfacial contact area and enhanced interfacial adhesion between $CaCO_3$ particles and PVC matrices.

Keywords: Calcium carbonate · PVC · Synthetic leather · Mechanical tests · Expanded layer

1 Introduction

Polyvinyl chloride (PVC) is now one of the most thermoplastic used and produced all over the world; it is used in many industrial applications. PVC synthetic leather is one of the industrial applications which PVC is an essential compound. Synthetic leather is generally known as coated textile surface with different layers, each layer gives its own physic-chemical properties to the final product.

To improve the properties of PVC synthetic leather, adding additives such as plasticizer, heat stabilizers, filler and lubricant, is important (Fernando and Thomas 2007).

It is well known that mechanical properties of polymeric material are affected by adding inorganic fillers, especially calcium carbonate (Liu et al. 2006).

S. Msahli and F. Debbabi (Eds.): CIRATM 2020, SPM 17, pp. 216–220, 2022.
https://doi.org/10.1007/978-3-031-08842-1_35

The effect of inorganic fillers on the microstructure and mechanical properties of PVC composites depends strongly on its amount and its particle size (Basilia et al. 2007).

Tear resistance, break strength and the elongation at break are the most important mechanical properties of PVC synthetic leather.

The effect of adding CaCO3 and varying their particle size was examined, the result was interpreted.

2 Materials and Methods

In this part, we explain the industrial process used to obtain the final coated textile within the company PLASTISS and in which step of this process we managed to vary the amount and particle size of the filler used.

2.1 Coating Process

The key point of the coating is the quality of the mixture composition which will be applied to the textile to obtain the synthetic leather: tow mixtures are prepared by stirring to form the superficial and internal layers of the final product. During the process, one mixture is applied to a paper sheet covered with silicon layer (thereafter called transfer paper) using a blade by controlling the desired thickness and go through an oven during 20 s at 200 °C. By this way, we obtain a plastic sheet with a given thickness sticking on the transfer paper. In a second step, the coating paste associated to the internal layer which is an expanded layer is applied to the product using a blade at a given thickness. After that, the knitted fabric is put in contact with this second layer and slightly pressed before going through an oven heated at 200 °C during 80 s. This step allows the reticulation of the second expanded layer and the transfer of the plastic system from the paper to the textile substrate. Finally, we obtain a micro porous structure made up of a first layer called superficial layer and a second cellular layer tethered to a textile substrate (Fig. 1).

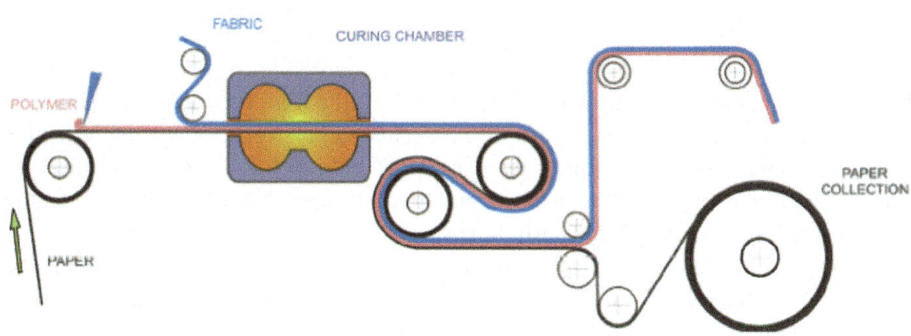

Fig. 1. Coating process

2.2 Materials

Dry PVC resin, diisononyl phthalate (DINP), mixed-metal heat stabilizer, puffing agent, calcium carbonate filler, pigment, transfer paper and polyester knitted fabric base were kindly provided by Plastiss Company (Monastir, Sayada, Tunisia). The ground micron size $CaCO_3$ particles were provided by SOFAP (Sfax-Tunisia).

2.3 Mechanical Properties

Break strength and elongation at break was carried out by a dynamometer according to ISO 1421 (December 2017), Tear resistance was measured according to ISO 4674-1 (April 2017).

3 Results and Discussions

3.1 Effect of $CACO_3$ Particle Size on the Mechanical Behaviour of the Expanded Layer of PVC Synthetic Leather

See Table 1.

Table 1. Different types of $CaCO_3$ used with different particle size

Filler type	Particle size(μ)
Ultra fine coated calcium carbonate made from extremely white marble:	
5MC	**2.5**
2AMC	**1.4**
Extremely fine natural calcium carbonate made from marble of exceptional whiteness:	
2M	**1.8**
95M	**0.9**

Figures 2 and 3 showed that for each type of $CaCO_3$, the finer the filler, the greater the improvement of mechanical properties of the expanded layer of PVC synthetic leather, it is also noted that the PVC synthetic leather expanded layer filled with fine $CaCO_3$ had higher elongation at break, tear resistance and break strength than PVC synthetic leather expanded layer filled with a large particle size.

It is known that mechanical properties of composites are influenced by the filler fraction and the interfacial adhesion between particles and matrix..Decreasing particle size results in dramatic increases in the specific surface area of inorganic particles, which leads to an increase in the interfacial contact area between the filler and PVC matrix. The increase in interfacial contact area would favor transfer of stress from the matrix to inorganic particles, thereby resulting in higher mechanical properties of the composite.

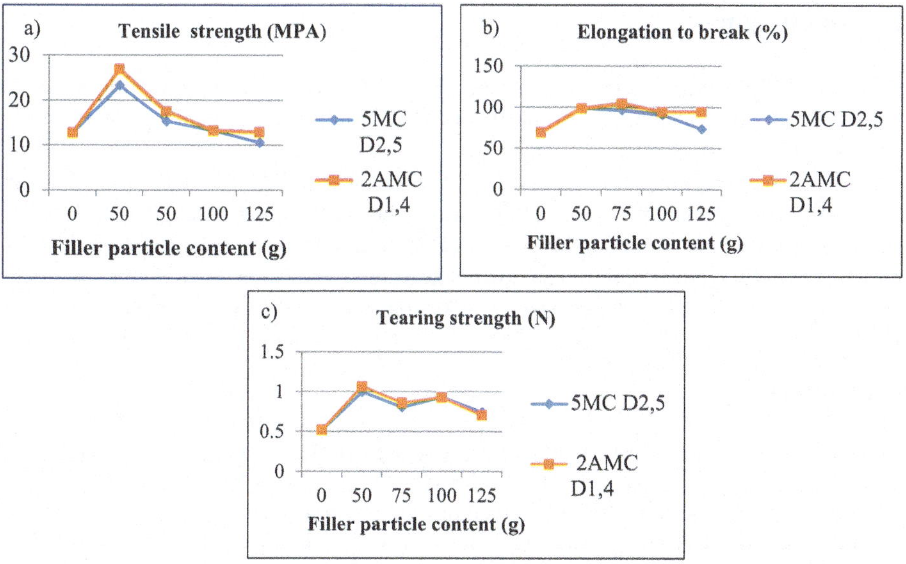

Fig. 2. Impact of different particle size calcium carbonate on the mechanical properties of PVC synthetic leather ($CaCO_3$ TYPE 1)

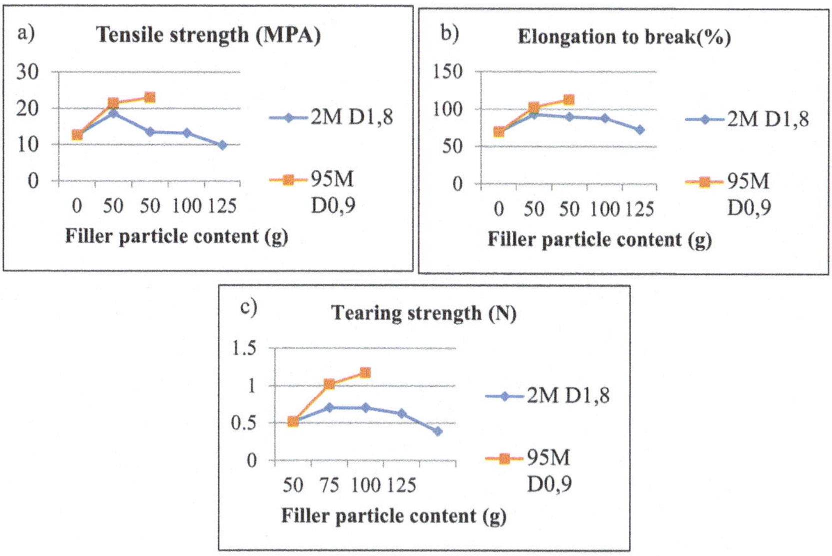

Fig. 3. Impact of different particle size calcium carbonate on the mechanical properties of PV synthetic leather (CaCO3 TYPE 2)

4 Conclusion

The aim of this study is to demonstrate that The effect of inorganic fillers on the microstructure and mechanical properties of PVC composites depends strongly on its amount and its particle size.

The mechanical tests showed that the addition of the $CaCO_3$ fillers could have an effect on the mechanical properties, the tear resistance, the break strength and the elongation at break of the expanded layer of PVC synthetic leather increased with decreasing $CaCO_3$ particle size, which was attributed to increased interfacial contact area and enhanced interfacial adhesion between $CaCO_3$ particles and PVC matrices.

References

Fernando, N.A.S., Thomas, N.: Effect of precipitated calcium carbonate on the mechanical properties of poly (vinyl chloride). J. Viny l Addit. Technol. **13**, 98–102 (2007)

Liu, P., Zhao, M., Guo, J.: Thermal stabilities of poly(vinyl chloride)/calcium carbonate (PVC/CACO3) composites. J. Macromol. Sci. Part B Phys. **45**, 1135–1140 (2006)

Basilia, B.A., Panganiban, M.E.G., Collado, A.A.V., Pesigan, M.O.D., De Yro, P.A.: Study on the functionality of nano-precipitated calcium carbonate as filler in thermoplastic. J. Solid Mech. Mater. Eng. **4**, 564–570 (2007)

Effect of the Calcium Carbonate Properties on the Mechanical Behaviour of the Superficial Layer of PVC Synthetic Leather

Mouna Stambouli[1(✉)], Walid Chaouech[1], Sondes Gargoubi[2], Riadh Zouari[1], Aweb Baccar[3], and Slah Msahli[1]

[1] Textile Engineering Laboratory of ISET Ksar Hellal, University of Monastir, Monastir, Tunisia
Mounastambouli@gmail.com

[2] Textile Engineering Laboratory of ISET Ksar Hellal, University of Monastir, Tunisia, Chimitex Plus, Sousse, Tunisia

[3] PLASTISS: Industry of Synthetic Leather Situated in Sayada, Monastir, Tunisia

Abstract. The main objective of this study was to study the relation between carbonate calcium and mechanical properties of PVC synthetic leather. Filler is an additive that is used for plastic material essentially to reduce costs.

Although, filler is primarily used to make PVC synthetic leather cheaper, they also influence a variety of parameters during production and technical properties of the final product.

Polyvinylchloride calcium carbonate composites (PVC/CaCO3) with different particle size and types were prepared.

The mechanical tests showed that the addition of the CaCO3 fillers could have an effect on the mechanical properties, the tear resistance, the break strength and the elongation at break of the PVC synthetic leather increased with decreasing CaCO3 particle size, which was attributed to increased interfacial contact area and enhanced interfacial adhesion between CaCO3 particles and PVC matrices.

Keywords: Calcium carbonate · PVC · Synthetic leather · Mechanical tests

1 Introduction

Polyvinyl chloride (PVC) is now one of the most thermoplastic used and produced all over the world; it is used in many industrial applications. PVC synthetic leather is one of the industrial applications which PVC is an essential compound. Synthetic leather is generally known as coated textile surface with different layers, each layer gives its own physic-chemical properties to the final product.

To improve the properties of PVC synthetic leather, adding additives such as plasticizer, heat stabilizers, filler and lubricants, is important (Fernando et al. 2007).

It is well known that mechanical properties of polymeric material are affected by adding inorganic fillers, especially calcium carbonate (Liu et al. 2006).

S. Msahli and F. Debbabi (Eds.): CIRATM 2020, SPM 17, pp. 221–226, 2022.
https://doi.org/10.1007/978-3-031-08842-1_37

The effect of inorganic fillers on the microstructure and mechanical properties of PVC composites depends strongly on its amount and its particle size (Basilia et al. 2007).

Tear resistance, break strength and the elongation at break are the most important mechanical properties of PVC synthetic leather.

The effect of adding CaCO3 and varying their particle size was examined, the result was interpreted.

2 Materials and Methods

In this part, we explain the industrial process used to obtain the final coated textile within the company PLASTISS and in which step of this process we managed to vary the amount and particle size of the filler used.

2.1 Coating Process

The key point of the coating is the quality of the mixture composition which will be applied to the textile to obtain the synthetic leather: tow mixtures are prepared by stirring to form the superficial and internal layers of the final product. During the process, one mixture is applied to a paper sheet covered with silicon layer (thereafter called transfer paper) using a blade by controlling the desired thickness and go through an oven during 20 s at 200 °C. By this way, we obtain a plastic sheet with a given thickness sticking on the transfer paper. In a second step, the coating paste associated to the internal layer which is an expanded layer is applied to the product using a blade at a given thickness. After that, the knitted fabric is put in contact with this second layer and slightly pressed before going through an oven heated at 200 °C during 80 s. This step allows the reticulation of the second expanded layer and the transfer of the plastic system from the paper to the textile substrate. Finally, we obtain a micro porous structure made up of a first layer called superficial layer and a second cellular layer tethered to a textile substrate (Fig. 1).

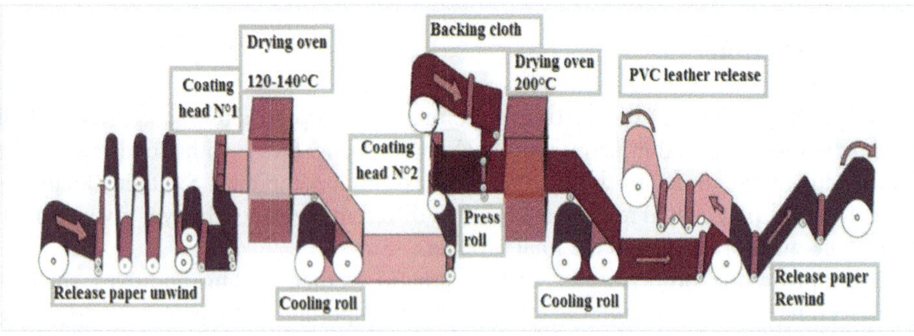

Fig. 1. Coating process

2.2 Materials

Dry PVC resin, diisononyl phthalate (DINP), mixed-metal heat stabilizer, puffing agent, calcium carbonate filler, pigment, transfer paper and polyester knitted fabric base were kindly provided by Plastiss Company (Monastir, Sayada, Tunisia). The ground micron size $CaCO_3$ particles were provided by SOFAP (Sfax-Tunisia).

2.3 Mechanical Properties

Break strength and elongation at break was carried out by a dynamometer according to ISO 1421 (December 2017), Tear resistance was measured according to ISO 4674-1 (April 2017).

3 Results and Discussions

3.1 Varying $CaCO_3$ Amount

Table 1. PVC synthetic leather superficial layer formulation.

Ingredients	Parts per hundred resin(phr)							
PVC	100	100	100	100	100	100	100	100
plasticizer	**70**	72	74	75	80	82	94	102
Stabilizer	**1.5**	1.5	1.5	1.5	1.5	1.5	1.5	1.5
CACO3 (sup 2)	**0**	5	10	15	25	50	75	100

Table 1 illustrated the formulation of the superficial layer of PVC synthetic leather, Fig. 2 showed that the addition of $CaCO_3$ gave rise to an increase in the elongation at break, the break strength and the tear resistance, the optimum of this increase was identified when we added 25 phr of $CaCO_3$ on the composition of superficial layer.

Elongation at break, strength breaks and tear resistance decreases, though, once the optimum at 25 phr filler is surpassed.

These results showed the relationship between filler level and mechanical properties of PVC synthetic leather (Table 2).

3.2 Varying CACO3 Particle Size

Figures 2 and 3 showed that for each type of $CaCO_3$, the finer the filler, the greater the improvement of mechanical properties, it is also noted that the PVC synthetic leather filled with fine $CaCO_3$ had higher elongation at break, tear resistance and break strength than PVC synthetic leather filled with a large particle size (Fig. 4).

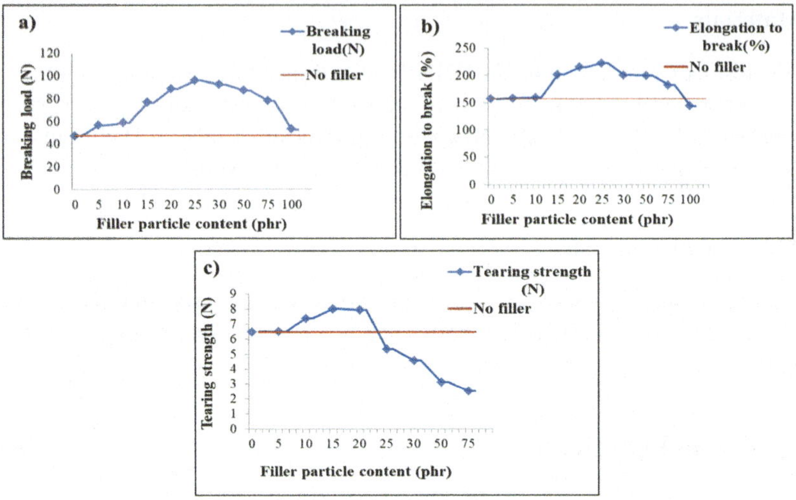

Fig. 2. Effect of varying $CaCO_3$ amount on the mechanical properties of PVC synthetic leather

Table 2. Different type of $CaCO_3$ used with different particle size

Filler type	Particle size(μ)
Ultra fine coated calcium carbonate made from extremely white marble:	
5MC	**2.5**
2AMC	**1.4**
Extremely fine natural calcium carbonate made from marble of exceptional whiteness:	
2M	**1.8**
95M	**0.9**

It is known that mechanical properties of composites are influenced by the filler fraction and the interfacial adhesion between particles and matrix. Decreasing particle size results in dramatic increases in the specific surface area of inorganic particles, which leads to an increase in the interfacial contact area between the filler and PVC matrix. The increase in interfacial contact area would favor transfer of stress from the matrix to inorganic particles, thereby resulting in higher mechanical properties of the composite.

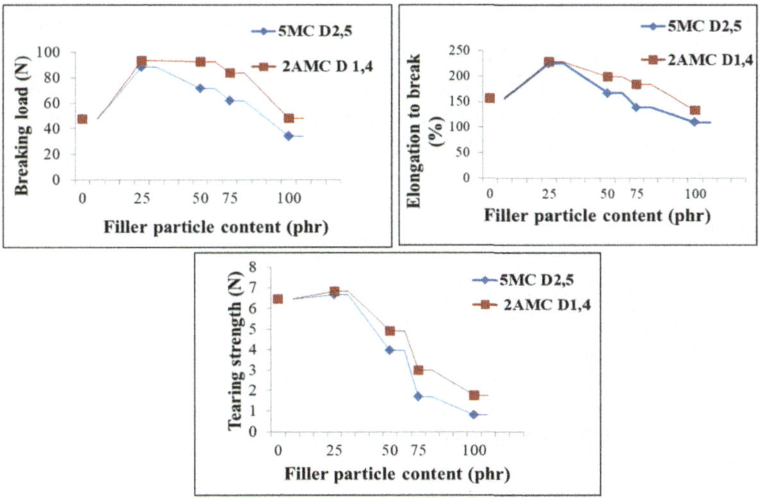

Fig. 3. Impact of different particle size calcium carbonate on the mechanical properties of PVC synthetic leather ($CaCO_3$ TYPE 1)

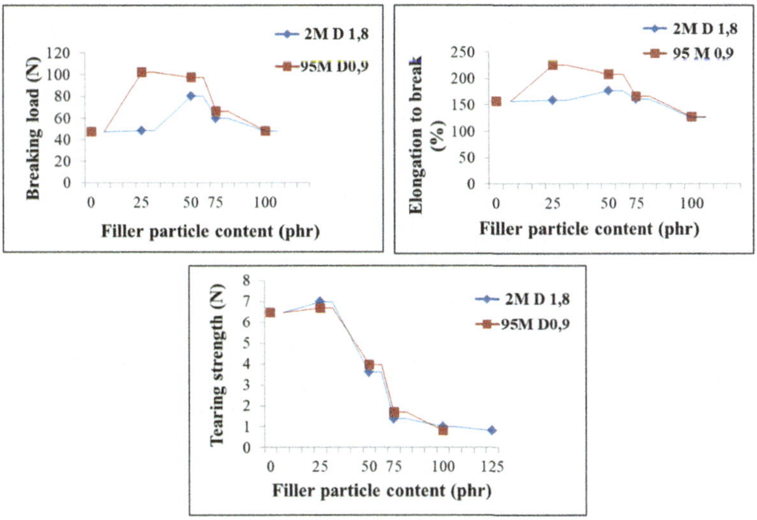

Fig. 4. Impact of different particle size calcium carbonate on the mechanical properties of PVC synthetic leather ($CaCO_3$ TYPE 2)

4 Conclusion

The aim of this study is to demonstrate that The effect of inorganic fillers on the microstructure and mechanical properties of PVC composites depends strongly on its amount and its particle size.

The mechanical tests showed that the addition of the CaCO3 fillers could have an effect on the mechanical properties, the tear resistance, the break strength and the elongation at break of the PVC synthetic leather increased with decreasing CaCO3 particle size, which was attributed to increased interfacial contact area and enhanced interfacial adhesion between CaCO3 particles and PVC matrices.

References

Fernando, N.A.S., et al.: Effect of precipitated calcium carbonate on the mechanical properties of poly (vinyl chloride). J. Viny l Addit. Technol. **13**, 98–102 (2007)

Liu, P., et al.: Thermal stabilities of poly(vinyl chloride)/calcium carbonate (PVC/CACO3) composites. J. Macromol. Sci. Part B Phys. **45**, 1135–1140 (2006)

Basilia, B.A., et al.: Study on the functionality of nano-precipitated calcium carbonate as filler in thermoplastic. J. Solid Mech. Mater. Eng. **4**, 564–570 (2007)

Usage Quality of Semi Processed Full Grain Leather

Antoneta Tomljenović(✉), Juro Živičnjak, Jadranka Akalović, and Franka Žuvela Bošnjak

Faculty of Textile Technology, University of Zagreb, Prilaz baruna Filipovića 28a, 10000 Zagreb, Croatia
antoneta.tomljenovic@ttf.unizg.hr

Abstract. Full grain refers to the strongest and most durable part of the cattle hide, consisting of the grain layer and the part of the underlying corium, and has not been sanded or buffed to remove any imperfections, that means it displays the more natural characteristics of leather. Leather quality changes with sampling location and were also affected by tanning and finishing processes utilized during production. Full grain leather, instead finishing, sometimes go only through an ironing process for the desired sheen is obtained. Therefore, with the purpose of analysing the influence of the leather processing level, and sampling location on their usage quality, in this paper, differently tanned semi processed cattle full grain leathers, sampled from the bend, shoulder and belly of side leather cuts were evaluated. Their applicability for footwear uppers were assessed by measuring leather thickness, mass per unit area, apparent density, tensile strength and percentage elongation at break, as well as flexibility using penetrometer and flexometer, all according to the standardised test methods.

Keywords: Full grain leather · Semi processed cattle leather · Leather testing · Usage quality · Footwear

1 Introduction

Full grain refers to the strongest and most durable part of the cattle hide, which is just below the hair, and has not been sanded or buffed to remove any imperfections, that means it displays the more natural characteristics of leather. The grain has densely packed collagen fibers that are finer. This results in a surface that is strong, durable, and can withstand tough use. This makes it good for footwear uppers. As full grain consisting of the grain layer and the part of the underlying corium, for shoe uppers can be split from 1.3 to 2 mm thick. The strength of the leather is dependent not only on its total thickness but on the proportion that is corium tissue and the frequency with which the corium fibers interweave and cross over each other. Three-dimensional weave of the fiber bundles is not random throughout the leather. There is, parallel to the grain surface, a predominating direction in which the majority of the fibers run. This directional run profoundly affects the physical properties of the leather. When a leather is stretched or

S. Msahli and F. Debbabi (Eds.): CIRATM 2020, SPM 17, pp. 227–233, 2022.
https://doi.org/10.1007/978-3-031-08842-1_38

compressed the angle of weave falls and the structure becomes more compact. One of the attractive features of leather is the line creasing of the outer grain surface that occurs when leather is curved grain inwards. The fineness of the surface folds also depends on uniformity of structure throughout the leather. The natural fiber weave changes with location on the cattle. Along the line of the backbone the leather is thickest and the weave most compact and dense. In the central bend region, which originally covered the back of the cattle, the fiber weave is particularly compact with the interweaving at a high angle. In the belly region, the weave is looser with fiber bundles running at a far lower angle of weave (Kite and Thomson 2011). Therefore, some parts of the same leather have different properties.

Leather quality is also affected by tanning and finishing processes utilized during production. The process of making full grain leather features three main stages: preparation, tanning and crusting. A fourth step may also be added where the leather is given a surface coating or finish. Full grain leathers, however, sometimes will skip this stage as it is not needed. Instead finishing, this leather will only go through an ironing process, which will use varying degrees of pressure and heat until the desired sheen is obtained.

Therefore, with the purpose of analyzing the influence of the leather processing level and sampling location on their properties, in this paper the quality of differently tanned (synthetic and chromium) semi processed cattle full grain leathers, sampled from the bend, shoulder and belly of side leather cuts, were evaluated and their applicability assessed.

2 Materials and Methods

The investigation was carried out on four semi processed full grain leather samples for footwear uppers: L1 – chrome free, synthetic tanned and re-tanned, drum dyed in cyan, waterproofed, slightly lubricated, vacuum dried, stretched and softened leather; L2 – chrome free, synthetic tanned and re-tanned, slightly lubricated, vacuum dried, stretched and softened, lining leather of surface treated with slightly beige pigmented casein and polyurethane polymerization binders; L3 – chromium tanned and re-tanned, drum dyed in black, waterproofed, slightly lubricated, vacuum dried, stretched and softened leather; L4 – chromium tanned and re-tanned, waterproofed, slightly lubricated, vacuum dried, stretched and softened wet blue leather. Three side of leather cuts from one batch for each leather (L1–L4) were tested.

After the sampling of bend (Bd), shoulder (Sh) and belly (Bl) from each side leather cut according to the EN ISO 2418 and conditioning according to the EN ISO 2419 at temperature: 23 ± 2 °C and air relative humidity: $50 \pm 5\%$, leather quality were assessed by determination of physical and mechanical properties – apparent density and mass per unit area according to the EN ISO 2420, thickness according to the EN ISO 2589, tensile strength and percentage elongation at break according to the EN ISO 3376; and usage properties – the dynamic water resistance of flexible leather by repeated linear compression equivalent to a 10% compression of the test piece using penetrometer (by measuring of the time taken for water to penetrate through the test piece and by determining of the percentage mass of water absorbed, all according to the EN ISO 5403-1) and dry and wet flex resistance of leather using flexometer (by visual examination using magnifier after 150000 dry and 50000 wet flex cycles according to the EN ISO 5402-1).

3 Results and Discussion

The results obtained by investigation are presented in Tables 1, 2 and 3. The average values of thickness, mass per unit area and apparent density calculated for sampled areas for all side leather cuts from one batch of leather samples L1–L4 were presented in Table 1. The average values of tensile strength and breaking elongation, calculated for each sampled area of the leather samples (L1–L4) from one batch, in the direction parallel to the backbone and perpendicular to the backbone was presented also in Table 1.

Table 1. Physical and mechanical properties of semi processed full grain leather samples

Leather sample	Sampling location	Thickness (mm)	Apparent density (kg/m^3)	Mass per unit area (g/m^2)	Tensile strength (MPa)		Breaking elongation (%)	
					↑	→	↑	→
L1	Bd	2.20	715.557	1572.275	22.45	15.66	84.78	108.44
	Sh	2.05	745.951	1528.646	22.05	18.85	70.86	81.64
	Bl	1.93	699.613	1351.483	23.48	20.67	78.47	98.51
L2	Bd	1.27	746.733	944.224	21.43	14.04	71.30	93.47
	Sh	1.30	764.317	996.559	19.85	20.73	69.50	95.13
	Bl	1.24	718.135	888.789	18.82	16.61	67.77	97.80
L3	Bd	1.64	768.423	1257.270	23.44	12.69	55.49	59.57
	Sh	1.63	765.476	1249.376	23.50	18.13	45.73	59.44
	Bl	1.72	767.900	1319.168	22.60	10.63	49.13	66.06
L4	Bd	1.33	725.411	959.629	19.81	15.34	66.44	66.24
	Sh	1.23	757.556	926.784	17.88	15.22	56.64	70.95
	Bl	1.24	707.608	875.696	23.40	15.32	59.31	79.38

Bd – bend, Sh – shoulder, Bl – belly, ↑ parallel to the backbone, → perpendicular to the backbone

The influence of sampling location, type of tanning agent used and the leather processing level on the results obtained were analised. It was found that thicker leathers (L1 and L3), intended for making of the outer parts of the footwear uppers, are also heavier and therefore of larger mass per unit area. Lining leathers (L2 and L4), intended for making of the inner parts of the footwear uppers, are thinner, lighter, and at the same time of lower mass per unit area. No significant differences in the obtained results of leather thickness, sampled from the bends and shoulders were found, which indicates the high quality of the leathers tested. In general, tested samples from the belly of the leather cuts were thinner, lighter and more uneven in structure.

Belly leather samples are, in general, also of less apparent density. Regardless of the leather type, in samples sampled from the shoulder of leather cuts, higher values of apparent density were determined. By evaluation of the apparent density of thicker full grain leathers, higher values of apparent density were observed in chromium tanned full grain leather. When compared the apparent density values of synthetic tanned leathers, higher apparent density values were found in L2 leather samples, which is associated with the structure of the processed leather grain surface.

Values of breaking strength are in average higher in the direction parallel to the backbone for all leather samples. No significant differences in obtained results within the same leather type, sampled from the bends, shoulders and belies were found in the direction parallel to the backbone. That also indicates the high quality of the leathers tested. Thicker leathers show higher values of breaking strength. But, at the same time in average, in thinner chromium tanned leather L3 was found higher breaking strength when compared with thicker synthetic tanned leather L1 in the direction parallel to the backbone. For thinner differently tanned leather samples that is not a case, which indicates the fact that the proportion of corium tissue in full thickness and type of tanning agent used affects the obtained values of breaking strength.

Breaking elongation determined in all tested leather samples are higher in the direction perpendicular to the backbone - generally in bellies, but the highest values were found in bends of samples L1 and L3. Higher average values of breaking elongation were found in synthetic tanned leathers when compared with the cromium tanned leathers. The percentage elongation at break of all leather samples is also affected by their thickness and apparent density.

After 150000 dry and 50000 wet flex cycles on flexometer no damage and cracks were not found at the grain surface of all leather samples tested in booth direction, but due to the looser weave in the belly the grain surface has a greater tendency to form coarse wrinkles. As an example, grain surface appearance of leather L1 is shown with magnification of 20× in Table 2.

The average values of penetration time and water absorption after repeated linear compression using penetrometer (method particularly suitable for leathers intended for footwear application), calculated for each sampled area of the leather samples (L1–L4) from one batch, in the direction parallel to the backbone and perpendicular to the backbone was presented in Table 3.

Table 2. Flex resistance of leather after repeated 150000 dry and 50000 wet flex cycles using flexometer – appearance of grain surface of leather sample L1 (magnification of 20x)

Leather sample	Sampling location	Dry sample		Wet sample	
		↑	→	↑	→
L1	Bd	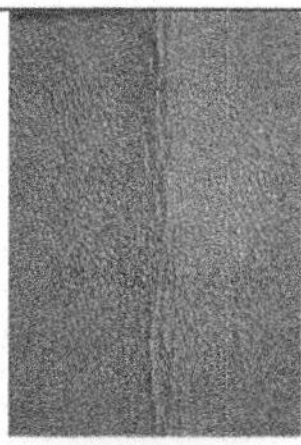	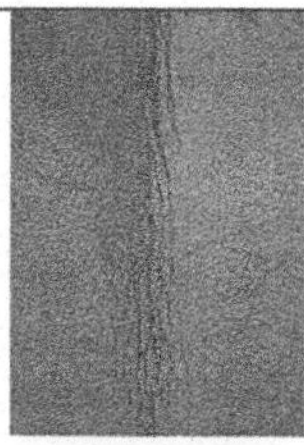	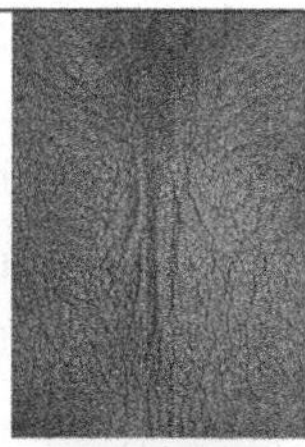	
	Sh	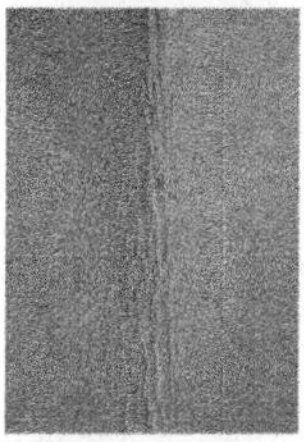	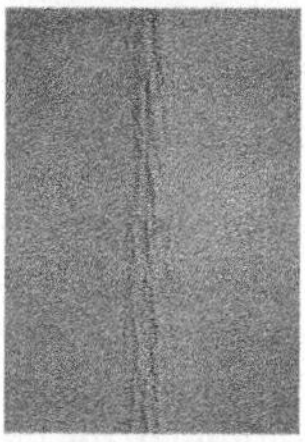	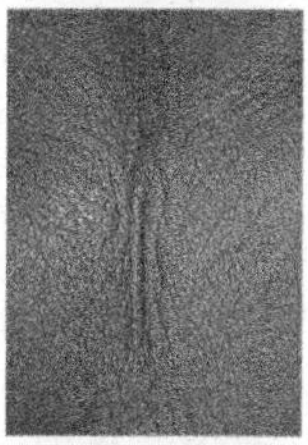	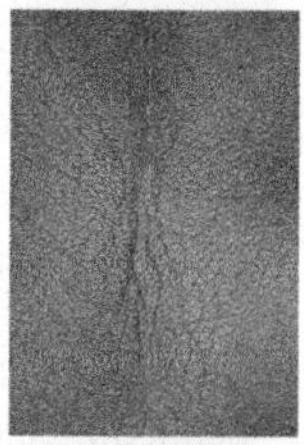
	Bl	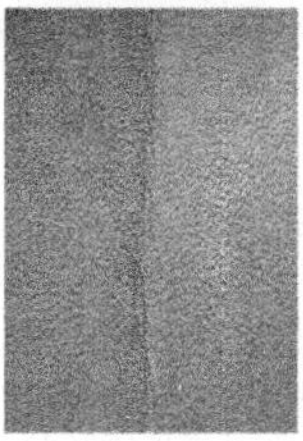	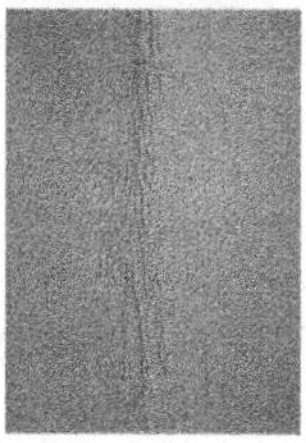	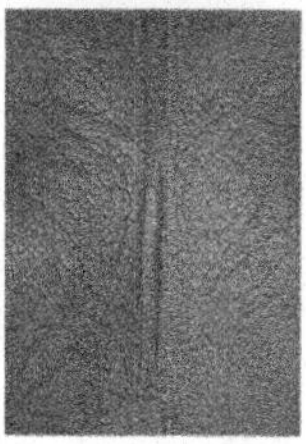	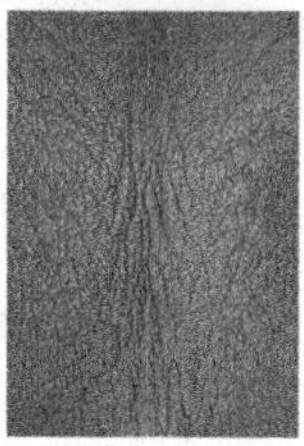

Bd – bend, Sh – shoulder, Bl – belly, ↑ parallel to the backbone, → perpendicular to the backbone

In semi processed full grain waterproofed thicker leather samples 1 and 3 (heavier and of larger mass per unit area) were evaluated water penetration time higher than 4 h. In non-waterproofed thinner lining leather L2 highest values of the percentage mass of water absorbed were determined. When compared waterproofed leathers (L1, L3 and L4) no significant differences in the results obtained for the direction parallel to the backbone and perpendicular to the backbone for each leather were found. In general, tested samples from the belly of the leather cuts were more uneven in structure and of lower values of the water penetration time.

Table 3. Water resistance of flexible leather after repeated linear compression

Leather sample	Sampling location	Penetration time (h:min)		Water absorption (%)	
		↑	→	↑	→
L1	Bd	5:14**	5:14**	9.96	10.36
	Sh	5:27**	6:00**	9.55	9.77
	Bl	3:48**	6:00**	9.64	10.74
L2	Bd	0:01**	0:01**	81.03	82.6
	Sh	0:01**	0:01**	64.17	64.73
	Bl	0:01**	0:01**	69.59	64.74
L3	Bd	4:14**	4:56**	9.62	11.21
	Sh	5:04**	4:59**	8.34	10.69
	Bl	4:52**	5:06**	8.85	8.97
L4	Bd	3:16*	3:18**	9.49	9.86
	Sh	3:23*	3:51*	10.07	9.76
	Bl	3:12*	2:37*	9.52	10.01

Bd – bend, Sh – shoulder, Bl – belly, ↑ parallel to the backbone, → perpendicular to the backbone, *droplet of water formed on the surface, **damp patch formed on the surface

4 Conclusion

Full grain is looked upon as the highest quality leather available. Since the outer layer is not removed, it develops a patina over time that can be pleasing to the eye. It was concluded that the full grain leather possesses different physical, mechanical and usage properties depending on the manufacturing process, sampling location and their macro- and microstructural characteristics. Differences observed in the tested properties of full grain leathers also define their applicability for footwear uppers. They should be especially taken into account when the specific parts of footwear uppers are cut – facing, tongue, collar, upper, quarter or vamp lining. Furthermore, three full grain leathers, used in this study, instead of finishing, is passed only through an ironing process which significantly reduces their production costs. Therefore, high usage quality in semi processed full grain leather samples evaluated confirm their applicability for footwear uppers.

Acknowledgment and Funding. This paper is funding by the Croatian science foundation within the Project IP-2016–06-5278 (Comfort and antimicrobial properties of textiles and footwear, principal investigator: Prof. Zenun Skenderi, Ph.D.).

References

Kite, M., Thomson, R.: Conservation of Leather. Routledge Taylor & Francis Group, New York (2011). ISBN 978-0-7506-4881-3

EN ISO 2418:2017 Leather - Chemical, physical and mechanical and fastness test - Sampling location

EN ISO 5402–1:2017 Leather - Determination of flex resistance - Part 1: Flexometer method

EN ISO 5403–1:2011 Leather - Determination of water resistance of flexible leather - Part 1: Repeated linear compression (penetrometer)

EN ISO 2420:2017 Leather - Physical and mechanical tests - Determination of apparent density and mass per unit area

EN ISO 3376:2020 Leather - Physical and mechanical tests - Determination of tensile strength and percentage elongation

EN ISO 2589:2016 Leather - Physical and mechanical tests - Determination of thickness

EN ISO 2419:2012 Leather - Physical and mechanical tests - Sample preparation and conditioning

Total Quality Management Practices in Nigerian Textile Industry

Idowu Jamiu Diyaolu(✉)

Family, Nutrition and Consumer Sciences, Obafemi Awolowo University, Ile-Ife, Nigeria
diyaolu@oauife.edu.ng

Abstract. The study examined Total Quality Management (TQM) practices in Nigerian textile firms. Eighteen (18) production managers and 227 production staff were purposively selected. Data was collected using a structured questionnaire and interview schedule. Percentages mean and analysis of variance were used for the analysis. The staff had a high perception of employee participation ($x = 3.28$) and product design ($x = 3.24$), while education and training ($x = 2.83$) had a low perception. The implementation of leadership ($F = 3.423$, $p < 0.05$) and supplier quality management ($F = 4.381$, $p < 0.05$) varied significantly with the location of the firms. Also, the implementation of leadership ($F = 3.764$, $p < 0.05$), evaluation ($F = 3.305$, $p < 0.05$) as well as recognition and reward ($F = 5.541$, $p < 0.05$) varied significantly with the ages of the firms. The study recommends the implementation of reward and recognition as well as training for staff in Nigerian textile firms.

Keywords: Total quality management · Production staff · Operations · Leadership · Product design

1 Introduction

Globally, Total Quality Management (TQM) has been observed to be a proven technique that guarantees the survival of business in a competitive environment (Sule et al. 2016; Hasan et al. 2013; Demirbag et al. 2006). The practice of TQM as a strategy is one of the most important technological innovations of the last decade which has integrated fundamental management techniques to improve the traditional practices in production and services (Gherardini et al. 2017; Besterfield et al. 2006; Bayazit 2003). TQM is a management philosophy that encourages customer satisfaction, continuous improvement and teamwork. It is the unrelenting pursuit of continuous improvement which is realized by accessing and utilizing the intensive knowledge and experience of managers and employees at all levels.

The concept of quality has been established from different perspectives ranging from the perception of values, conformance to requirement, fitness to use and meeting customer's expectations. The quality of a product or service makes it consistently meet or exceed customers' expectations (Russell and Taylor 2006). There is an ever-increasing

S. Msahli and F. Debbabi (Eds.): CIRATM 2020, SPM 17, pp. 234–242, 2022.
https://doi.org/10.1007/978-3-031-08842-1_39

demand for quality products and this global revolution had forced the industry to invest substantial resources in adopting and implementing total quality management strategies.

TQM is mostly needed in the textile industry. In Nigeria, the textile industry used to be a major employment generating sector (Textile Report 2005). The industry is linked with the clothing, furnishing, cordage, bagging, chemical and petrochemical industries. The production of high-quality textiles and apparel employs a complex process requiring high-quality inputs, technology, processing and delivery to markets. There is need to implement TQM in policy statements and quality manuals (Hasan et al. 2013). Poor leadership management commitment is always considered one of the reasons for the failure of TQM efforts. The aspect of sourcing for materials, supplier quality management, is an important aspect of TQM.

Today, many firms in the Nigerian textile sector have gone out of operations with thousands of workers rendered jobless, leading to spending over $2billion on imported textile goods (Federal Ministry of Industry, Trade and Investment 2014). In addition, the perception of consumers is that Nigerian textile products are inferior in quality and expensive relative to imported textiles. This resulted to low patronage of local products (Umar and Nwachukwu 2007). If TQM is given priority, especially as it has been effective in enhancing quality, this trend may be reversed. There are significant relationships between TQM, competitive advantage and business excellence (Gherardini et al. 2017). It is, therefore, necessary to examine the practice of TQM among Nigerian firms and identify areas where there are needs for improvement to engender growth and development in the industry. The study examined TQM practices in the Nigerian textile industry, identified relationships between TQM operations vis-a-vis the location and age of the firms.

2 Methodology

In Nigeria, resilient textile firms are located in Lagos, Kano and Kaduna states. These states were purposively selected for the study, consisting of 18 firms. Eighteen (18) production managers and 227 production staff were purposively selected. The production staff are those involved in the various operations such as weaving, printing, dyeing and finishing processes. A structured questionnaire and interview schedule were used for data collection. Eleven TQM constructs were examined which included leadership, supplier quality management, vision and plan statement, process control and improvement, product design, quality system improvement, employee participation, recognition and reward, education and training as well as customer focus (Hasan et al. 2013). Percentages mean and analysis of variance were used for the analysis.

3 Results

Table 1 showed the demographic characteristics of the production staff. A higher percentage (84.6%) were males with (15.4%) females. The wide variation in male and female participation may be due to the stereotypical assignment of certain jobs to women in Nigeria as it is believed that women are to keep the home and avoid strenuous jobs (Kolawole and Fasina 2009). This is contrary to the report that, as far back as 2003, about 43,000 employees (90% female) were operating in the garments sector in Lesotho (OECD 2005). Females should, however, be encouraged to participate in textile production. About 69.6% were married while 29.5% were single. Also, 22.9% had HND. The highest academic qualification among the management staff in the textile firms is MSc. It has been observed that higher education institutes for studies in textile majors have reduced drastically (Lottersberger 2012). The level of operations among the firms in Nigeria will be better with more management staff possessing higher educational qualification like Ph.D, with other professional qualifications.

Table 1. Demographic and socio-economic characteristics of production staff

Characteristics	Frequency	Percentage
Age in years		
<20	2	0.01
20–30	59	25.99
31–40	58	25.55
41–50	85	37.44
51–60	23	10.13
>60	-	-
Gender		
Male	192	84.6
Female	35	15.4
Marital status		
Single	67	29.5
Married	158	69.6
Divorced	2	0.01
Widowed	-	-
Educational status		
No formal education	5	2.2
First school leaving certificate	19	8.4
Secondary school certificate	51	22.5
Ordinary National Diploma (OND)	40	17.6
Higher National Diploma (HND)	52	22.9

(continued)

Table 1. (*continued*)

Characteristics	Frequency	Percentage
Bachelor of Science (BSc)	55	24.2
Masters of Science (MSc)	5	2.2
Doctor of Philosophy (PhD)	-	-

Table 2 showed the perception of TQM among staff in textile firms. The result revealed that process control and improvement has the highest mean of 3.99. This shows that the firms use statistical packages for process control, implements various inspections effectively and also keep the firms clean and neat. Employee participation (mean = 3.28) also showed that the employee were very devoted to the success of the firms and are keenly concerned with quality-related matters. Furthermore, the result on leadership (mean = 3.23) implies that top management keenly participates in quality management and improvement process and focus on product quality rather than yields. However, the low response was observed under recognition and reward (mean = 2.74) as well as education and training (mean = 2.83).

The implication of this is that employees in Nigerian textile firms would be more encouraged with rewards in salary and better promotion plans as this will boost their commitment to the work. The employee would also want resources to be made available for training on how to use quality administration tools. The rewarding system improves the loyalty of employees and their disposition to work (Atakan and Ay 2009).

Table 2. Perception of production staff on total quality management in the textile firms

TQM constructs	SA (%)	A (%)	D (%)	SD (%)	Mean
Leadership					
Top management participate in quality improvement	35.2	61.7	0.9	2.2	
They empower employees to solve quality issues	37.9	6.4	0.4	5.3	
They focus on product quality rather than yields	29.5	5.1	5.3	10.1	3.23
Supplier quality management					
The firm has long-term relations with suppliers	36.1	55.9	1.3	6.6	
The firm has information about supplier performance	19.8	69.6	2.2	7.9	
The firm always gives feedback on product performance	37.9	54.2	0.9	7.0	3.21
Vision and plan statement					
Our company has a clear long term vision statement	45.4	47.1	0.4	7.0	

(*continued*)

Table 2. (*continued*)

TQM constructs	SA (%)	A (%)	D (%)	SD (%)	Mean
Our company has a clear short-term business plan	11.9	37.0	15.0	36.1	
Various policies correspond to the employees	22.0	56.8	4.4	16.7	2.93
Evaluation					
Benchmarking is used comprehensively in our firm	23.3	61.7	2.6	12.3	
Quality related data are used for evaluation	21.6	62.1	2.2	14.1	
The aim of the evaluation is for improvement	30.4	59.0	3.1	7.5	3.09
Process control and improvement					
The firm is kept neat and clean at all times	40.1	51.5	1.8	6.6	
Firm implements various inspections effectively	32.6	55.9	0.4	11.0	
Our firm uses statistical packages for process control	23.8	63.9	2.6	9.7	3.19
Product design					
Design engineers have some shop floor experiences	21.1	69.2	2.2	7.5	
Customer s are thoroughly considered in a new design	38.8	57.3	4.0	0.0	
New product designs are thoroughly reviewed	34.8	62.1	0.9	2.2	3.24
Quality system improvement					
Firm uses ISO 9000 instruction for quality	36.6	48.9	1.3	12.8	
Our firm has a sensible quality manual	28.6	65.6	2.2	3.5	
Our firm has clear implementation directions	22.9	70.0	0.0	7.0	3.19
Employee participation					
Employees are very devoted to the success of the firm	46.3	49.8	0.0	4.0	
Reporting work problems is encouraged in the firm	39.2	53.3	1.3	6.2	
The workforce is concerned with quality-related behaviour	27.3	59.0	1.8	11.9	3.28
Recognition and reward					
The industry has a promotion plan for employee	22.5	43.2	8.8	25.6	
Excellent suggestions are economically rewarded	15.9	40.5	9.3	34.4	
Position promotions are based on work quality	23.3	46.3	11.5	18.9	2.74

(*continued*)

Table 2. (*continued*)

TQM constructs	SA (%)	A (%)	D (%)	SD (%)	Mean
Education and training					
Employees are encouraged to further education	15.9	45.4	5.3	33.5	
Employees are trained on quality administration tools	22.0	57.3	3.5	17.2	
Resources are available for employee training	20.3	46.7	6.6	26.4	2.83
Customer focus					
Quality-related customer complaints are treated	40.5	54.2	1.3	4.0	
The firm conducts market research to improve products	34.4	55.9	2.6	7.0	
The firm provides a warranty on products to customers	22.5	56.8	2.2	18.5	3.18

Table 3 showed the Analysis of Variance of significant difference in TQM implementation by age of the firms. It is revealed that the implementation of leadership ($F = 3.764$, $p < 0.05$), evaluation ($F = 3.305$, $p < 0.05$) product design ($F = 5.177$, $p < 0.05$), quality system improvement ($F = 3.764$, $p < 0.05$), employee participation ($F = 3.387$, $p < 0.05$) as well as recognition and reward ($F = 5.541$, $p < 0.05$) varied significantly by the ages of the firms. Factors, such as the size of the firms can affect TQM implementation. There is a significant difference between the size of the establishment and the difficulties experienced in the TQM process. According to Talapatra et al. (2020) TQM implementation in Bangladeshi textile sector depends on structural, strategic, contextual and human resource-enabling factors. Nigerian textile firms have to educate the employee about Total Quality Management, Statistical Process Control (SPC), Evaluation system and Employee participation in quality-related decisions (Hasan et al. 2013).

Table 3. Table of analysis of variance showing significant difference in TQM implementation by age of firms

TQM	<20 years	20–39 years	≥40 years	F - Statistics	df	P
Leadership	3.18 ± 0.55	3.17 ± 0.52	3.36 ± 0.42	3.764	2	.025
Supplier quality management	3.1 4 ± 0.41	3.17 ± 0.46	3.28 ± 0.48	1.594	2	.205
Vision and plan statement	2.96 ± 0.41	2.92 ± 0.52	2.92 ± 0.42	0.186	2	.830
Evaluation	3.00 ± 0.62	3.16 ± 0.44	2.99 ± 0.55	3.305	2	.039

(*continued*)

Table 3. *(continued)*

TQM	<20 years	20–39 years	≥40 years	F - Statistics	df	P
Process control/improvement	3.08 ± 0.59	3.16 ± 0.55	3.29 ± 0.49	1.941	2	.146
Product designs	3.43 ± 0.32	3.18 ± 0.40	3.31 ± 0.44	5.177	2	.006
Quality system improvement	3.23 ± 0.47	3.11 ± 0.42	3.30 ± 0.53	4.051	2	.019
Employee participation	3.11 ± 0.51	3.24 ± 0.48	3.38 ± 0.49	3.387	2	.036
Recognition and reward	3.02 ± 0.51	2.60 ± 0.84	2.91 ± 0.66	5.541	2	.004
Education and training	3.05 ± 0.58	2.80 ± 0.69	2.85 ± 0.60	1.373	2	.255
Customer focus	3.24 ± 0.44	3.16 ± 0.51	3.21 ± 0.56	0.386	2	.680
Total TQM	103.26 ± 10.49	101.06 ± 11.99	104.51 ± 11.07	2.272	2	.105

Table 4 showed the Analysis of Variance of significant difference in TQM implementation by location of the firms. It is shown that the implementation of leadership ($F = 3.423$, $p < 0.05$), supplier quality management ($F = 4.381$, $p < 0.05$) and quality system improvement ($F = 8.525$, $p < 0.05$) varied significantly by the location of the firms. It has been observed that the adoption of total quality management by SMEs in Nigeria, including textiles manufacture, is subject to many factors and affects different performance outcomes (Sule et al. 2016). In Nigeria, the firms are located in the North (Kano and Kaduna) and south (Lagos). It is noteworthy to mention that there is a clear demarcation in the clothing traditions among these two settings which might have influenced the variations.

Table 4. Analysis of variance showing significant difference in TQM implementation by location of firms

TQM	Kaduna	Kano	Lagos	F - Statistics	df	P
Leadership	2.94 ± 0.81	3.24 ± 0.50	3.28 ± 0.42	3.423	2	.034
Supplier quality management	2.964 ± 0.74	3.30 ± 0.43	3.18 ± 0.43	4.381	2	.014
Vision and plan statement	2.77 ± 0.81	2.98 ± 0.45	2.92 ± 0.44	1.454	2	.236
Evaluation	3.08 ± 0.56	3.10 ± 0.49	3.07 ± 0.51	0.66	2	.936
Process control/improvement	3.00 ± 0.83	3.18 ± 0.47	3.24 ± 0.54	1.433	2	.241
Product designs	3.19 ± 0.72	3.25 ± 0.38	3.26 ± 0.39	0.210	2	.811

(continued)

Table 4. *(continued)*

TQM	Kaduna	Kano	Lagos	F - Statistics	df	P
Quality system improvement	2.79 ± 050	3.15 ± 0.44	3.28 ± 0.47	8.525	2	.000
Employee participation	3.29 ± 0.71	3.26 ± 0.46	3.29 ± 0.49	0.110	2	.895
Recognition and reward	2,44 ± 1.03	2.66 ± 0.87	2.85 ± 0.63	2.896	2	.057
Education and training	2.56 ± 1,03	2.91 ± 0.64	2.81 ± 0.59	2.106	2	.124
Customer focus	3.06 ± 0.77	3.26 ± 0.43	3.14 ± 0.54	1.826	2	.164
Total TQM	96.25 ± 16.65	102.87 ± 11.45	103.03 ± 10.79	2.524	2	.082

4 Conclusion

From the foregoing, there is evidence of TQM practices in Nigerian textile firms. However, there is a need for improvement in education and training as well as recognition and reward. The variation in ages and location of textile firms in Nigeria significantly influence TQM implementation. Most of the firms in Nigeria are located in the northern and southern regions. Since most of the constructs in TQM implementation were observed, it is believed that the firm should be doing well contrary to the observation of closure and low employment generation. The study, however, recommends a specific study on each of the areas of TQM construct, for example, supplier quality management, to identify specific problems facing the operation. Reward and training in the firms are strongly recommended.

References

Atakan, T., Ay, O.: Total quality management practices in textile and confection corporations. Trakia J. Sci. **14**(2), 20–23 (2009)

Bayazit, O.: Total quality management (TQM) practices in Turkish manufacturing organizations. Total Qual. Manag. Mag. **15**(5), 345–350 (2003)

Besterfield, D.H., Besterfield-Michna, C., Besterfield, G.H., Besterfield-Sacre, M.: Total Quality Management, 3rd edn, pp. 34–56. Prentice-Hall of India, New Delhi (2006). ISBN-81-203-2883-3

Demirbag, M., Tatoglu, E., Tekinkus, M., Zaim, S.: An analysis of the relationship between total quality management implementation and organizational performance: evidence from Turkish SMEs. J. Manuf. Technol. Manag. **17**(6), 829–847 (2006)

Federal Ministry of Industry, Trade and Investment: The Nigeria Industrial Revolution Plan: Textiles and Garments. Document NIRP\2\1\7. Abuja, Nigeria (2014)

Gherardini, F., Renzi, C., Leali, F.: A systematic user-centred framework for engineering product design in small-and medium-sized enterprises (SMEs). Int. J. Adv. Manuf. Technol. **91**(5–8), 1723–1746 (2017)

Hasan, I.U., Sohail, M.M., Piracha, J.L., Ahmad, K.: Implementation status of TQM practices in textile and apparel industrial organization: a case study from Faisalabad, Pakistan. Br. J. Econ. Manag. Trade **3**(3), 201–223 (2013)

Kolawole, S.O., Fasina, O.A.: Gender sensitivity and women empowerment in Nigeria. J. Res. Natl. Dev. **7**(1), 3–5 (2009)

Lottersberger, A.: Design, innovation and competitiveness in the textile industry upstream design driven innovation. Ph.D. thesis, Politecnico Di Milano School of Design, Italy (2012)
OECD: Trade and Structural Adjustment Policies in Selected Developing Countries, p. 45. OECD Publication, USA (2005)
Russell, R.S., Taylor, B.W.: Operations Management, 4th edn, p. 45. Prentice-Hall Inc., New Jersey (2006). ISBN-81-203-2383-1
Sule, J.G., Ogbadu, E.E., Nafiu, A.T.: Factor analysis of total quality management adoption by SMES in Nigeria. Facta Universitatis Ser. Econ. Organ. 13(4), 365–377 (2016). ISSN 2406-050X
Talapatra, S., Uddin, M.K., Antony, J., Gupta, S., Cudney, E.A.: An empirical study to investigate the effects of critical factors on TQM implementation in the garment industry in Bangladesh. Int. J. Qual. Reliab. Manag. **37**(9/10), 1209–1232 (2020). https://doi.org/10.1108/IJQRM-06-2018-0145. ISSN 0265-671X
Textile Report: Industry Report. Agusto and Co. Limited, Colombia, South America, p. 34 (2005)
Umar, A.A., Nwachukwu, J.C.: Interface of quality related-costs with automation in nigerian industries. Cont. J. Eng. Sci. **1**, 45–52 (2007)

Development of an Application for Planning and Selecting Subcontractors Using the WSM Method

Lahdhiri Mourad(✉), Jmali Mohamed, and Babay Amel

Laboratory of Textile Engineering (LGTex), University of Monastir, Monastir, Tunisia
lahdhirimrad@yahoo.fr

Abstract. The purpose of the management of supply chain activities is to maximize customer value, to achieve sustainable competitive benefits, and to organize the material and the information flows in the supply chain in order to provide the highest degree of customer satisfaction at the lowest possible cost. This paper aims to present an application based on the WSM method to find the best selection of the subcontractors in the supply chain system.

Keywords: Apparel industry · Supply chain · WSM · Subcontractor selection

1 Introduction

Subcontractors categorization, selection and performance evaluation are decisions of strategic importance to companies. Several studies are interested to this topic, one of them is focused the best way on supplier selection and evaluation using the model multiple criteria decision-making (MCDM) according the concept TOPSIS (Chen et al. 2006). Bayrak et al. and Maurizo. B et al. have used a fuzzy supplier selection algorithm (Bayrak et al. 2007; Maurizio et al. 2010). Felix et al. have developed their research according to the model fuzzy AHP (Analytic Hierarchy Process), in fact they combined the fuzzy method with the AHP method to clarify the fuzzy. As a result, the used model is proved simple, less time taking and having a less computational expense (Felix et al. 2008). In addition, Sharon et al. have described a decision model that incorporates decision maker's subjective evaluations and applies fuzzy arithmetic operators to manipulate and quantify these assessments (Sharon et al. 2009). In this research, our aim is to facilitate the planning task, the aim of which is to add an application, which allows automatic selection of the subcontractor. We have applied the WSM method since it helps to make a decision and to evaluate several options in situations where no possibility is perfect.

Therefore, we developed an application in which the WSM method was used to choose the subcontractors in an apparel supply chain.

S. Msahli and F. Debbabi (Eds.): CIRATM 2020, SPM 17, pp. 243–247, 2022.
https://doi.org/10.1007/978-3-031-08842-1_40

2 Materials and Methods

2.1 WSM Method

The Weighted Sum Model (WSM) is probably the most commonly used approach, especially in single dimensional problems. The assumption that governs this model is the additive utility assumption (Adriyendi 2015).

2.1.1 WSM Process

Step 1: Construct the decision matrix X: as mentioned in Table 1.

Table 1. Decision matrix

	C1	C2	C3	…	Cn
A1	X11	X_{12}	X_{13}	…	X_{1n}
A2	X11	X_{22}	X_{23}	…	X_{2n}
A3	X31	X_{32}	X_{33}	…	X_{3n}
.	…	…	…	…	…
An	Xn1	X_{n2}	X_{n3}	…	X_{nn}

Where: A_i are the alternative, Ci are the criteria

Step 2: Construct the normalized decision matrix Z: as mentioned in Table 2.

Table 2. Normalized decision matrix

	C1	C2	C3	…	Cm
A1	Y11	Y_{12}	Y_{13}	…	Y_{1m}
A2	Y11	Y_{22}	Y_{23}	…	Y_{2m}
A3	Y31	Y_{32}	Y_{33}	…	Y_{3m}
.	…	…	…	…	…
An	Yn1	Y_{n2}	Y_{n3}	…	Y_{nm}

For beneficial attribute or alternative:

$$\mathrm{Y_{ij}} = \frac{x_{ij}}{x_{ij\,\max}} \tag{1}$$

For non-beneficial attribute or alternative:

$$\mathrm{Y_{ij}} = \frac{x_{ij\,\min}}{x_{ij}} \tag{2}$$

Step 3: Construct the weighted normalized decision matrix Z′.
In this step, multiplying the weighted vector W of criteria with the normalized matrix.

$$\mathrm{W} = \begin{bmatrix} \mathrm{W_1} \\ \mathrm{W_2} \\ \mathrm{W_3} \\ \ldots \\ \mathrm{W_n} \end{bmatrix} = \begin{bmatrix} \mathrm{W_1 * Y_{11}\ W_1 * Y_{12} \ldots W_1 * Y_{1m}} \\ \mathrm{W_2 * Y_{21}\ W_1 * Y_{22} \ldots W_1 * Y_{2m}} \\ \mathrm{W_3 * Y_{31}\ W_1 * Y_{32} \ldots W_1 * Y_{3m}} \\ \ldots \quad \ldots \quad \ldots\ldots \\ \mathrm{W_n * Y_{n1}\ W_1 * Y_{n2} \ldots W_1 * Y_{nm}} \end{bmatrix}$$

Step 4: Calculate the score S for WSM:
Calculate the score for each alternative I using the following formula;

$$S_i = \sum\nolimits_{j=1}^{m} W_i * Y_{ij} \tag{3}$$

3 Results and Discussion

In this study, we developed a production management application in a manufacturing denim products company and we integrated a WSM decision support tool in our application.

The Fig. 1 represents our application in which we integrated the WSM method. To evaluate the WSM method, a dataset composed of 12 production orders was used. Then, this data was tested both in a real case and in our application. It should be noted that these production orders, which are given to well-defined subcontractors, are selected from those, which have been successful in the choice made. We present in Table 3 the rank of the solution adopted by the company in the list of solutions determined by the WSM method.

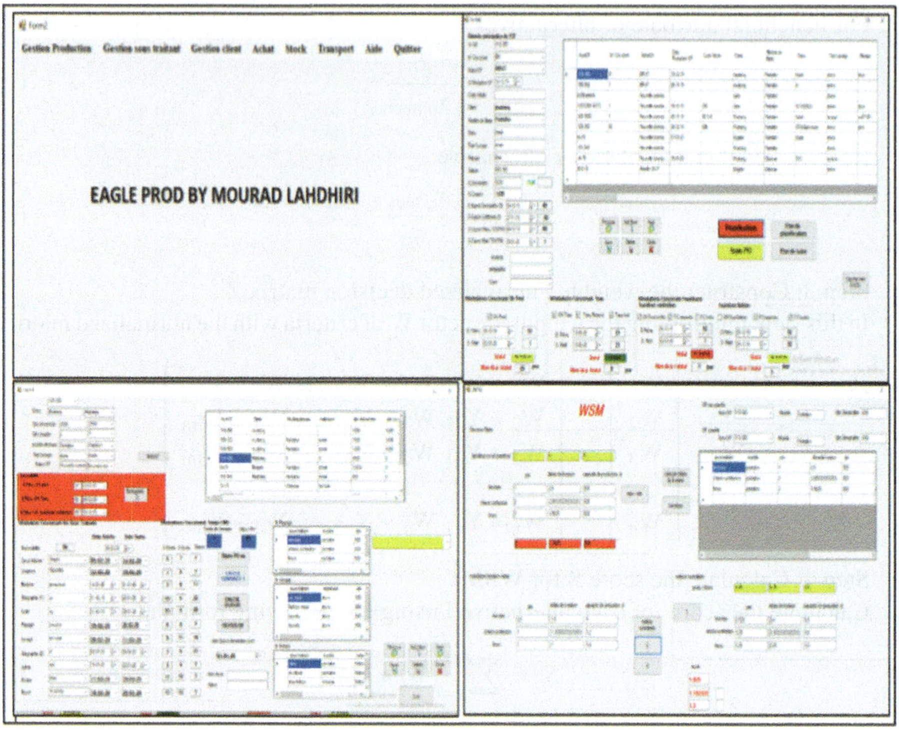

Fig. 1. Production management application

Table 3. Evaluation test

N° PO	Quantity (pieces)	The selected Subcontractors by company	The selected Subcontractors by WSM	The rank of subcontractors selected
1	1080	S1	S1	1
2	2000	S1	S2	2
3	1350	S5	S5	1
4	1500	S7	S7	1
5	800	S8	S8	1
6	3400	S3	S3	1
7	1600	S4	S4	1
8	1560	S2	S2	1
9	2302	S5	S5	1
10	7600	S3	S3	1
11	3400	S6	S6	1
12	790	S7	S7	1

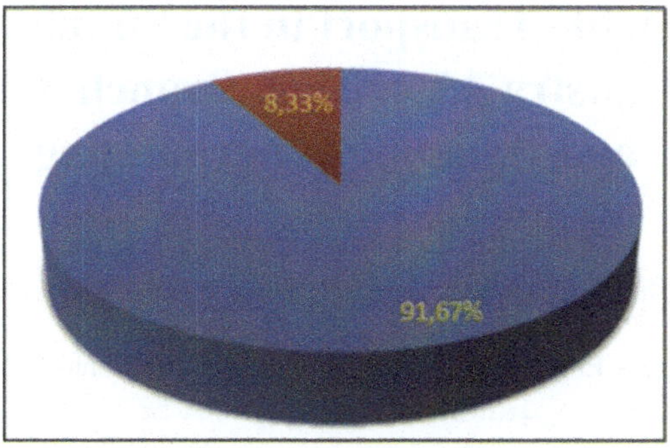

Fig. 2. Percentage of coincidence for the WSM method with the choice of the company

From the Fig. 2, we can conclude that the results found are very acceptable. In fact the percentage of coincidence for WSM method with the choice of the company is equal to 91, 67%.

4 Conclusion

In this paper, it has been proved that our application is effective for the selection of the best subcontractors; it represents a computerized decision support tool. In fact, we found a coincidence with the result of the company with a percentage equal to 91, 67%. Our application is performant to choose the best subcontractors.

References

Adriyendi, M.: Multi-attribute decision making using simple additive weighting and weighted product in food choice. Int. J. Inf. Eng. Electron. Bus. 8–14 (2015)

Chen, C.T., et al.: A fuzzy approach for supplier evaluation and selection in supply chain management. Int. J. Prod. Econ. **102**, 289–301 (2006)

Chen, F.T.S., et al.: Global supplier selection: a fuzzy-AHP approach. Int. J. Prod. Res. **46**(14), 3825–3857 (2008)

Bayrak, M.Y., et al.: A fuzzy approach method for supplier selection. Product. Plann. Control. **18**(1), 54–63 (2007)

Maurizio, B., et al.: From traditional purchasing to supplier management: a fuzzy logic based approach to supplier selection. Int. J. Logist. Res. App. Lead. J. Supply. Chain. Manag. **5**, 235–255 (2010)

Ordoobadi, S.M., et al.: Development of a supplier selection model using fuzzy logic. Supply Chain Manag. Int. J. **14**(4), 314–327 (2009)

Sustainable Transport in the Supply Chain Using Dijkstra Algorithm, Branch & Bound, and Dynamic Programming

Lahdhiri Mourad(✉), Jmali Mohamed, and Babay Amel

Laboratory of Textile Engineering (LGTex), University of Monastir, Monastir, Tunisia
lahdhirimrad@yahoo.fr

Abstract. Finding the optimum path represent several solutions to solve shortest path search problems. It allows determining a shorter way to get from one city to another, knowing the road network of a region. Specifically, it calculates shorter paths from a source in an oriented graph weighted by positive real. In this paper, we utilize the three methods to optimize the transportation of technicians from the company to each subcontractor and make it sustainable.

Keywords: Apparel industry · Supply chain · Dijkstra algorithm · Branch and bound · Dynamic programming

1 Introduction

The path search problem involves finding the optimum path between the present location and the destination under given conditions. Currently, these problems arise in networks such as the highway system, railroads, logistics, and communication networks and cover a wide range of applications (Noto and Sato 2000). Grassin and Minoux have developed their research according to the Danzig algorithm to choose the shortest path (Grassin and Minoux 1973). Thomas and Michael have used dynamic programming to solve the longest path's problem (Thomas, Michael). In the research of Gondran, he used linear algebra to find the best path with the minimum distance (Gondran 1975). HU used integer programming and network flows to solve and reduce the longest road problem (Hu 1969). Kaufmann and Desbazeille, in their research, described the method of the critical path (Kaufmann and Desbazeille 1964). In the work of Grinold, he used the steepest ascent for large-scale linear programs (Grinold 1972). In this research, we tried to optimize technicians' transportation from the company to each subcontractor to make it sustainable. We applied the Dijkstra algorithm, Branch & bound and dynamic programming methods to find the shortest path between the company and the furthest subcontractors corresponding to the minimum distance and cost.

S. Msahli and F. Debbabi (Eds.): CIRATM 2020, SPM 17, pp. 248–257, 2022.
https://doi.org/10.1007/978-3-031-08842-1_41

2 Materials and Methods

2.1 Dijkstra Algorithm

The Dijkstra algorithm conceived by E. W. Dijkstra in 1956 (Edsger 1971). It is an algorithm for finding the shortest paths between nodes in a graph that may represent, for example, road networks and is used in the field of transport management (Xu et al. 2007).

The Dijkstra algorithm:

```
INITIALISATION:
For each  x ∈ S do δs (x) =∞
  δs(x)=0
      X=S
      E= ∅
Treatment:
While X ≠∅ DO
 Select in the list X the vertex x with δs(x) minimum          Remove the vertex x from the list X
 Add the vertex x to the list E
     FOR EACH y ∈ V+ (x) ∩X DO
        IF δs (y) >  δs (x) + l (x, y) THEN
            δs (y) takes the value  δs (x) + l (x, y)
                            p (y) = x
                        END IF
          END FOR
  END WHILE
```

2.2 Branch & Bound Method

Branch and bound (BB, B&B, or B&B) is an algorithm design paradigm for discrete and combinatorial optimization problems and mathematical optimization.

2.2.1 Branch & Bound Algorithm

1. Using a heuristic, find a solution xh to the optimization problem. Store its value, B = f(xh). (If no heuristic is available, set B to infinity.) B will denote the best solution found so far and be used as an upper bound for candidate solutions.
2. Initialize a queue to hold a partial solution with none of the problem's variables.
3. Loop until the queue is empty:
4. Take a node N off the queue.
5. If N represents a single candidate solution x and $f(x) < B$, then x is the best solution so far. Record it and set $B \leftarrow f(x)$.
6. Else, Branch on N to produce new nodes Ni. For each of these:
7. If bound (Ni) > B, do nothing; since the lower bound on this node is greater than the upper bound of the problem, it will never lead to the optimal solution and can be discarded.
8. Else, store Ni on the queue.

2.3 Dynamic Programming

Dynamic programming is both a mathematical optimization method and a computer programming method.

- Regard a tour to be a simple path that starts and ends at vertex 1.
- Every tour consists of an edge $<1, k>$ for some $k \in V - \{1\}$ and a path from k to vertex 1. The path from vertex k to vertex 1 goes through each vertex in $V - \{1, k\}$ exactly once.
- Let g(i, S) be the shortest path starting at vertex i, going through all vertices in S, and terminating at vertex 1.
- $G(1, V - \{1\})$ is the length of an optimal salesperson tour.
- From the principle of optimality it follows that:
- $g(1, V - \{1\}) = \min \{c1k + g(k, V - \{1, k\})\}$ (1)
- $2 \leq k \leq n$
- Generalizing from (1) we obtain
- $g(i, S) = \min j \in S \{cij + g(j, S - \{j\})\}$ (2)
- Formula (2) may be solved for $g(1, V - \{1\})$ if we know $g(k, V - \{1,k\})$ for all choices of k.
- The g values may be obtained by using (2).
- Clearly, $g(i, \emptyset) = ci,1$, $1 \leq i \leq n$. Hence, we may use (2) to obtain g(i, S) for all S of size 1.
- Then we can obtain g(i, S) for S, which $|S| = 2$.
- Then, we can obtain g(i, S) for S, which $|S| = 3$, etc.
- When $|S| < n - 1$, the value of i and S for which g(i, S) is needed are such that $i \neq 1$; $1 \notin S$ and $i \notin S$

3 Numerical Example

In this numerical example, we applied the dynamic programming, Dijkstra's algorithm, and Branch and bound to do a turn from node 1 with a minimum distance. Therefore, we define our networks; each node represent subcontractor of sewing (Fig. 1):

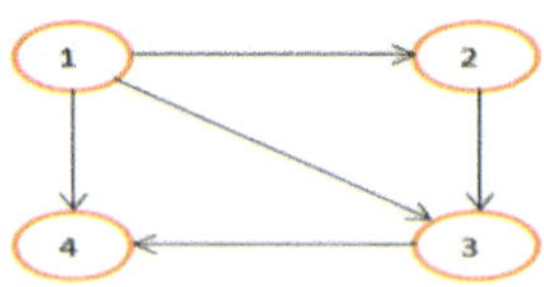

Fig. 1. Road network of subcontractors

The matrix of distance is mentioned in Table 1.

Table 1. Matrix of distance

		NODES			
NODES		1	2	3	4
	1	0	10	15	20
	2	5	0	9	10
	3	6	13	0	12
	4	8	8	9	0

3.1 Application of Dynamic Programming

We applied the dynamic programming based on the Formula:

$$g\,(i,\ S) = \min_{j \in S}\left\{c_{ij} + g(j,\ S - \{j\})\right\} \tag{1}$$

Our objective is to determine the minimum turn from 1, so we construct the tree level per level in the first step as mentioned in Fig. 2 and Table 2.

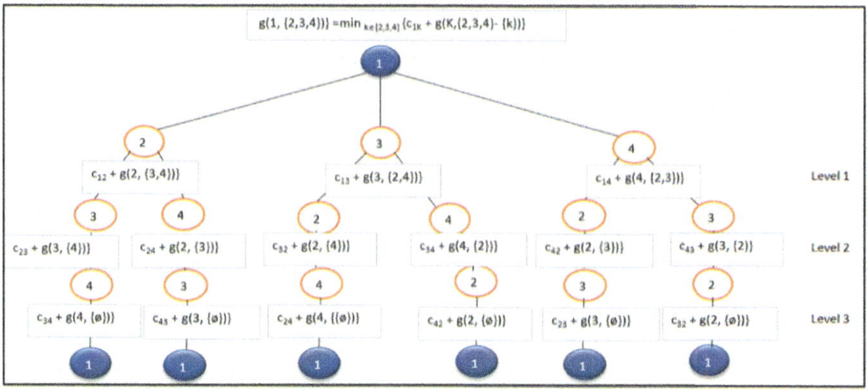

Fig. 2. The tree of dynamic programming

From the Table 2, $g(1, \{2,3,4\})\} = \min_{k \in [2,3,4]} \{c_{12} + g(2,\{3,4\}\}, c_{13} + g(3,\{2,4\}), c_{14} + g(4,\{2,3\}\}\} = 35$ km, So the minimum distance for a tour 1 = 35 km, In addition, the optimum path is:1–2–4–3–1.

Table 2. Application of dynamic programming level per level

Level 0	Level 1	Level 2	Level 3
g(1, {2,3,4})} = min k ∈ {2,3,4] {c1K + g(K,{2,3,4}- {k})}	c12 + g(2, {3,4})} = 35	c23 + g (3, {4})} = 29	c34 + g(4, {ø})} = 20
	c13 + g(3, {2,4})} = 40	c24 + g(2, {3})} = 25	c43 + g(3, {ø})} = 15
	c14 + g(4, {2,3})} = 43	c32 + g (2, {4})} = 31	c24 + g(4, {{ø)} = 18
		c34 + g(4, {2})} = 25	c42 + g(2, {ø})} = 13
		c42 + g(2, {3})} = 23	c23 + g(3, {ø})} = 15
		c43 + g (3, {2)} = 27	c32 + g(2, {ø})} = 18

3.2 Application of Branch and Bound

Step 1: Reduce the matrix of distance: Eliminate the minimum value on the row and eliminate the minimum value on the column as mentioned in Table 3.

Table 3. Reduced matrix

		NODES			
NODES		1	2	3	4
	1	∞	0	1	5
	2	5	∞	0	0
	3	0	7	∞	1
	4	0	0	5	∞

Total reduced = 29 + 9 = 38 = C1.

Step 2: calculate the distance for each node level per level:

Level 1:

From 1 to 2: C2 = C [1, 2] + C1 + reduced value; C [1, 2] = 0; 1 rows = ∞ and 2 column = ∞ (Table 4)

Table 4. Reduced matrix

		NODES			
NODES		1	2	3	4
	1	∞	∞	∞	∞
	2	∞	∞	0	0
	3	0	∞	∞	1
	4	0	∞	5	∞

Reduced value = 0; C2 = C [1, 2] + C1 + reduced value = 0 + 38 + 0 = 38.

From 1 to 3: C3 = C [1, 3] + C1 + reduced value; C [1, 3] = 1; 1 rows = ∞ and 3 column = ∞ (Table 5)

Table 5. Reduced matrix

		NODES			
NODES		1	2	3	4
	1	∞	∞	∞	∞
	2	0	∞	∞	0
	3	∞	7	∞	1
	4	0	0	∞	∞

Reduced value = 0; C3 = C [1, 3] + C1 + reduced value = 1 + 38 + 0 = 39; so we choose the minimum value;

From 1 to 4: C4 = C [1, 4] + C1 + reduced value; C [1, 4] = 5; 1 rows = ∞ and 4 column = ∞ (Table 6)

Table 6. Reduced matrix

		NODES			
NODES		1	2	3	4
	1	∞	∞	∞	∞
	2	0	∞	0	∞
	3	0	7	∞	∞
	4	∞	∞	5	∞

Reduced value = 0; C4 = C [1, 4] + C1 + reduced value = 5 + 38 + 0 = 43;

For level 1, we choose the minimum value for each node. In our case, C2 is the minimum value.

Level 2:
From 2 to 3: C3 = C [2, 3] + C2 + reduced value; C [2, 3] = 0; 2 rows = ∞ and 3 column = ∞ (Table 7)

Table 7. Reduced matrix

		NODES			
NODES		1	2	3	4
	1	∞	∞	∞	∞
	2	∞	∞	∞	∞
	3	∞	∞	∞	1
	4	0	∞	∞	∞

Reduced value = 1; C3 = C [2, 3] + C2 + reduced value = 0 + 38 + 1 = 39.

From 2 to 4: C4 = C [2, 4] + C2 + reduced value; C [2, 4] = 1; 2 rows = ∞ and 4 column = ∞ (Table 8)

Table 8. Reduced matrix

		NODES			
NODES		1	2	3	4
	1	∞	∞	∞	∞
	2	∞	∞	∞	∞
	3	0	∞	∞	∞
	4	∞	∞	5	∞

Reduced value = 0; C4 = C [2, 4] + C2 + reduced value = 1 + 38 + 5 = 43,

For level 2, we choose the minimum value for each node. In our case, C3 is the minimum value.

Level 3:
From 3 to 4: C4 = C [3, 4] + C3 + reduced value; C [3, 4] = 0; 3 rows = ∞ and 4 column = ∞ (Table 9)

Table 9. Reduced matrix.

		NODES			
NODES		1	2	3	4
	1	∞	∞	∞	∞
	2	∞	∞	∞	∞
	3	∞	∞	∞	∞
	4	0	∞	∞	∞

Reduced value = 0; C4 = C [3, 4] + C3 + reduced value = 0 + 39 + 0 = 39,

So the optimal path for a tour from 1: 1–2–3–4–1 with a minimum distance equal to 39 km.

3.3 Application of Dijkstra Algorithm

We applied the Dijkstra algorithm to the following Table 10:

Table 10. Results obtained by applying the Dijkstra algorithm

1	2	3	4	Chosen node
1(0)	∞	∞	∞	1 (0)
	2(10)	∞	4(20)	2 (10)
		3(9)	4(20)	3(9)
			4 (20)	4(20)

So the optimal path for a tour from 1: 1–2–3–4–1 with a minimum distance equal to 39 km.

4 Results and Discussion

In this study, three methods were used: dynamic programming, Dijkstra algorithm, and B&B to optimize technicians' transportation from the company to each subcontractor and make it sustainable. The studied network contains seven nodes from node A to node H. We define the distance between every two nodes in Kilometer as mentioned in Fig. 3. Node A represents the manufacturing denim products company, and the node H represents the furthest subcontractor.

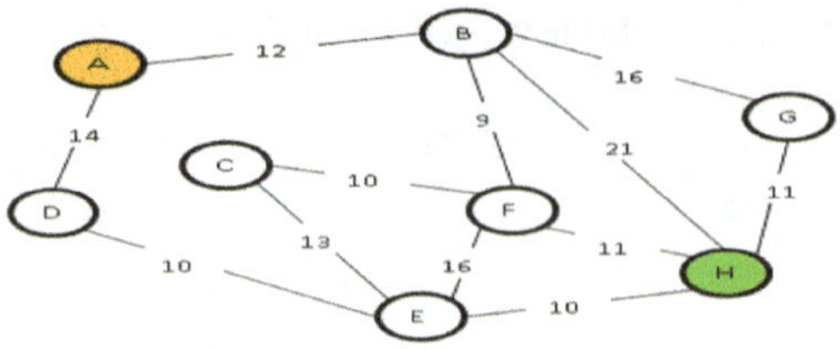

Fig. 3. Road network of subcontractors

The result of applying the three methods is mentioned in the following table:

Table 11. Results obtained by applying the Dijkstra algorithm, B&B, and Dynamic Programming

Path	Dijkstra's algorithm		Branch and bound		Dynamic programming	
	Distance (km)	Cost (dt)	Distance (km)	Cost (dt)	Distance (km)	Cost (dt)
A–E	33	25	37	25,5	24	23,5
A–F	45	28	40	27	21	25,9
A–C	50	29	31	26	31	26
A–G	43	26	45	27	28	25,5
A–H	33	26,5	39	27	32	26

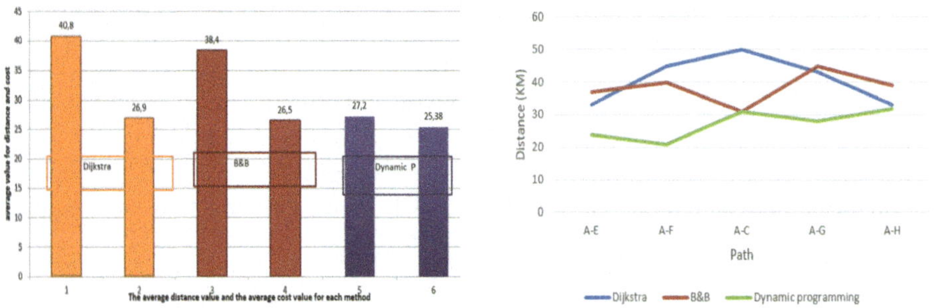

(a) (b)

Fig. 4. Effects of Dijkstra method, B&B, and dynamic programming: (a) the effect on the average value for the cost and the distance, (b) the effect on the distance.

Table 11 and Fig. 4 show that we had an optimization in terms of distance and cost for each path when we applied dynamic programming. This method allows a minimum average value in the distance equal to 27.2 km and cost equal to 25 DT than the other methods. We have an optimization at a distance with a percentage of 33% compared to the Dijkstra method and 29% compared to the B&B method. We have an optimization at the cost with a value of 4% compared to the Dijkstra method.

5 Conclusion

This paper has proved that dynamic programming is an effective method in determining the shortest path. Indeed this method offers an optimization at a distance with percentage values of 33% and 29% compared to Dijkstra and B&B methods, respectively, and optimization at the cost level with a value of 4% compared to the Dijkstra method.

References

Edsger, W.: A short introduction to the art of programming. Technische Hogeschool, Eindhoven (1971)

Hu, T.: Integer Programming and Network Flows. Addison-Wesley Pub Co. ISBN:452 (1969)

Grassin, J., Minoux, M.: Variations on a DANTZIG algorithm: application to find the shortest paths in large networks. IT Oper Res. **1**, 53–62 (1973)

Grinold, R.C.: Steepest ascent for large scale linear programs. SIAM J. **14**(3), 447–464 (1972)

Gondran, M.: Algèbre linéaire et cheminement dans un graphe, Revue française d'automatique, d'informatique et de recherche opérationnelle. Recherche opérationnelle, tome 9. **1**(9), 77–99 (1975)

Kaufmann, A., Desbazeille, G.: La méthode du chemin critique, Dunod, 170 (1964)

Noto, M., Sato, H.: A method for the shortest path search by extended Dijkstra algorithm. In: SMC 2000 Conference Proceedings, 2000 IEEE International Conference on Systems, Nashville USA (2000)

Xu, M., et al.: An improved Dijkstra shortest path algorithm for sparse network. Appl. Math. Comput. **185**, 247–254 (2007)

Thomas, H., Michael, S.: Determining all optimal and near-optimal solutions when solving shortest path problems by dynamic programming. Oper. Res. **32**(6), 1381–1384 (1984)

Advance in Polymers and Textile Biomaterials

Comparative Study of Dyeing Performances of Advanced Polyesters Filaments

Marwa Souissi[1,2](✉), Ramzi Khiari[1,3,4], Mounir Zaag[5], Nizar Meksi[1,2], and Hatem Dhaouadi[1]

[1] Faculty of Sciences of Monastir, Laboratory of «Chimie de l'Environnement et des Procédés Propres» (LCE2P - LR21ES04), University of Monastir, 5019 Monastir, Tunisia
souissi.marwa20@yahoo.com
[2] University of Monastir, National Engineering School of Monastir, 5019 Monastir, Tunisia
[3] Higher Institute of Technological Studies (ISET) of Ksar-Hellal, 5070 Ksar-Hellal, Tunisia
[4] University of Grenoble Alpes, CNRS, Grenoble INP, LGP2, 38000 Grenoble, France
[5] Société Industrielle des Textiles (SITEX), 5070 Ksar-Hellal, Tunisia

Abstract. This paper is devoted to study the chemical, physical and mechanical properties of various advanced polyesters and their influence on dyeing performances. Four kinds of filaments are used and compared: two PET monofilaments with different geometries were investigated and namely: **(A)** circular and **(B)** cross-sectional shapes, **(C)** PTT monofilaments with tetrachannel shapes, and **(D)** bicomponent PET/PTT filaments. The influence of dyeing parameters (temperature, time and carrier concentration) on dyeing performances (color strength, CIELab coordinates) was investigated and optimized using Box-Behnken experimental design. Comparing results showed that PET/PTT bicomponent filaments **(D)** have a lower percentage of crystallinity and a lower value of glass transition temperature which gives them an excellent dyestuff affinity compared to the other monofilaments.

Keywords: Polyester filaments · Bicomponent filaments · Dyeing · Box-Behnken experimental design

1 Introduction

In the last decades, considerable attention has been paid to improve the performances of synthetic fibers, expand their field areas, and facilitate their uses (Bansal et al. 2016). Among the recent advanced synthetic polymers, a new 100% PTT filament with tetrachannels cross section has been produced (Hernandez et al. 2016). This kind of filaments with special sections have many advantages as they readily wick away perspiration and allow air to flow more quickly due to their larger surface area than circular cross section fibers while ensuring a pleasant touch and a feeling of comfort for the wearer (Oh 2006). Moreover, bicomponent polyester filaments, such as polyethylene terephthalate/polytrimethylene terephthalate (PET/PTT) filaments, have been gaining importance recently due to their great extensibility and their excellent elastic recovery (Wenjing et al.

S. Msahli and F. Debbabi (Eds.): CIRATM 2020, SPM 17, pp. 261–267, 2022.
https://doi.org/10.1007/978-3-031-08842-1_42

2010; Sihaiet al. 2011). These bicomponent filaments are extruded from the same spinneret, adjacent and arranged side by side. In this context, this work presents an original study by establishing a comparative analysis of chemical and mechanical properties, and dyeing performances between four innovative mono and bicomponent filaments: three different geometries shapes of monofilaments were investigated: (A) circular (PET), (B) cross-sectional (PET) and (C) tetra-channels (PTT) shapes, and (D) bi-component PET/PTT filaments. The first part, the mainly characteristics of examined filaments using different tools such as SEM, DSC and XRD analysis were discussed. The second part, an adequate dyeing process for each filament was developed and optimized using response surface methodology (RSM). Finally, the dyeing performances were evaluated and discussed.

2 Materials and Methods

2.1 Textile Support

Based on each of studied multifilament yarns, jersey knits are made with circular knitting machine type Tricolab gauge 12 (Sodemat, France). The obtained fabrics are used to optimize dyeing process and evaluate dyeing performances.

2.2 Dyeing Procedure and Color Measurement

Knitted fabrics were dyed with CI Disperse Yellow 211 (concentration of 1.2%), a liquor-to-fiber ratio equal to 10:1 and a pH value of dyebath equal to 7. Colors of dyed samples were measured using a spectrophotometer Spectraflash 600 Plus (Datacolor, USA).

2.3 Experimental Design

Box-Behnken experimental design allows to analyze the effect of the three studied factors (dyeing temperature, dyeing time, and quantity of carrier added to the dye bath) and to deduce the optimal dyeing conditions maximizing dyeing performances. The analyzed response is the color yield (*K/S*) of the dyed samples. Factors and levels used in the experimental design are listed in Table 1.

Table 1. Factors and levels used in this study.

Factors	Units	Variation levels		
		Low level (−1)	Middle level (0)	High level (+1)
Dyeing temperature	°C	110	120	130
Dyeing time	min	25	50	75
Concentration of carrier	g/L	0	6	12

3 Results and Discussions

3.1 Morphological Analysis of Studied Filaments

Table 2 shows the different SEM catches of jersey knits made for the four studied multifilaments yarns as well as lengthwise views of these multifilaments. Obtained micrographs confirm the geometry shape of monofilaments. In fact it is clear that filaments (A) have a conventional circular section, filaments (B) have a section in the form of a cross and filaments (C) have a flat section which consists of four adjacent channels hence the name tetrachannels. Concerning bicomponent filaments (D), it can be seen that there are two spun filaments side by side with unequal proportions (in our case 60% PET and 40% PTT).

Table 2. Scanning electron micrographs of studied filaments.

Filaments	Configuration of filaments in the knitted fabrics	Lengthwise view of filaments
(A)		
(B)		
(C)		
(D)	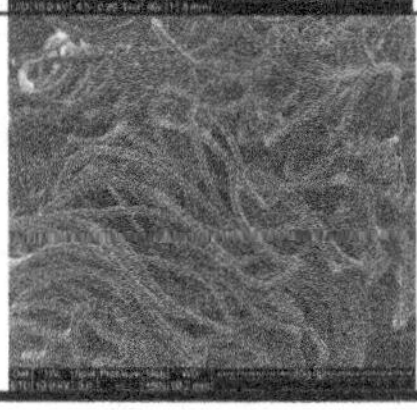	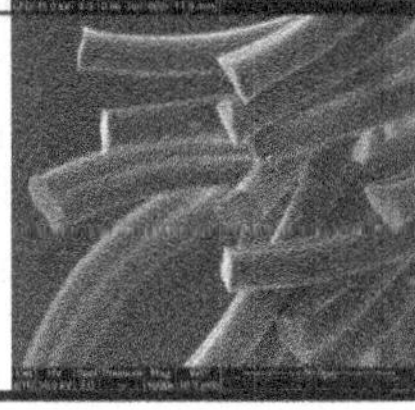

3.2 DSC and DRX Analysis of Studied Filaments

Obtained results show that melting temperatures in the case of PET filaments with different sectional shapes (circular and cross-sectional shapes) were relatively close with values equal to 249 °C and 253 °C, respectively. For the monofilament (C) which is composed of 100% PTT, the melting temperature is equal to 231 °C. Concerning bi-component filaments (D), DSC analysis shows the existence of two principal peaks. The first peak is located at a temperature equal to 248 °C. It corresponds to the melting point of the polyethylene terephthalate (PET). The second important peak is observed at 222 °C, and corresponds to the melting point of (PTT).It can be concluded that the bicomponent filaments (D) have the lowest value of (T_g) (equal to 49 °C) compared to others monofilaments, which proves that their dye affinity is quite important. While the other monofilaments which have fairly large values of (T_g) equal to 52 °C (C), 62 °C (B) and 62 °C (A), respectively, are much more difficult to dye. From DRX analysis, it can be seen that in the case of bicomponent filaments (D), the crystallinity is equal to 57% which is less important than crystallinity values of monofilaments. So filaments (D) have more amorphous zones which make them more suitable for dyeing and this also give them a great elasticity which is attributed to the arrangement and orientation of polymers in the chain (Table 3).

Table 3. Results of DSC and XRD analysis obtained for studied filaments.

Filaments	DSC analysis			XRD analysis						
	T_m (°C)	T_c (°C)	T_g (°C)	$2\theta_1$(°)	$2\theta_2$(°)	$2\theta_3$ (°)	Cr (%)	a (Å)	b (Å)	c (Å)
(A)	249	199	62	16.9	23.4	25.7	60.0	4.85	6.03	12.81
(B)	253	197	62	17.6	22.7	25.5	58.2	4.84	5.90	13.13
(C)	231	179	52	17.4	22.5	25.7	61.9	4.80	5.90	13.13
(D)	222 (PTT) 248 (PET)	185 (PTT) 163 (PET)	37 (PTT) 62 (PET)	18.8	23.4	25.5	56.6	4.84	6.08	13.40

Where: T_m, T_c and T_g are the melting temperature, the crystallization temperature, and the glass transition temperature, respectively; θ_1, θ_2, and θ_3 are the peak diffraction angles; Cr is the degree of crystallinity; a, b and c are the unit cell parameters.

4 Evaluation of Dyeing Performances

The objective is to develop an adequate dyeing process for studied filaments. The use of the Box-Behnken experimental design allowed us to analyze the effect of dyeing conditions (dyeing temperature, dyeing duration and concentration of carrier added to the dye bath) on the analyzed response (*K/S*) and interactions between them.

4.1 Study of the Effects of Dyeing Parameters

Figure 1 shows the main effects plots of the studied parameters. Obtained results showed that dyeing temperature has considerable effect on the dyeability of all studied filaments with CI Disperse Yellow 211 dye. However, the present study promotes a more economical dyeing process in terms of temperature by adding a phthalamide-based non-toxic carrier; it allows to reduce dyeing temperature by 15 °C (120 °C instead of 135 °C) while obtaining the same results from point of view (*K/S*).The dyeing duration has also a great influence on the obtained values of *(K/S)*. The optimum dyeing duration required for the disperse dye molecules to insert into the amorphous areas of the filaments with special sections (50 min for filaments (B), (C) and (D)) is lower compared to the case of the conventional PET monofilament (A) with circular section (75 min). In addition, obtained results showed that in the case of the filament (A), the adding of carrier agent in the dye bath has a negative effect on the color strength of dyed samples. For the other monofilaments (B) and (C) having special morphologies, the carrier concentration has a great effect for the dyeability of samples. Concerning the PET/PTT bicomponent filaments (D), it is observed that the necessary carrier concentration does not exceed 6 g/L and thus a higher *(K/S)* value is obtained. Beyond this value, the adding of carrier has a little effect on the *(K/S)* response.

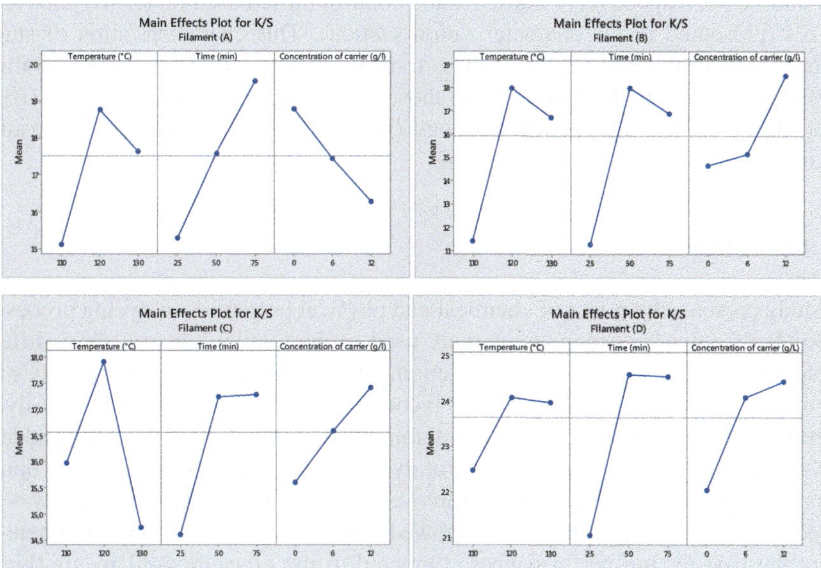

Fig. 1. Mains effects plots for color yield (*K/S*) of studied filaments dyed with CI Disperse Yellow 211.

4.2 Colors of Dyed Filaments Under Optimum Conditions

Table 4 shows the optimal dyeing conditions, predicted and experimental values of color strength (*K/S*). It also shows the CMC color differences (in D65/10°) between samples dyed under these optimal dyeing conditions. In fact, each dyeing was repeated three times in three different pots to verify the reproducibility of dyeing under these optimal conditions. All the CMC color differences obtained are less than 1; this proves the good dyeing reproducibility of studied filaments.

Table 4. Optimal dyeing conditions and obtained values of *(K/S)* and color differences (in D65/10°) of studied filaments dyed using CI Disperse Yellow 211.

Filaments	Optimal dyeing conditions			Predicted values of K/S	Experimental values	
	Temperature (°C)	Time (min)	Conc. of carrier (g/L)		K/S	ΔECMC(2:1)
(A)	122.95	64.34	10.69	19.40	19.10 ± 0.08	0.14
(B)	115.00	57.85	12.00	21.43	21.14 ± 0.22	0.72
(C)	110.00	50.00	12.00	19.92	19.80 ± 0.31	0.75
(D)	113.26	50.58	8.78	25.28	24.76 ± 0.12	0.66

Obtained results showed that bicomponent filaments (D) have (*K/S*) value equal to 24.76 which is higher than those of monofilaments (A), (B) and (C) (equal to 19.10, 21.14 and 19.80, respectively).These results confirm those based on (DSC) and (XRD) analyzes (presented in the characterization section). This characterization of studied filaments showed that the values of glass transition temperature (T_g) and cristallinity rate of filament (D) are much lower than those of the other monofilaments (A), (B) and (C). So, the PET/PTT bicomponent filaments (D) have larger amorphous zones, resulting in better dye penetration and a darker hue.

5 Conclusion

This study presents the effect of chemical and physical properties in dyeing process for four kinds of polyesters filaments largely used in the textile industry: three different monofilaments with (A) PET circular section, (B) PET cross-sectional shape, (C) PTT tetrachannel section, and (D) PET/PTT bicomponent filaments. The effect of dyeing conditions (dyeing temperature, dyeing duration and concentration of carrier added to the dye bath) on the color strength *(K/S)* of dyed samples was analyzed and the optimal dyeing conditions were determined. Moreover, the reproducibility of dyeing was also verified and proved. Finally, the results showed that PET/PTT bicomponent filaments(D) present the best dyeing performances compared to the other monofilaments (highest value of (*K/S*)) and more ecological dyeing process (saves of time, energy and auxiliary products).

References

Bansal, S., Raichurkar, P.: Review on the manufacturing processes of polyester-PET and nylon-6 filament yarn. Int. J. Textile Eng. Proces. **2**, 23–28 (2016)

Hernandez, I,A,. Hietpas, G.D,. Howell, J.M,. Schultze, C.: Poly(trimethylene terephthalate) tetrachannel cross-section staple. US 6,458,455B1 (2016)

Oh, T.: Studies on melt spinning process of hollow polyethylene terephthalate fibers. Polymer Eng. Sci. **46**, 609–616 (2006)

Wenjing, L., Jin, L., Mei, W.: Crimp and tensile properties of PTT/PET self-crimping staple fiber, Synth. Fiber China. 1, 27–31 (2010)

Sihai, C., Shanyuan, W.: Latent-crimp behavior of PET/PTT elastomultiester and a concise interpretation. J. Macromol. Sci. Part B Phys. **50**, 1447–1459 (2011)

Infrared Spectra for Alfa Fibers Treated with Thymol

Arwa Turki(✉), Asma El Oudiani, Slah Msahli, and Faouzi Sakli

Laboratory of Textile Engineering, University of Monastir, Monastir, Tunisia
turkiarwa@hotmail.fr

Abstract. In this work, we attempt to analyze the hydrogen bond network of cellulose by comparing the 2nd derivative infrared spectra of Alfa fiber treated with different concentrations of thymol. It was proved, from this study, that the 2nd derivative infrared spectrum is an effective method to gain further insight on OH bond system of cellulosic materials and to clarify some grey areas in terms of infrared band assignment. The careful analysis of 2nd derivative infrared spectra of untreated and thymol treated Alfa fibers revealed an increase in band intensity related to both intermolecular and intramolecular hydrogen bonds and a decrease in bands related to free hydroxyl groups.

Keywords: FTIR spectrum · Cellulose · Alfa fiber · Thymol · Hydrogen bonds

1 Introduction

Nowadays, development of eco-friendly materials gives rise to a new generation of packaging made with cellulosic fibers. To enhance some antimicrobial properties of these packaging, fibers were treated with essential oils such as lemongrass (Cybopogon citratus), rosemary pepper (Lippia sidoides), basil (Ocimum gratissimum) oregano, Zataria multiflora…(Bastos et al. 2016; Kristo et al. 2008; Liakos et al. 2015, 2016; Shakeri et al. 2011). Among these great variety of essential oils, thymol (also known as 2-isopropyl-5-methylphenol, IPMP) is widely used in food packaging as a nature substance with strong bactericidal and anti-fungal properties (Galotto et al. 2016; Leite et al. 2010; Milovanivic et al. 2015; Zamani et al., 2015). Extracted from Thymus vulgaris (thyme) and some other varieties of plants, thymol is a "phenolic" compound with the chemical composition of $C_{10}H_{14}O$. Thymol molecule has a hydroxyl group attached to a benzene-type ring. Thanks to this aromatic ring, thymol is very polar and reacts irreversibly with many materials such as cellulosic fibers. Indeed, as cellulose is composed of hydroxyl groups, thymol may irreversibly form stable ethers with these –OH groups.

Early studies on glucose and polysaccharides showed that cellulose is a polysaccharide comprising β (1 → 4) linked D-glucopyranose units. Cellulose formula is (C 6 H 10 O 5) n, and it is a homopolymer with a several hundred to many thousands of units. Crystallographic study of D- glucose and cellobiose demonstrated that the D-glucose residues have the 4C1 chair conformation (Atalla and VanderHart 1984; Meyer and Misch 1937). This specific conformation and glycosidic linkages of cellulose units

S. Msahli and F. Debbabi (Eds.): CIRATM 2020, SPM 17, pp. 268–275, 2022.
https://doi.org/10.1007/978-3-031-08842-1_43

result in two chemically different chain ends: the first end comprises an anomeric carbon atom involved in a glycosidic linkage, whereas in the second chain end, the anomeric carbon atom is free (see Fig. 1).

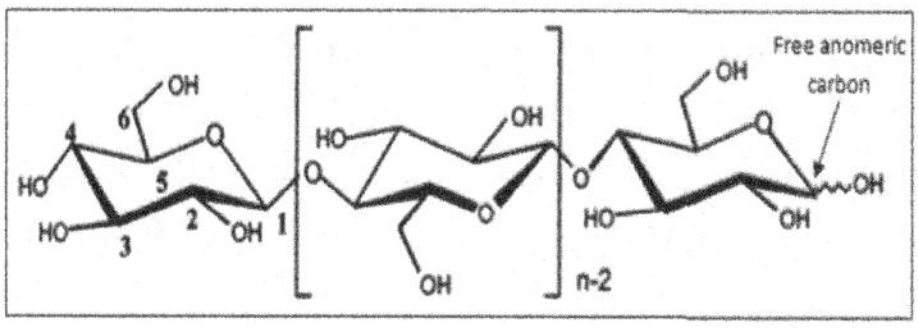

Fig. 1. Representation of cellulose chain.

In this work, FTIR spectroscopy was used to fully analyze the way of incorporation of thymol molecules into cellulosic structures. Through a careful study of infrared spectra for untreated and thymol treated cellulosic fibers, we have tried to analyze and properly understand the way of incorporation of thymol into cellulosic structure. Besides, this work is an attempt to clarify some uncertainties regarding infrared assignments of hydrogen bonds.

2 Materials and Methods

2.1 Extraction of Ultimate Alfa Fibers

The Alfa plant is formed by several circular bundles of filaments. To extract individual fibers from these filaments, we have used mixed treatment involving NaOH (sodium hydroxide) and H_2O_2 (hydrogen peroxide) since it gives the best separation results as demonstrated by Ben Marzoug et al. (2010). In this method, Alfa filaments were placed in a bath solution containing 25 g/L of stabilizing agent, 3 g/L of wetting agent, 30 g/L NaOH (sodium hydroxide, 98%) and 3 5mL/L H_2O_2 (hydrogen peroxide, 35%) at a temperature of 120 °C during 90 min. All chemical products were purchased from Chimitex, Tunisia. The treatments were made with a dyeing machine (Mathis) under pressure and during continuous mixing. After each treatment, individual fibers were rinsed with tap water and dried at room temperature. The obtained individual Alfa fibers are characterized by a length between 0.2 and 3 mm, a diameter of 8 μm, a density of 1.2 and a crystallinity of 80%. Furthermore, it has been proved that this fiber is rich in cellulose and contains a very low lignin content (<1%) (Ghali et al. 2006).

2.2 Thymol Treatment of Alfa Fibers

Thymol (2-Isopropyl-5-methylphenol, IPMP), a white crystalline substance, was purchased from Sigma-Aldrich and meets analytical specification of Ph. Eur., BP, NF. It has the following characteristics: CAS number: 89-83-8; chemical formula: $C_{10}H_{14}O$; molecular weight MW: 150.22; density: 0.965 g/mL at 25 °C (M.); purity: 99–101%.

Ultimate Alfa fibers were treated by impregnation in thymol solutions with different concentrations (2, 4, 6, 8 and 10% w/w). The temperature of all treatments was set at 50 °C and their duration was varied (2 min and 30 min). Fibers' drying was achieved in an oven at a temperature of 40 °C during 60 min.

2.3 Infrared Analyses

Attenuated total reflection (ATR) is a technique used in conjunction with infrared spectroscopy that enables samples to be passively examined in the solid or liquid state without any preparation. ATR-FTIR is used in many fields since it is a nondestructive, direct and rapid method. A spectrophotometer of type PerkinElmer UATR (Single Reflection Diamond) with an attenuated total reflectance (ATR) attachment was used to undertake infrared spectra analysis. 100 scans were taken per sample with a resolution is 2 cm^{-1}. The infrared spectra were recorded in the range of 4000–500 cm^{-1}.

2.4 Preprocessing of IR Spectra

2.4.1 Spectra Normalization

Normalization helps give all samples an equal impact on the model. Without normalization, some samples may have such severe multiplicative scaling effects that they will not be significant contributors to the variance and, as a result, will not be considered important by many multivariate techniques.

In this work, we have used the multiplicative signal correction (MSC) method to normalize infrared spectra of which the basic concept is to remove non-linearities in the data caused by scatter from particulates in the sample.

2.4.2 Noise Removal and Baseline Correction

Noise represents random fluctuations around the signal that may originate from the instruments or environmental laboratory conditions. In this work, we have used the Fast Fourier Transform (FFT) algorithm available in Origin 6.0. Besides, spectra were baseline corrected with a segmented linear baseline linking the following frequencies: 380, 1185, 1515, 2790, 3660 cm^{-1}. Processing of spectra, baseline corrections, and derivative spectra were generated using software resolution program (Origin 6.0).

3 Results and Discussion

Figure 2a represents the normalized FTIR spectra relating to the different samples of treated Alfa fibers. Only the region of interest was retained for further analysis. This region corresponds to the stretch vibration of OH bonds between 2800 and 3800 cm^{-1}. It is clear from Fig. 2b, that the OH band area and intensity is altered with the treatment of fibers. Figure 2c represents the 2nd derivative infrared spectrum for the region of 2800–3800 cm^{-1}. This figure shows many peaks that correspond to different mode of OH bending namely those related to interchain, intrachain and free OH bonds (Fig. 2d).

Bands at 2850 cm^{-1} and 2917 cm^{-1} (Fig. 3) are assigned to symmetric and asymmetric CH stretching in aromatic methoxyl groups and in methyl and methylene groups of side chains (El oudiani et al. 2017; Pandey and Pitman 2003; Schwanninger et al. 2004). Since cellulose molecules are free of these types of groups, the increase of intensity of the bands at 2850 cm^{-1} and 2917cm^{-1} gradually with concentrations may be explained by the incorporation of thymol molecules containing methyl groups into cellulosic chains. This is to ensure the compatibility between thymol and cellulose molecules.

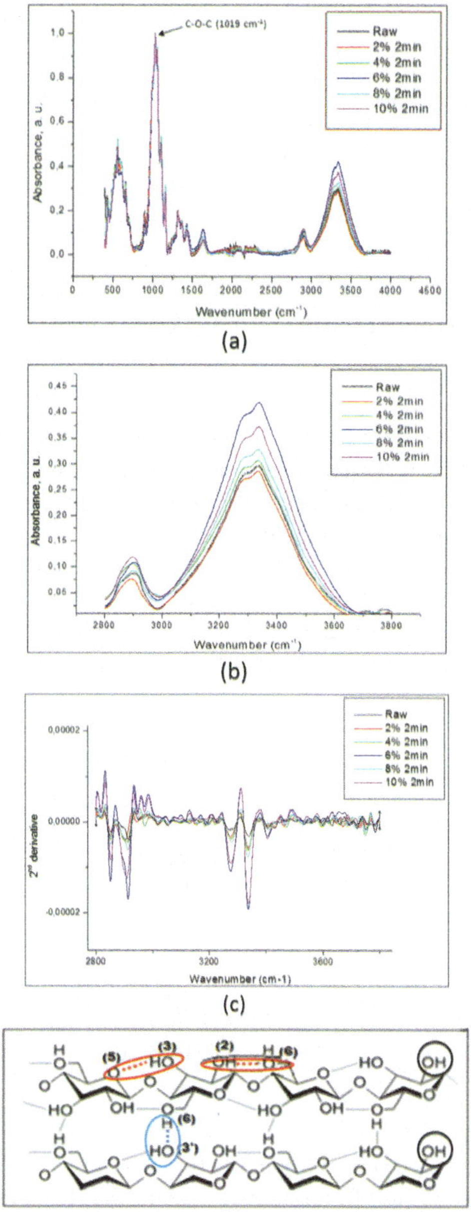

Fig. 2. FTIR spectra of Alfa fibers treated with different concentrations of thymol, (a)Normalized FTIR spectra of treated Alfa fibers; (b) 2800–3800 cm^{-1} OH band region, (c) 2nd derivative spectra for OH band region; (d) Representation of inter (blue), intra(red) and free (black) hydrogen bonds in cellulose.

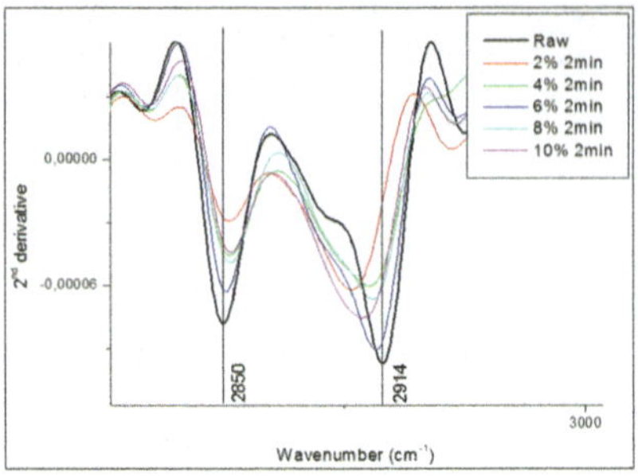

Fig. 3. Variation of 2850 cm^{-1} and 2916 cm^{-1} FTIR bands in the 2nd derivative spectra with thymol concentration.

Figure 4 presents the 3200–3400 cm^{-1} region of hydrogen bond stretching. The bands at 3273cm^{-1} and 3337 cm^{-1} that correspond to intramolecular OH bonds, O(6)H…O(2) and O(3)H…O(5) respectively, show a notable increase in intensity when the fiber is subjected to thymol treatment.

Thus, the incorporation of thymol into fibers leads to the formation of new intramolecular hydrogen bonds because it was proven that these intramolecular hydrogen bonds are quickly developed to provide more stabilization when the structure is subjected to a certain change (Nishiyama et al. 2003, 2008).

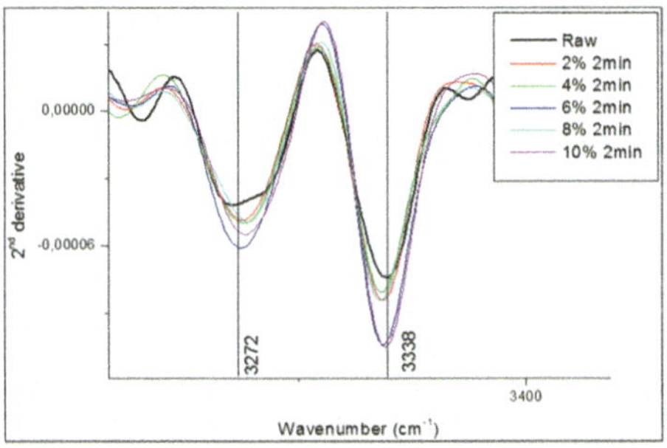

Fig. 4. Variation of 3272 cm^{-1} and 3338 cm^{-1} FTIR bands in the 2nd derivative spectra with thymol concentration.

In this work, we can argue that the bands between 3400 and 3500 cm^{-1} are due to the intermolecular hydrogen bonds formed between cellulose chains and thymol molecules. As shown in Fig. 5, bands at 3466 cm^{-1} and 3430 cm^{-1} do not exist for untreated fiber, they only appear with a relatively high intensity once the fiber is subjected to thymol treatment.

It must be pointed out that red and green peaks that correspond to low thymol concentrations (2% and 4%) are the most intense ones at 3466 cm^{-1} and there is no significant effect of high concentrations for this particular wavenumber. However, opposite effect is observed for lower wavenumbers.

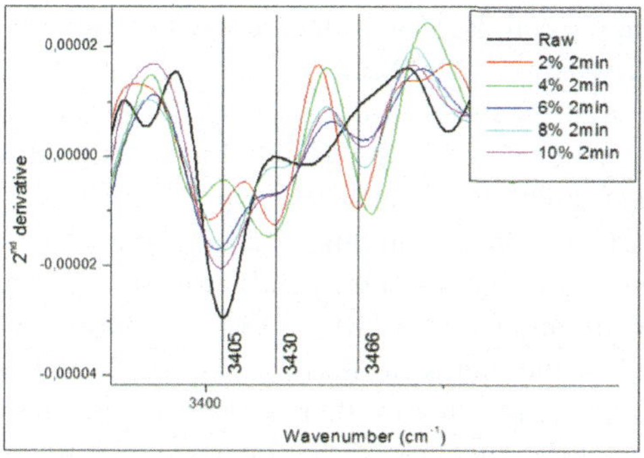

Fig. 5. Variation of 3405 cm^{-1}, 3430 cm^{-1} and 3466 cm^{-1} FTIR bands in the 2nd derivative spectra with thymol concentration

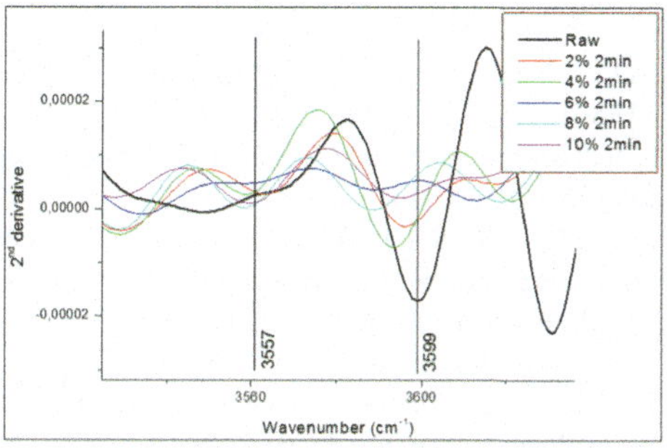

Fig. 6. Variation of 3569 cm^{-1} and 3599 cm^{-1} FTIR bands in the 2nd derivative spectra with thymol concentration.

Thus, bands at 3466 cm^{-1} corresponding to weak intermolecular hydrogen bonds are formed for low thymol concentrations, whereas bands at 3405 cm^{-1} may be assigned to strong intermolecular H bonds formed for high thymol concentrations (Lee et al. 2015; Fernandes et al. 2011; Thomas et al. 2013).

Peaks in the range of 3500–3600 cm^{-1} are generally attributed to stretching vibration of the free OH groups (Kondo 1997, 2005; Schwanninger et al. 2004; Watanbe et al. 2006; Silveerstein et al. 1981).

As shown in Fig. 6, primary alcohols appear at around 3590 cm^{-1} (OH(2) and OH(3)) and secondary ones at 3560 cm^{-1} (OH(6)). Reducing of these bands by fiber treatment may be explained by the decrease in the number of free OH groups that interfere with thymol molecules to form intermolecular hydrogen bonds. Thus, we can conclude that the OH stretching bands in the range of 3500–3600 cm^{-1} (free OH groups) are replaced by bands around 3400–3500 cm^{-1} (intermolecular OH bonds).

4 Conclusions

FTIR spectroscopy is a widely used technique to analyze fiber structure. It was proved that the 2nd derivative Infrared spectrum is a useful method for identifying the least changes taking place in the cellulosic supramolecular structure. In the present work, we focus on the region of 2800 cm^{-1}–3800 cm^{-1} which corresponds to OH group vibration. The 2nd derivative infrared spectrum comprises many different peaks corresponding to intermolecular, intramolecular and free OH bonding. We noticed an increase in the intensity of the bands at 2850 cm^{-1} and 2917 cm^{-1} which may be explained by the incorporation of thymol molecules containing methyl groups into cellulosic chains. The same increase in intensity was observed for bands related to intramolecular (3272cm^{-1} and 3338 cm^{-1}) and intermolecular (3405, 3430 and 3466 cm^{-1}) hydrogen bonds. However, the intensity of the bands in the range of 3500–3600 cm^{-1} corresponding to free hydroxyl groups is reduced with fiber treatment. This may be explained by the decrease in the number of free OH groups that interfere with thymol molecules to form intermolecular hydrogen bonds.

References

Atalla, R.H., VanderHart, D.L.: Native cellulose: a composite of two distinct crystalline forms. Science **223**, 283–285 (1984)

Bastos, M.S.R., et al.: Physical and mechanical testing of essential oil-embedded cellulose ester films. Polym. Testing **49**, 156–161 (2016)

Ben Marzoug, I., Sakli, F., Roudesli, S.: Separation of ultimate and technical esparto grass fibres: comparison between extraction methods. J. Text. Inst. **101**(12), 1050–1056 (2010)

El Oudiani, A., Msahli, S., Sakli, F.: In-depth study of agave fiber structure using Fourier transform infrared spectroscopy. Carbohyd. Polym. **164**, 242–248 (2017)

Galotto, M., López De Dicastillo, C., Torres, A., Guarda, A.: Thymol: use in antimicrobial packaging. Antimicro. Food Package **2016**, 553–562 (2016)

Ghali, L., Zidi, M., Roudesli, S.: Physical and mechanical characterization of technical esparto (ALFA) fibres. J. Appl. Sci. **6**, 2450–2455 (2006)

Kondo, T.: The assignment of IR absorption bands due to free hydroxyl groups in cellulose. Cellulose **4**(4), 281–292 (1997). https://doi.org/10.1023/A:1018448109214

Kondo, T.: Hydrogen bonds in cellulose and cellulose derivatives. In: Dumitriu S (ed.) Polysaccharides, CRC Press, New York (2005). ISBN 3-540-37102-8

Kristo, E., Koutsoumanis, K.P., Biliaderis, C.G.: Thermal, mechanical and water vapor barrier properties of sodium caseinate films containing antimicrobials and their inhibitory action on Listeria monocytogenes. Food Hydrocolloids **22**(3), 373–386 (2008)

Lee, C.M., Kubicki, J.D., Fan, B., Zhong, L., Jarvis, M.C., Kim, S.H.: Hydrogen-bonding network and OH stretch vibration of cellulose: comparison of computational modeling with polarized IR and SFG spectra. J. Phys. Chem. **119**(49), 15138–15149 (2015)

Leite de Souza, E., Carneiro de Barros, J., Vasconcelos de Oliveira, C.E., Lúcia da Conceição, M.: Influence of Origanum vulgare L. essential oil on enterotoxin production, membrane permeability and surface characteristics of Staphylococcus aureus. Int. J. Food Microbiol. **137**, 308–311 (2010)

Liakos, I.L., et al.: Fibrous wound dressings encapsulating essential oils as natural antimicrobial agents. J. Mater. Chem. **3**, 1583–1589 (2015)

Liakos, I.L., et al.: Antimicrobial lemongrass essential oil—copper ferrite cellulose acetate nanocapsules. Molecules **21**, 520–530 (2016)

Meyer, K.H., Misch, L.: Positions des atomes dans le nouveau modèle spatial de la cellulose. Helv. Chim. Acta **20**(1), 232–244 (1937)

Milovanovic, S., Stamenic, M., Markovic, D., Ivanovic, J., Zizovic, I.: Supercritical impregnation of cellulose acetate with thymol. J. Supercrit. Fluids **97**, 107–115 (2015)

Nishiyama, Y., Sugiyama, J., Chanzy, H., Langan, P.: Crystal structure and hydrogen bonding system in cellulose 1(alpha), from synchrotron X-ray and neutron fiber diffraction. J. Am. Chem. Soc. **125**, 14300–14306 (2003)

Nishiyama, Y., Johnson, G.P., French, A.D., Forsyth, V.T., Langan, P.: Neutron crystallography, molecular dynamics, and quantum mechanics studies of the nature of hydrogen bonding in cellulose I-beta. Biomacromol **9**, 3133–3140 (2008)

Pandey, K.K., Pitman, A.J.: FTIR studies of the changes in wood chemistry following decay by brown-rot and white-rot fungi. Int. Biodeterior. Biodegradation **52**(3), 151–160 (2003)

Schwanninger, M., Rodrigues, J.C., Pereira, H., Hinterstoisser, B.: Effects of short- time virbratory ball milling on the shape of FTIR spectra of wood and cellulose. Vib. Spectrosc. **36**, 23–40 (2004)

Shakeri, M.S., Shahidi, F., Toosi, S.B., Bhrami, A.: Antimicrobial activity of Zataria multiflora Boiss. Essential oil incorporated with whey protein based films on pathogenic and probiotic bacteria. Int. J. Food Sci. Technol, **46**, 3, p. 549 -554 (2011)

Silverstein, R.M., Bassler, G.C., Morrill, T.C.: Spectrometric Identification of Organic Compounds, 4th edn. John Wiley & Sons, New York (1981)

Thomas, L.H., et al.: Structure of cellulose microfibrils in primary cell walls from collenchyma. Plant Physiol. **161**, 465–476 (2013)

Watanbe, A., Morita, S., Ozaki, Y.: Temperature-dependent structural changes in hydrogen bonds in microcrystalline cellulose studied by infrared and near-infrared spectroscopy with perturbation correlation moving-window two-dimensional correlation analysis. Appl. Spectrosc. **60**, 611–618 (2006)

Zamani, Z., Alipour, D., Moghimi, H.R., Mortazavi, S.A.R., Saffary, M.: Development and evaluation of thymol microparticles using cellulose derivatives as controlled release dosage form Iran. J. Pharma. Res **14**(4), 1031–1040 (2015)

Physical, Chemical and Surface Properties of Alkali-Treated Kenaf Fiber

Yosr Ben Mlik(✉), Mounir Jaoudi, Foued Khoffi, and Slah Msahli

Textile Engineering Laboratory, University of Monastir, Monastir, Tunisia
yosrbenmlik@gmail.com

Abstract. This paper deals with the effect of alkaline treatment on the kenaf fiber physico-chemical and mechanical properties. The kenaf stems was treated by sodium hydroxide solution (NaOH) at different conditions. The chemical extraction allows the elimination of non cellulosic materials but also destroy the cellulosic chains at high extraction conditions (temperature, concentration and treatment duration). The results an improvement on the mechanical properties especially for tenacity and this also proved by the cristallinity index that was 57% for crude fiber pass to 72% for treated fibers. The results show that the chemical structure of the extracted fiber is the same with a decrease of the amount of non cellulosic material on the fibers.

Keywords: Kenaf · Alkali treatment · Fiber mechanical properties · DRX · FTIR

1 Introduction

Cellulosic fibers have recently attracted the attention of scientists because they are the most abundant renewable and biodegradable resources. That's why scientists worldwide have begun to show interest in exploiting the full potential of natural fibers and their uses (Mylsamy 2011; Chattopadhyay et al. 2012). There are plenty of renewable resources obtainable from the plant kingdom (Ashish et al. 2015) and a vast resource for different natural fibers such as Jute, Banana, Coir, etc., which are abundantly available in many parts of the world. However, there are still a number of other vegetable fibers which have not been used as textile fibers. One of the abundant sources of natural fiber is kenaf plant. Therefore, this study aims to study the effect of chemical treatment on the kenaf fibers morphological, mechanical and chemical properties.

2 Material and Methods

2.1 Sampling

Kenaf (Hibiscus Cannabinus L.) used in this study were cultivated in experimental field in semi-arid region in Tunisia. Kenaf stalks are air-dried and stored under standardized conditions. To extract fibers and the kenaf stem was cut at a length of 10 cm.

S. Msahli and F. Debbabi (Eds.): CIRATM 2020, SPM 17, pp. 276–282, 2022.
https://doi.org/10.1007/978-3-031-08842-1_44

2.2 Extraction Method

The chemical extraction of vegetable fibers is generally performed using sodium hydroxide. The extraction was carried out in a closed bath containing kenaf stem in NaOH solution at liquor ratio of 1/40. The extraction was carried in different conditions of time, temperature and NaOH concentration as shown in the Table 1.

Table 1. Summary of the chemical extraction conditions

Parameters	Min	Max	Step
NaOH concentration (N)	0,5	1,5	0,5
Treatment time (min)	120	240	60
Temperature (°C)	80	120	20

At the end of the chemical process, the treated fibers were thoroughly washed in warm hot water to remove dissolved substances. Then the fibers were neutralized with 10 ml/L acetic acid (Song et al. 2006; Columbus et al. 1994; Guzman et al. 1982; Henriksson et al. 1998).

2.3 Fibers Characterization

2.3.1 Morphological Properties

In order to study the morphology of the extracted fibers and to evaluate changes in the surface due to the chemical treatment, crude fibers and treated fibers were analyzed by scanning electron microscopy (SEM) Hitachi S-2360N.

2.3.2 Finesse Measurement

The linear density was determined using the gravimetric method according to the standard NF G07-007.

2.3.3 Mechanical Properties

The mechanical properties were determined according to the NF G07-002, 50 tests were carried out to determine the maximum strength that a fiber can support and the corresponding strength and elongation.

2.3.4 Chemical Structure

FTIR spectra of the kenaf fiber samples were performed. The spectra were recorded over the range 400–4000 cm^{-1}. The IR spectroscopy was applied to determine the functional groups and chemical structure of the kenaf fibers. Infrared radiation will excite molecular vibrations within a material; the frequencies of these vibrations, and hence the absorption peaks in the spectrum, are characteristic of the chemical composition of the specimen specimen (Morrison,et al. 1976; Gardette 1996; Messaoud 2011).

2.3.5 Crystallinity Index

The degree of crystallinity, an average property, is the fraction of the crystalline content in the sample under consideration. Different methods to calculate the degree of crystallinity of cellulose from X-ray diffraction spectra have been published (Dallel 2012). The method used in this work to calculate the crystallinity index was based on calculating the ratio of intensity at $2\Theta = 22{,}8°$ and $2\Theta = 18{,}5°$ which gives the ratio of crystallinity to amorphous fractions (Bansal et al. 2010).

$$CrI = {}^{100 \times (I_{200} - I_{Am})}\big/_{I_{200}}$$

where:
CrI: Crystallinity index
I_{Am}: the minimum in intensity above baseline at $2\Theta = 18{,}5°$
I_{200}: the maximum in intensity above base line at $2\Theta = 22{,}8°$
2Θ: the diffraction angle

3 Results and Discussion

3.1 Morphological and Physical Properties

Figure 1 features the evolution of kenaf fiber surface at different stages of alkali treatment and the degree of cellulose ultimate fiber separation from the rest of lignocellulosic material existing in the natural fiber. It is clear that the fibers are degraded after prolonged exposure to high concentrations of sodium hydroxide.

For low NaOH treatment condition (T = 80 °C, C = 1N for 180 mn), precipitation of the dissolved lignin on the fiber surface still exist on the fiber surface. When increasing the treatment conditions, it allows cleaning more the fiber surface; So, separation performance improves. For higher NaOH treatment conditions (T = 100 °C, C = 1,5N for 120 mn), a larger removal of the non-cellulosic materials was observed and the cellulose ultimate fibers looked better defined.

Figure 2 presents the linear density of all obtained fibers. We notice that the linear density varied depending on the fiber extraction parameters. After alkali treatment, there is a reduction in the fiber mass per unit length, which can be explained by the removal of foreign material from the fiber surface. The finest fibers are obtained when working at high extraction conditions. The linear density passes from 21,76tex to 10,31tex and this not due only to the removal of non cellulosic materials, but also to the separation of the ultimate fibers as shown in the SEM images Fig. 1.

Table 2 presents a summary of the mechanical properties of the extracted fibers. The results show that the fibers tenacity increased with the duration of the treatment and the concentration and this is due to the removal of impurities and the arrangement of the cellulosic chain. Compared to the untreated fiber, the strain of the treated fiber decreases this could be explained by the rearrangement of the macromolecular chains which block intramolecular movement. But we noted that fibers obtained by sever chemical extraction conditions present low tenacity and strain compared to crude fiber this could be explained by the fact that the chemical extraction attacks not only gummy materials (lignin and

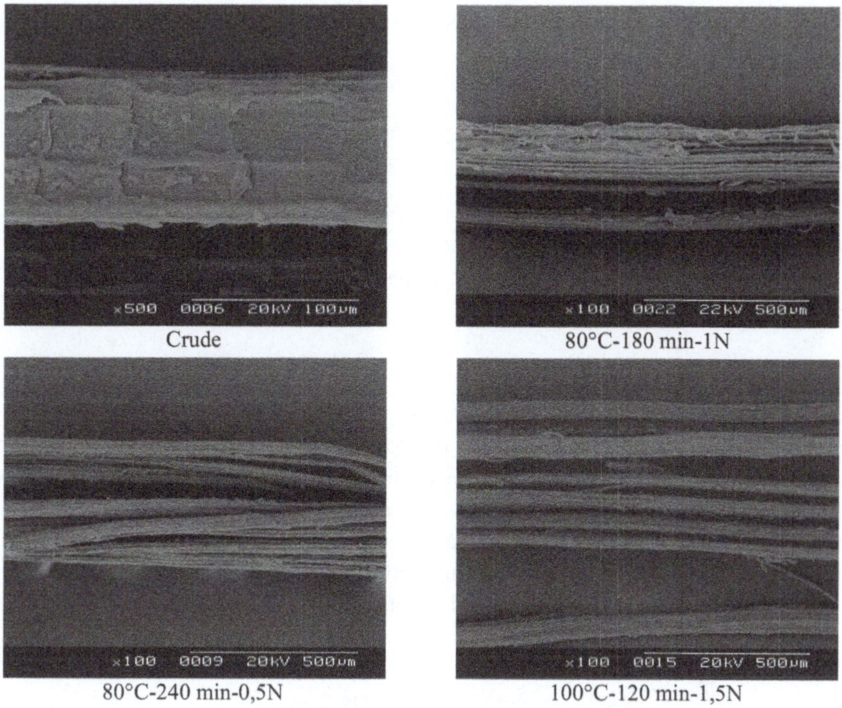

Fig. 1. SEM of kenaf fiber surface at different extraction conditions

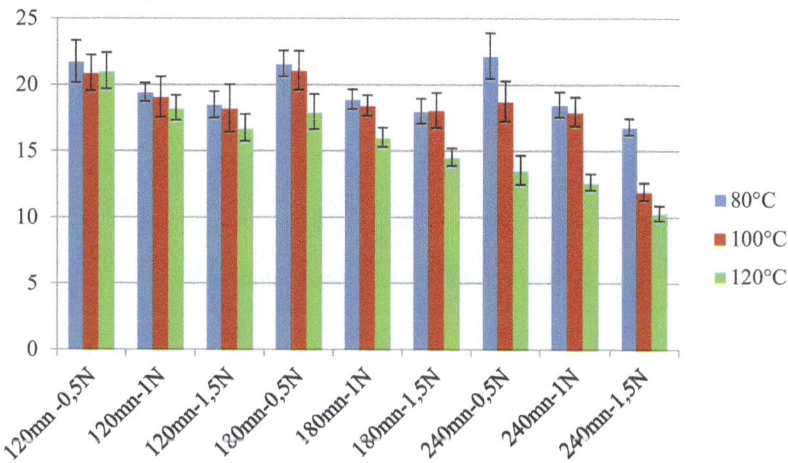

Fig. 2. Linear density (tex) of chemically extracted kenaf fibers

hemicelluloses) but also the cellulosic chain which make the fiber structure weakest (Samad et al. 2002). This result is proved through the SEM images where we noticed a destroyed surface with non cellulosic substance for chemical treated fiber in Fig. 1.

Table 2. hysical and mechanical properties of the extracted fibers

		Crude	Chemical extraction		
			80 °C	100 °C	120 °C
Tenacity (cN/tex)	Average	14,19	13,02–21,43	10,46–17,04	8,65 - 15,52
	CV (%)	39,08	35,10	44,06	46,37
Strain (%)	Average	7,28	2,35–3,90	2,42–3,35	1,92 - 2,91
	CV (%)	31,25	37,02	53,66	50,67

4 Chemical Structure

The FTIR allows us to follow the changes in fiber chemical constitution after the different treatment done. When examining the spectra in Fig. 2, the same absorption bands with significant differences in intensity and width at half height was observed but there is no appearance of new bands or disappearance of other bands and this proved that the chemical treatment allows the elimination of non cellulosic materials without modifying the fiber chemical structure (Garside and Wyeth 2003; Sgriccia et al. 2008).

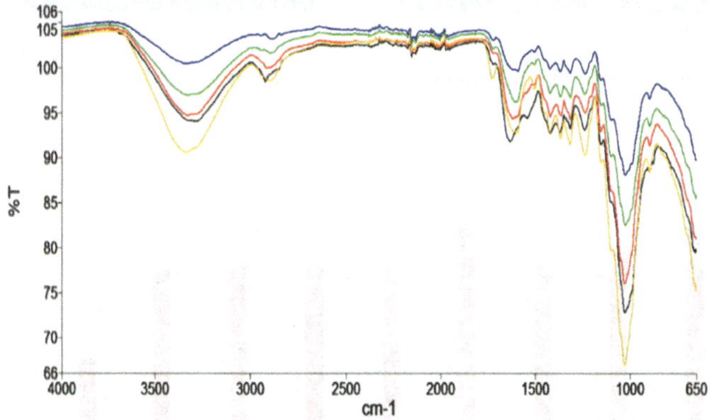

Fig. 3. FTIR spectra for treated and untreated fibers

4.1 Crystallinity Index

The results of the crystallinity index are summarized in Fig. 4 which presents the crystallinity index of the chemically treated fiber. Figure 4 allows showing the differences of crude fibers and the enzyme treated fibers.

The crystallinity index for treated fibers increases relative to that of crude fibers. This can be explained by the fact that the combined action of sodium hydroxide, temperature and time has an impact on the fiber structure and on the organization of the macromolecular chains which means that the increase of the crystallinity index is not only caused by the removal of the amorphous content (lignin and hemicelluloses) but also due to a

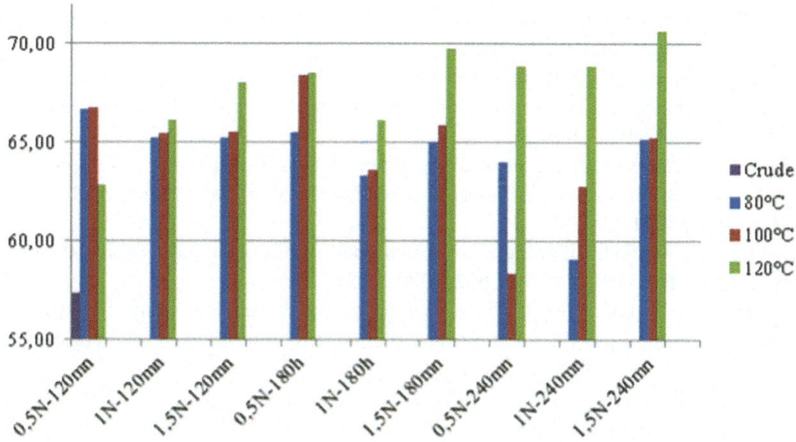

Fig. 4. Crystallinity index (%) of crude and chemically treated fiber

new reorganisation in the fiber structure. The crystallinity index for the crude fiber is 57,39% and this index reaches a value of 71,93% for the chemically treated fiber (with an increase of 25% compared to crude fiber).

5 Conclusion

The purpose of this study was to investigate the effect of the chemical extraction method on the kenaf fiber extracted from the kenaf stem. Fibers were extracted by using sodium hydroxide. The obtained fibers were characterized and the obtained results are compared to the crude fiber manually extracted. Physic-chemical and mechanical properties of all extracted fibers were studied. The extracted fiber have better mechanical properties compared crude fiber especially for tenacity, but for high concentration of NAOH, the cellulose chains are attacked which damage the fiber and lead to decrease the fiber mechanical properties. This results is proven through the crystallinity index that increases to reach a maximum value and then decreases proving the appearance of new amorphous area. To conclude chemical extraction allows having fibers with improved properties compared to crude fiber and this by controlling the extraction condition to eliminate only the non cellulosic materials without destroying the cellulose structure.

References

Chattopadhyay, et al.: Agave americana/Anew source of textile fiber. Colourage **6**, 33–36 (2012)

Columbus, et al:. Kenaf separation: Annual report-crop in U.S. Cotton Ginning Laboratory. – Stoneville (1994)

Mohammed, D.: Evaluation du potentiel textile des fibres d'alfa (Stipa Tenacissima L.): Caractérisation physico-chimique de la fibre au fil, Université de Haute Alsace, LPMT. - haute-alsace (2012)

Gardette J.L. :Caractérisation des polymères par spectrométrie optique in technique de l'ingénieur (1996):

Garside, P., Wyeth, P.: Identification of cellulosic fibers by FTIR spectroscopy: thread and single fiber analysis by attenuated total reflectance. Stud. Conserv. **51**, 269–275 (2003)
De, G., et al.: Kenaf and Pinapple fibers for nonwoven fabrics. NSTA Technol. J. 77–78 (1996)
Henriksson, et al.: Chemical/Physical retting of flax using detergent and oxalic acid at high pH. Text. Res. J. **76**, 942–947 (1998)
Ashish, H., et al.: Agave Americana leaf fibers. Fibers **3**, 64–75 (2015)
Mylsamy, K.: Studies on Agave Americana fiber reinfoorced composite Materials, Ph.D. Thesis / Anna University (2011)
Samad, M.A., et al.: Mechanical properties of kenaf fibers (hibiscus cannabinus) and their spinning quality [Revue]. Pak. J. Biol. Sci. **5**, 662–674 (2002)
Messaoud, M. : Fonctionnalisation anti-bactérienne passive ou Active de supports textiles par voie Sol-Gel ou photochimique: l'association du TiO2 et de la chimie douce, Ph.D. Thesis, Grenoble (2011)
Morrison, et al.: Evaluation of chemically retted kenaf using chemical, histochemical and microspectrophotometric analysis. Text. Res. J. **66**, 651–656 (1976)
Sgriccia, N., et al.: Characterisatization of natural fiber surfaces and natural fiber composites [Revue]. Composite Part A Appl. Sci. Manuf. **39**, 1632–1637 (2008)
Song, et al.: Chemical and biological retting of kenaf fibers. Text. Res. J. **76**, 751–756 (2006)

Morphological and Properties Characterization of Melt-Spun Poly(Lactic Acid)/Cellulose Nanowhiskers Fibers: Effect of Filler Content

Tassadit Aouat[1,2] and Mustapha Kaci[1(✉)]

[1] Laboratoire des Matériaux Polymères Avancés (LMPA), Université de Bejaia, 06000 Bejaia, Algeria
kacimu@yahoo.fr

[2] Faculté des Sciences et de la Technologie, Université Yahia Farès, 26000 Médéa, Algeria

Abstract. The production of poly(lactic acid) (PLA)/cellulose nanowhiskers (CNW) bionanocomposites represents an efficient route to enlarge their application in many industrial fields, with the possibility to control the properties by filler content adjustment. In this work, fibers yarns of PLA and PLA/CNW filled at 1 and 3 wt.% were prepared by melt spinning in the presence of PLA-*grafted*-maleic anhydride (PLA-*g*-MA) used as the compatibilizer. The study aimed at investigating the influence of filler content on the morphology and properties of PLA/CNW bionanocomposites fibers. The results showed that adding only 1 wt% of CNW in PLA in presence of the compatibilizer improved the morphology and the thermal stability of the bionanocomposite fibers than those filled with 3 wt%, while the tensile properties were almost comparable to the neat PLA.

Keywords: Poly(lactic acid) · Cellulose nanowhiskers · Bionanocomposites · Melt spinning · Fibers

1 Introduction

In recent years, much attention has been paid to biopolymers as an answer to the environmental issues and to the depletion of fossil resources (Dehouche et al. 2019). However, the application of biopolymers is often limited by their poorly mechanical and thermal properties. In order to be competitive to petroleum based polymers, they need to be modified (Luiz de Paula et al. 2011). Nanoscale reinforcements have strong promise in designing ecofriendly green bionanocomposites combining (natural/bio) fibers with biodegradable polymers (Valentini et al. 2019). In this respect, the attraction for cellulose nanowhiskers (CNW) as reinforcing fillers in the polymer matrices has largely increased due to the unique combination of their impressive mechanical properties coupled with their high aspect ratio. Indeed, CNW offer many advantages such as high reactivity, renewability, biodegradability and natural abundance (Aouat et al. 2018). Poly(lactic acid) (PLA) is one of the most representative bio-based and biodegradable polymers (Nampoothiri et al. 2020). However, some of PLA properties, like flexural properties, gas permeability and heat distortion temperature are too low for widespread applications. Therefore, the

S. Msahli and F. Debbabi (Eds.): CIRATM 2020, SPM 17, pp. 283–288, 2022.
https://doi.org/10.1007/978-3-031-08842-1_45

development of PLA/CNW bionanocomposites could be one of the most efficient routes to improve the properties of the biomaterials by adjusting the suitable filler content. In the field of textiles, the electrospinning process is the most used ones to produce polymer fibers (Park et al. 2007). Nevertheless, the need to develop new processing techniques, as an extension of conventional plastics industry, remains an important challenge. In this paper, melt spinning process was used to elaborate neat PLA and PLA/CNW bionanocomposite fibers at 1 and 3 wt% in the presence of PLA-*grafted*-maleic anhydride (PLA-*g*-MA) used as the compatibilizer at 7 wt%. Morphology, thermal stability and tensile properties of the fibers were investigated with respect to filler content ratio.

2 Experimental

2.1 Materials

PLA used was fiber-grade resin 6202D and supplied by Nature Works LLC. According to the manufacturer, the main physical characteristics of the polymer are as follows: d = 1.24g/cm^3, T_g = 60 °C and T_m ~ 160–170 °C. Microcrystalline Cellulose (MCC) was supplied by Sigma-Aldrich under the trade name Avicel PH 101and used as a raw material for extracting cellulose nanowhiskers (CNW) (Aouat et al. 2018). Sulfuric acid 95–97% from Sigma-Aldrich was used for CNW extraction step. PLA-*grafted*-maleic anhydride (PLA-*g*-MA) (~3% of MA) used as the compatibilizer for the PLA bionanocomposite fibers, was prepared in the Materia Nova laboratory (Mons, Belgium) by reactive extrusion in a Leistritz twin-screw extruder (L/D = 50).

2.2 Preparation of PLA Fiber Yarns by Melt Spinning

Neat PLA and PLA/CNW bionanocomposite fiber yarns were manufactured in two steps. However prior to processing, both CNW and PLA were dried at 60 °C for 12 h to remove the moisture. The first step was the transformation of PLA and PLA bionanocomposites to pellets using a Thermo-Haake co-rotating intermeshing twin-screw extruder (L/D = 25) at 1 and 3 wt% of CNW. The screw speed was kept at 100 rpm, while the temperature profile in the barrel was 160–190 °C. The extruded materials were directly granulated and designed as PLA, PLA/CNW1 and PLA/CNW3. Fiber yarns were elaborated by melt spinning machine, Model Spinboy I, manufactured by Busschaert Engineering. The fiber yarns were covered with an appropriate spin finish (oil/water emulsion composed of a lubricant, an antistatic agent, a cohesion agent and surfactants), which provides the fibers cohesion along the process, rolled up on two heated rolls with varying speeds (S_{R1} and S_{R2}) in order to ensure the optimal drawing.

2.3 Characterization Techniques

2.3.1 Scanning Electron Microscopy (SEM)

A QUANTA 200 FEG (FEI Company) Scanning Electron Microscope (SEM) was used to investigate the external as well as the fracture surface morphology of CNW fibers. The fibers surface was coated with a carbon thin layer before SEM observations.

2.3.2 Tensile Measurements

Tensile measurements of fibers were performed by a MTS machine following the standard NF EN ISO 2062. The cell force used was 1 KN with a loading speed of 200 mm/min and a distance of 200 mm between grips. All tensile tests were carried out on specimens previously stored for at least 48 h at 20 ± 2 °C under 50 ± 3% RH. The values represent an average of five replicates.

2.3.3 Thermogravimetric Analysis (TGA)

TGA analysis was performed on a TGA Setsys instrument (Setaram, France). About 10 mg of samples were heated from 30 to 900 °C. The heating rate was set at 10 °C/min under the nitrogen atmosphere and the flow rate of 100 ml/min.

3 Results and Discussion

3.1 Morphological Analysis of PLA and PLA Bionanocomposite Fibers

The morphological investigation of the neat PLA and PLA bionanocomposite fibers was performed by SEM. In this respect, Fig. 1a, b and c shows the external surface of neat PLA and PLA bionanocomposite fibers at 1and 3 wt% of CNW, respectively. Figure 2a, b and c show the morphology in cross sectional area of the corresponding samples. In Fig. 1a, the external surface of neat PLA fibers is smooth and regular with a good circular-shape. Moreover, no surface defects are observable on the cross-sectional area of neat PLA fibers as shown in Fig. 2a. For the PLA bionanocomposite fibers, the addition of CNW affects the morphological surface as illustrated in Fig. 1b and c in comparison with that of neat PLA (Fig. 1a). Further, the effect is more pronounced at 3wt%. Indeed in Fig. 2b, small filler aggregates are visible on the cross-sectional area of PLA/PLA-*g*-MA/CNW1 whose size significantly increases at 3wt% (Fig. 2c) displaying lower fibers yarns regularity and a bulging surface. Process conditions could probably be responsible for the poor dispersion of the cellulosic fillers in PLA matrix at 3 wt%, due to their natural tendency to re-agglomerate and to form strong hydrogen bonds as the water sublimate.

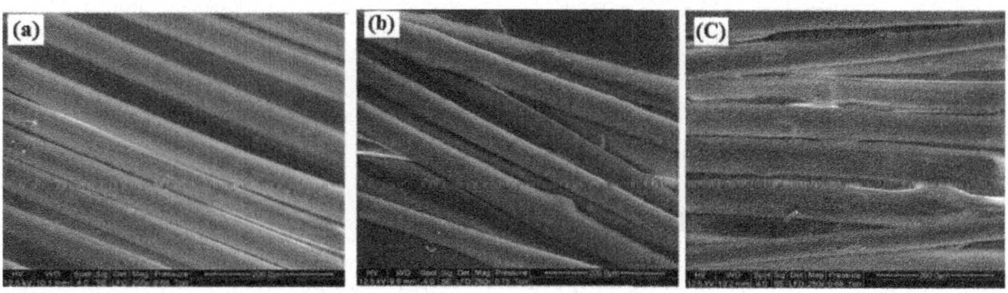

Fig. 1. SEM images of the external surface fibers of: **a)** neat PLA and PLA/PLA-*g*-MA/CNW bionanocomposites **b)**: 1 wt% and **c)**: 3 wt%.

Fig. 2. SEM images of the cross-sectional surface fibers of: **a)** neat PLA and PLA/PLA-*g*-MA/CNW bionanocomposites **b)**: 1 wt% and **c)**: 3 wt%.

3.2 Tensile Properties

The tensile properties of neat PLA fibers and those of PLA/CNW bionanocomposites at 1 and 3 wt% filler content are provided in Table 1. For the PLA/PLA-*g*-MA/CNW1 fibers, the results indicate an increase in elongation at break by almost 18% compared to that of neat PLA, whereas Young's modulus and tensile strength at break show almost comparable values within the experimental errors. However for PLA/CNW3 fibers, it is observed that the tensile properties are lower than those filled at 1 wt% and neat PLA due probably to the melt spun process. This may result from the fibers arrangements in the suspensions, which are frozen during the hot-pressing stage due to the polymer melt viscosity (Perisin et al. 2010). This drawback may be prevented by using electrospinning process. Indeed, (Zhou et al. 2013) reported that the preparation of CNW based PLA fibers by electrospun process produces fibers of better mechanical performances compared to the melt-spun process. Furthermore, the use of solvent in the electrospinning process leads to better possibility for CNW to form a percolating network.

Table 1. Values of Young's modulus, tensile strength at break and (%) elongation at break of neat PLA and PLA/CNW bionanocomposite fibers.

Samples	Young's modulus (MPa)	Tensile strength at break (MPa)	Elongation at break (%)
PLA	2510 ± 158	92.2 ± 5.4	77.7 ± 7.3
PLA/PLA-g-MA/CNW1	2334 ± 99	83.9 ± 3.8	91.6 ± 4.9
PLA/PLA-g-MA/CNW3	930 ± 92	22.2 ± 1.9	19.0 ± 3.5

3.3 Thermal Stability by TGA

TGA data, i.e., onset degradation temperature (T_{onset}), temperature at maximum degradation rate (T_{mrd}) and residue at 500 °C of neat PLA and PLA/CNW bionanocomposites fibers are provided in Table 2. It is noticed that the addition of CNW to PLA matrix results in an increase in the thermal stability of the bionanocomposite fibers, in particular at filler content of 1 wt%. Indeed T_{onset} increases by 15 °C compared to neat PLA. This is most likely due to the restriction of the mobility of polymer chains and suppression of the decomposition as a result of the homogeneous dispersion of CNW in the polymer matrix. Similar trend is also observed with T_{mrd} after adding CNW. This behavior is likely due to a better dispersion of CNW at 1 wt% in the PLA matrix with the compatibilizer compared to 3 wt%.

Table 2. TGA data of neat PLA and PLA/CNW bionanocomposite fibers.

Samples	T_{onset} (°C)	T_{mrd} (°C)	Char at 500 °C (%)
PLA	311	362	0.6
PLA/ PLA-*g*-MA/CNW1	326	369	0.1
PLA/PLA-*g*-MA/CNW3	320	367	1.3

4 Conclusion

The effect of filler content on the morphology and properties of PLA and PLA/CNW bionanocomposite fibers prepared by melt-spinning process at 1 and 3 wt% were investigated by several techniques such as SEM, TGA and tensile measurements. It can be concluded that overall, the melt spinning of cellulosic PLA fibers was suitably performed at 1 and 3 wt.%. All results indicated clearly that better fibers performances were achieved by adding CNW at 1 wt% in PLA compared to 3 wt%. Indeed a better morphological surface was observed at 1 wt% than 3 wt% in PLA matrix and almost 18% increase in elongation at break was obtained compared to neat PLA. In addition, the onset degradation temperature was increased by 15 °C at 1 wt%, which indicates a better thermal stability. Finally, the study showed that CNW is a promising reinforcement for biodegradable polymers such as PLA. However, important challenges need to be solved as the extension of the study to other compatibilizers aiming to improve the processing characteristics and fibers properties.

References

Aouat, T., et al.: Morphological, mechanical, and thermal characterization of poly(lactic acid)/cellulose multifilament fibers prepared by melt spinning. Adv. Polym. Technol. **37**, 21779–21713 (2018). ISSN: 07306679

Dehouche, N., Kaci, M., Kaid, N.: Thermo-mechanical recycling effects on morphology and properties of ethylene vinyl acetate copolymer/olive husk flour composites. Int. J. Plast. Technol. **23**(2), 246–252 (2019). https://doi.org/10.1007/s12588-019-09256-1

Luiz de Paula, E., et al.: Influence of cellulose nanowhiskers on the hydrolytic degradation behavior of poly(D, L-lactide). Polym. Degrad. StabiL **96**, 1631–1638 (2011). ISSN: 0141-3910

Nampoothiri, K.M., et al. : An overview of the recent developments in polylactide (PLA) research. Bioresour. Technol. **1010**, 8493 – 8501 (2020). ISSN: 0960-8524

Park, W., et al. : Electrospinning of poly(ethylene oxide) with bacterial cellulose whiskers. Macromol. Symp. **249–250**, 289–294 (2007). ISSN: 10221360

Peresin, M.S., et al.: Nanofiber composites of polyvinyl alcohol and cellulose nanocrystals: manufacture and characterization. Biomacromolecules **11**, 674–681 (2010). ISSN: 1525-7797

Valentini, F., et al.: Polyhydroxyalcanoates/fibrillated nanocellulose composites for additive manufacturing. J. Polym. Environ. **27**, 1333–1341 (2019). ISSN: 1566-2543

Zhou, C., et al.: Electrospun bio-nanocomposite scaffolds for bone tissue engineering by cellulose nanocrystals reinforcing maleic anhydride grafted PLA. Appl. Mater. Interf. **5**, 3847–3854 (2013). ISSN: 1944-8244

Morphological Characterization and Thermal Stability of poly(3-hydroxybutyrate-co-3-hydroxyhexanoate)/Microcrystalline Cellulose Biocomposites

Nadjet Dehouche[1(✉)], Mustapha Kaci[1], Rili Rosa[1], and Stéphane Bruzaud[2]

[1] Laboratoire des Matériaux Polymères Avancés (LMPA), Université de Bejaia, 06000 Bejaia, Algeria
dehouche_nadjet@yahoo.fr

[2] Institut de Recherche Dupuy de Lôme (IRDL), UMR CNRS 6027, Université de Bretagne-Sud, Rue Saint Maudé, 56321 Lorient Cedex, France

Abstract. In this paper, biocomposites samples based on poly(3-hydroxybutyrate-co-3-hydroxyhexanoate) (PHBHHx)/cellulose microcrystalline (MCC) extracted from olive husk flour were prepared in a twin-screw extruder at loading rates of 10, 20 and 30 wt%. Morphology, thermal stability and water uptake of PHBHHx biocomposites were investigated with respect with filler content. The study revealed that adding MCC to PHBHHx reduces the thermal stability of the biocomposite samples compared with neat polymer; this is however, more pronounced at higher filler content. This is due to filler aggregates as observed by scanning electron microscopy, inducing heterogeneities and defects within the polymer matrix. Further, water uptake increases with filler content ratio.

Keywords: poly(3-hydroxybutyrate-co-3-hydroxyhexanoate) · Cellulose microcrystalline · Biocomposites · Water uptake · Thermal stability

1 Introduction

The growing ecological and environmental consciousness has driven efforts for development of new innovative materials for various end-use applications (Blanco et al. 2020; Sutana et al. 2020). In particular, biopolymers can be widely applied in many technological fields like packaging, tissue engineering, agriculture and also in additive manufacturing processes (Valentini et al. 2019). In this regard, polyhydroxyalkanoates (PHAs), generated by microorganisms, have highly attracted attention in various fields due to their physico-mechanical properties, which are comparable to those of polyolefins or PET (Tarrahi et al. 2020). However, the application of PHAs to some industrial fields is rather limited due mainly to its slow crystallization rate, high crystallinity and high thermal sensitivity. Moreover, these PHA exhibit also a brittleness and high production costs (Aouat et al. 2019).

S. Msahli and F. Debbabi (Eds.): CIRATM 2020, SPM 17, pp. 289–294, 2022.
https://doi.org/10.1007/978-3-031-08842-1_46

Among the PHAs family, PHBHHx copolymer is considered as the most potential biopolymers to substitute traditional commodity plastics. It is fully biodegradable and biocompatible. Due to its ductile nature and wider processing window compared with poly(3-hydroxybutyrate), PHBHHx has a broad range of applications in various industries such as biomedical sector including tissue engineering, bio-implant patches, drug delivery, surgery, wound dressing and scaffolds, agriculture in the controlled release of insecticides and also flexible packaging (Dehouche et al. 2020). Together with the development of biocomposite materials, the addition of cellulosic fillers to PHAs could be one of the most promising routes to further improve their properties (Murphy and Collins 2018). Indeed, cellulose due to its abundant availability, renewability, biodegradability, high strength and stiffness, could replace advantageously layered silicates, carbon nanomaterials and other metallic oxide fillers (De Paula et al. 2011). Indeed, natural fiber-reinforced biocomposites have been used for many applications and accepted by consumers, especially in the automotive industry (interior products such as door panels, seat backs, headliners, package trays, dashboards, and interior parts), packaging and building industries, where a high load carrying capacity is not required (Asim et al. 2018). Therefore, in this work, the main objective was to investigate the influence of MCC loading rates on the morphology, thermal stability and water uptake of PHBHHx biocomposites prepared by melt compounding. The filler content ratio was 10, 20 and 30 wt%. The properties performances of the PHBHHx/MCC biocomposites have been discussed with respect to the neat polymer.

2 Experimental Part

2.1 Materials Used

PHBHHx containing 11% of hydroxyhexanoate was provided in granular form and manufactured by Kaneka Corporation (Osaka, Japan) under the grade Aonilex X151A. According to the manufacturer, the polymer has a density = 1.19 gcm^{-3}, a melting temperature (T_m) = 130 °C, a glass transition temperature (T_g) = −0.2 °C and a crystallinity index (X_c) = 34%. Microcrystalline cellulose (MCC) was extracted from olive husk flour, which has been collected in the region of Bejaia in Algeria. The detailed extraction procedure of MCC is described in literature (Dos Santos et al. 2016). The size fraction, selected after sifting, has a maximum average diameter of 50 μm. This is more appropriate to have a good dispersion of the filler within the thermoplastic matrix (Hassaini et al. 2017). The particle shape of MCC is rod-like with various sizes ranged from a few microns ~5 to 50 μm as illustrated in Fig. 1.

2.2 Sample Preparation

Prior to processing, all materials were dried in a vacuum oven at 70 °C for 12 h to minimize both the moisture effect and subsequently the hydrolysis of PHBHHx and MCC. The preparation of PHBHHx/MCC biocomposites was carried out in a twin-screw micro-compounder, Model DSM Xplore (version 1.0 - 2005) at 145 °C and 50 rpm. MCC was incorporated into PHBHHx at 10, 20 and 30 wt%. The residence time was kept at 6 min.

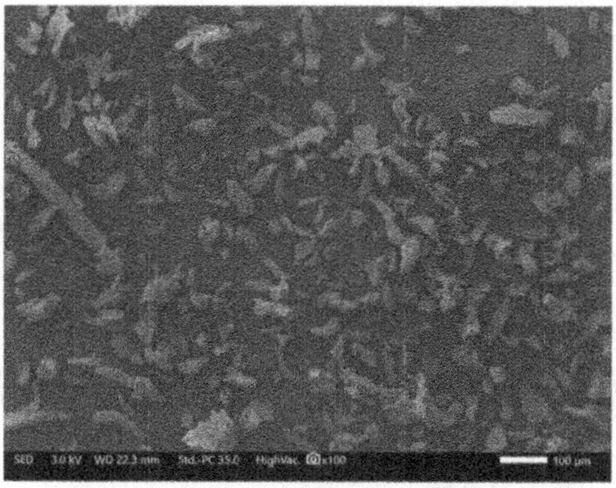

Fig. 1. Rod-like shape of MCC. 100×

2.3 Technical Characterization

TGA experiments were carried out in a thermal analyzer (Setaram TGDTA92-10) using a heating rate of 20 °C min^{-1} under nitrogen atmosphere in the temperature range starting from 20 to 600 °C.

The sample weight was approximately 10 mg.

The water absorption was determined by measuring the difference between the weight at the initial time of the test and the constant final weight of the sample after 24 h of immersion in distilled water according to ASTM D570-98(2018) standard method. The samples in rectangular shape have the following dimensions: 20 × 5 × 0.5 mm.

The surface morphology of the biocomposite samples at various filler contents was observed by a Jeol JSM-6031 scanning electron microscope. The neck region for the broken biocomposite samples fractured in liquid nitrogen is parallel to the draw direction in order to reveal the internal morphology. Prior to any observation, the fractured surface was coated with a thin gold layer by means of a Polaron sputtering apparatus.

3 Results and Discussion

Figure 2a, b and c shows the SEM micrographs of the fractured surface of PHBHHx/MCC biocomposite filled at 10, 20 and 30 wt%, respectively. In all figures, it is observed an heterogeneous and rough surface, which exhibits a phase separation due to the lack of interfacial adhesion between the filler and the matrix (Hassaini et al. 2016). Indeed, MCC particles are pulled out from the polymer matrix. This phenomenon is more pronounced at a higher filler content where the number and size of aggregates increase with increasing the MCC content resulting in more surface defects and voids.

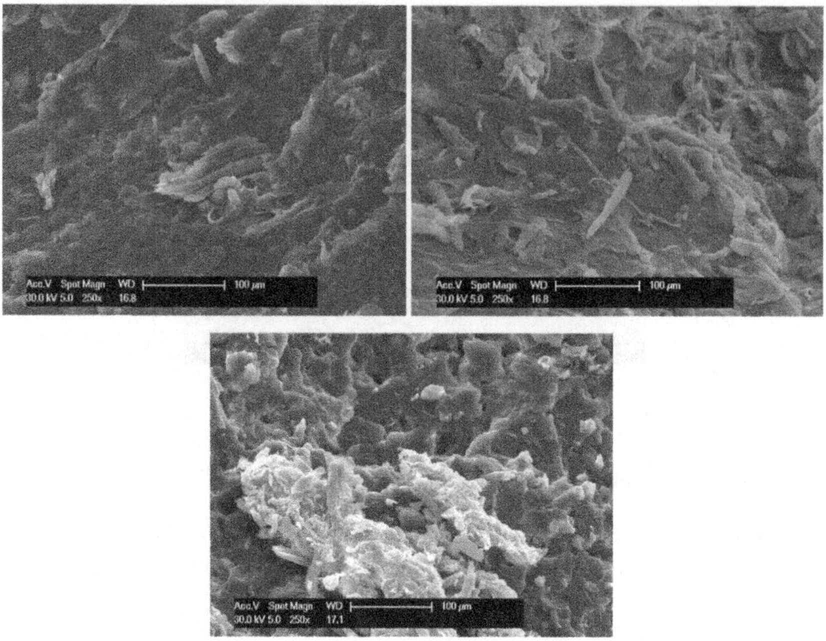

Fig. 2. SEM micrographs of the fractured surface of PHBHHx/MCC at 10 (a), 20 (b) and 30 wt% (c) biocomposites. 250×.

Figure 3 shows the TGA thermograms of neat PHBHHx and PHBHHx biocomposites at various filler content ratios. It is clear that the neat PHBHHx exhibits one single degradation step starting from almost 220 °C. According to the literature (Giubilini et al. 2020), the thermal degradation process of PHBHHx occurs by a random chain scission reaction with a six-membered ring ester intermediate, transforming eventually the ester groups into olefinic and carboxylic acid groups. However, the addition of MCC to PHBHHx decreases the onset degradation temperature of the biocomposites compared to the neat polymer. This is more pronounced at a higher filler content ratio. This behavior is likely due to the bad dispersion of MCC within the PHBHHx matrix in agreement with the morphological data. Indeed, the tendency of MCC particles to agglomerate, induces heterogeneities and defects within the polymer matrix besides the lower thermal stability of the MCC (Hassaini et al. 2017). Moreover from Fig. 3, TGA thermograms of the PHBHHx/MCC biocomposites show two main degradation phases in which the first step is due to the polymer matrix, while the second one results from the MCC degradation.

Figure 4 shows the water uptake for the neat PHBHHx and its biocomposites at 10, 20 and 30 wt% filler content determined after 24 h of immersion in distilled water. It is observed that the water absorption increases with increasing the filler content. This is well expected given that MCC is hydrophilic. Moreover, the bad dispersion of MCC may also be responsible for the increase in water uptake in agreement with the morphological observations shown in Fig. 2b and c.

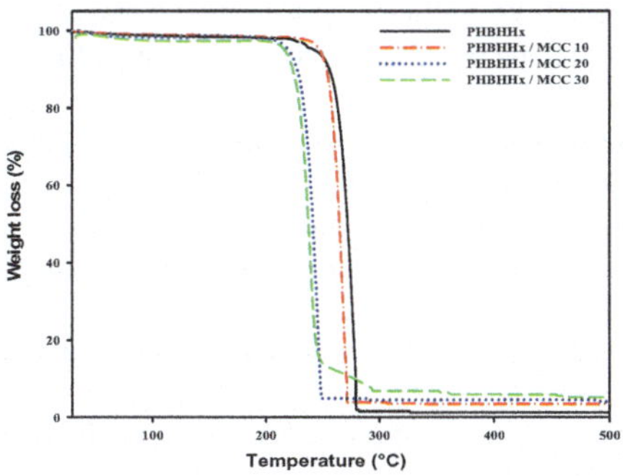

Fig. 3. TGA thermograms of PHBHHx and PHBHHx biocomposites.

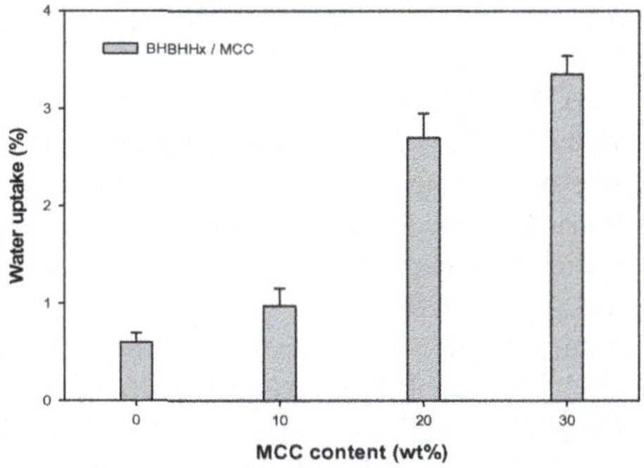

Fig. 4. Water uptake of PHBHHx and PHBHHx biocomposites after 24 h of immersion.

4 Conclusion

From the study, it was shown that the incorporation of MCC to PHBHHx at 10, 20 and 30 wt% led to a phase separation morphology with surface defects and roughness, being however more pronounced at a higher filler content. Further, a decrease in the thermal stability of the PHBHHx/MCC biocomposites was also observed. Water uptake of the biocomposites also increases with the filler content. Regarding the above results, it appears clearly the necessity to carry out the MCC surface modification for better dispersion of the MCC in PHBHHx matrix, thus enhancing strong interactions between them.

References

Aouat, T., et al.: Investigation on the durability of PLA Bionanocomposite fibers under hygrothermal conditions. Front. Mater. **6,** 323 (2019). ISSN 2296-8016

Asim, M., et al.: Propriétés thermiques, physiques et inflammabilité des composites hybrides phénoliques de fibres de kénaf/ananas traités au silane. Comp. Struct. **202**, 1330–1338 (2018)

Blanco, I., et al.: Analyse du cycle de vie dans le secteur des polymères: un examen complet des expériences d'application à l'échelle italienne. Polymers **12**, 1212 (2020)

Dehouche, N., et al.: Effects of various surface treatments on Aloe Vera fibers used as reinforcement in poly (3-hydroxybutyrate-co-3-hydroxyhexanoate)(PHBHHx) biocomposites. Polym. Degrad. Stab. **175**, 109131 (2020). ISSN: 0141-3910

De Paula, E.L., et al.: Influence of cellulose nanowhiskers on the hydrolytic degradation behavior of poly (d, l-lactide). Polym. Degrad. Stab. **96**, 1631–1638 (2011)

Dos Santos, F.A., et al.: The use of cellulose nanofillers in obtaining polymer nanocomposites: properties, processing, and applications. Mater. Sci. Appl. **7**, 257 (2016)

Giubilini, A., et al.: Composites biodégradables nanocellulose-poly (3-hydroxybutyrate-co-3-hydroxyhexanoate) par impression 3D par modélisation par dépôt fondu. ACS Sustain. Chem. Eng. **8**, 10292–10302 (2020)

Hassaini, L., et al.: The effects of PHBV-g-MA compatibilizer on morphology and properties of poly (3-hydroxybutyrate-Co-3-hydroxyvalerate)/olive husk flour composites. J. Adhes. Sci. Technol. **30**, 2061–2080 (2016)

Hassaini, L., et al.: Valorization of olive husk flour as a filler for biocomposites based on poly (3-hydroxybutyrate-co-3-hydroxyvalerate): effects of silane treatment. Polym. Test. **59**, 430–440 (2017). ISSN : 0142-9418

Murphy, C., Collins, M.N.: Filaments biocomposites d'acide polylactique renforcés de cellulose microcristalline pour l'impression 3D. Polym. Comp. **39**, 1311–1320 (2018)

Sultana, T., et al.: Études sur les propriétés mécaniques, thermiques et morphologiques des composites polymères biodégradables renforcés de nano cellulose à coque de noix de bétel. Compos. Sci. **4**, 83 (2020)

Tarrahi, R., et al.: Polyhydroxyalkanoates (PHA): From production to nanoarchitecture. Int. J. Biol. Macromol. **146**, 596–619 (2020)

Valentini, F., et al.: Composites de polyhydroxyalcanoates/nanocellulose fibrillée pour la fabrication additive. J. Polym. Env. **27**, 1333–1341 (2019)

Characterization of Tunisian Bleached Hair

Echhida Sayahi(✉), Taoufik Harizi, Slah Msahli, and Faouzi Sakli

Textile Engineering Laboratory, University of Monastir, Monastir, Tunisia
Chahida.sayahi@yahoo.fr

Abstract. The aim of this study was to determine the effect of bleaching on hair properties. The use of the SEM has demonstrated that the cuticles scales of bleached hair were lifted. Tensile properties on wet state were largely affected by this treatment. A significant variation of the denaturation temperature and enthalpie was also notified. After bleaching, the cysteic acid has largely increased, in contrast, a significant decrease of cystine was observed.

Keywords: Bleached hair · Mechanical properties · Thermal properties · Amino acid analysis

1 Introduction

All people want beautiful hair. The demand for products that enhance the appearance of these surfaces has created a huge cosmetic industry. For many years, especially in the second half of the twentieth century, scientists have focused on the physical and chemical properties of hair to constantly develop products that change the shine, color, softness and aesthetics of the whole of the hair.

On a daily basis, various treatments are performed on the hair. Washing with a shampoo is practiced to keep hair and scalp clean. Brushing, drying with a dryer and the use of hot plates are also performed. The hair is impaired by exposure to the sun's ultraviolet rays. In addition, chemical treatments such as permanent waving, whitening and hair coloring can significantly change the appearance of hair. They are used by many people who want to change their natural style to a desired style and color.

Much research has been done on hair damage from chemical treatments (bleaching and permanent waving) and grooming, including shampooing and combing (David et al. 1978; Nogueira et al. 2004; Fernandes and Cavaco-Paulo 2012; Wortman and Souren 1987. Inoue et al. 2007; Robbins and Kamath 2007). Beyak et al. (1969) established that both bleaching and perming can result in losses of up to 20% in mechanical properties. Robbins and Kelly (1969) studied the amino acid composition of chemically altered hair and found changes from both bleaching and permanent waving.

In this paper, we were interested to determine the effect of bleaching in hair. The changes in hair properties, introduced as a result of this treatment, were measured using various assessment methods such as tensile strength, thermal behavior, and scanning electron microscope surface analysis.

S. Msahli and F. Debbabi (Eds.): CIRATM 2020, SPM 17, pp. 295–300, 2022.
https://doi.org/10.1007/978-3-031-08842-1_47

2 Materials and Methods

2.1 Materials

The fibers used for the analysis are dark brown hair of a woman of 30 years of age.

The Bleaching treatment was done with a commercial preparation based on an alkaline solution containing 9% of hydrogen peroxide. The bleaching procedure is done following the manufacturer's instructions for use. The bleaching preparation was applied for 10 min and was repeated seven times.

2.2 Methods

2.2.1 Morphological Properties

The fiber surface change after bleaching treatment was examined using the SEM. We have used a SEM S360 (Zeiss NTS GmbH, Oberkochen).

2.2.2 Mechanical Properties

The tensile measurements were performed using a Miniature Tensile Tester Model 675 (MTT675) and a Fiber Dimensional Analysis Unit Model 765 (FDAS765) of Dia-Stron, UK. About 50 single fibers were tested for each sample at a stretching rate of 20 mm/min and a gauge force of 1.5 gmf, as initial condition. The measurements were performed in wet conditions. Prior to loading in the circular cassette, the samples were immersed in distilled water for 120 min to allow them wetting. During the measurements the cassettes were also filled with distilled water to ensure the 100% humidity content during the measurement. Mechanical properties were also measured in dry conditions at 65% $\pm$ 4 RH and 21 °C $\pm$ 2 °C.

2.2.3 Thermal Analysis

All DSC-experiments were performed on a power-compensated instrument (DSC-7, Perkin Elmer). The samples (3–5 mg) were cut and put in sealed aluminum crucibles with a small hole in the lid. The dynamic heating was carried out at a temperature range from 30 to 400 °C with a heating rate of 5 °C/min. An empty crucible was used as a reference.

2.2.4 Amino Acid Analysis

The amino acid composition of hair fibers was determined by High-Performance Liquid Chromatography (HPLC). Hair samples were hydrolyzed using HCl 6 M for 24 h at 110 °C in a nitrogen atmosphere.

3 Results and Discussion

3.1 Morphological Change

The morphological changes of the hair surface after bleaching were observed through SEM (Fig. 1). Figure 1(a) shows the representative micrographs of the untreated hair fibers used in this study. The untreated hair had a relatively complete and uniform appearance of the cuticle scales and the edge of the cuticle scale was clearly separated. On the other hand, the cuticle scales of the bleached hair were irregular and lifted. Some pitting and breaking were observed.

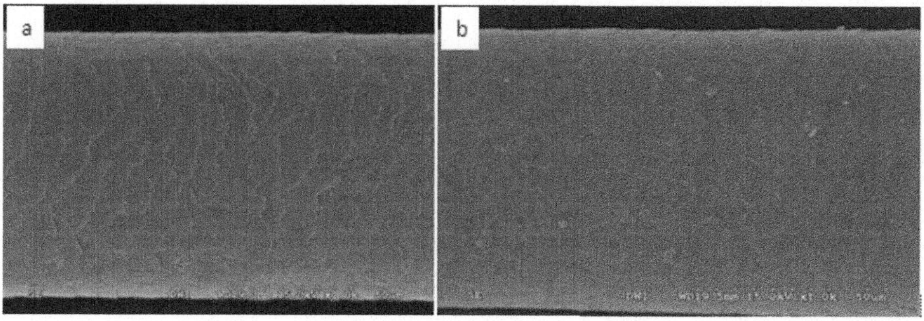

Fig. 1. Scanning electron microscopy images of the untreated hair (a) and the bleached hair (b)

3.2 Effect of Bleaching in Hair Mechanical Properties

Figure 2 shows the mechanical properties for bleached hair samples. An increase in the breaking elongation was observed as a general trend in the bleached samples compared to the control. The breaking extension increase by increasing the time of treatment (number of bleaching cycles). Indeed, the bleaching treatment breakdown the disulfide bonds. This breaking action ensures the sliding capacity of the polypeptide chains due to the decrease of the covalent bonds (S-S bonds) which ensure the stability of the fiber.

The same figure demonstrated a significant decrease at wet state ($p << 0,01$) of the elastic modulus, the elastic modulus has decreased by 25% after seven bleaching cycles of 10 min each. The hair becomes more flexible at 100% RH this is due, on the one hand, to the chemical degradation of the fiber following the oxidation of the cystine. On the other hand, the bleaching leads to the reduction of 18-MEA, which is one of lipids which ensure the cohesion of scales hair (Masukawa et al. 2004). Some scales are destroyed by bleaching. The decrease of the latter makes the fiber more porous and more hydrophilic and subsequently less resistant.

The data presented in Fig. 2 showed also a significant reduction of the breaking stress according to the number of bleaching cycles.

The whole mechanical properties of the bleached hair showed that this treatment has damaged the fiber. The bleaching has a great influence on the mechanical properties of the fiber especially in the wet state with a no significant alteration at 65% RH. Indeed, disulfide bonds contribute greatly to the resistance of keratin fibers at 100% moisture; it decreases in a manner virtually linear with the cystine content. On the other hand, the dry fiber strength is not significantly affected by the breakdown of covalent cross links, largely dependent on the main chain length and inter-chain hydrogen bonds (Alexander et al. 1951). Furthermore, the bleaching treatment conducted to the increase of the rate of cysteic acid, this reaction increases the ionic interactions that allow to consolidate the fiber in dry state and which are eliminated in the presence of water, hence the slight decrease in mechanical properties at 65% RH compared to the variation of these properties in wet state after this treatment.

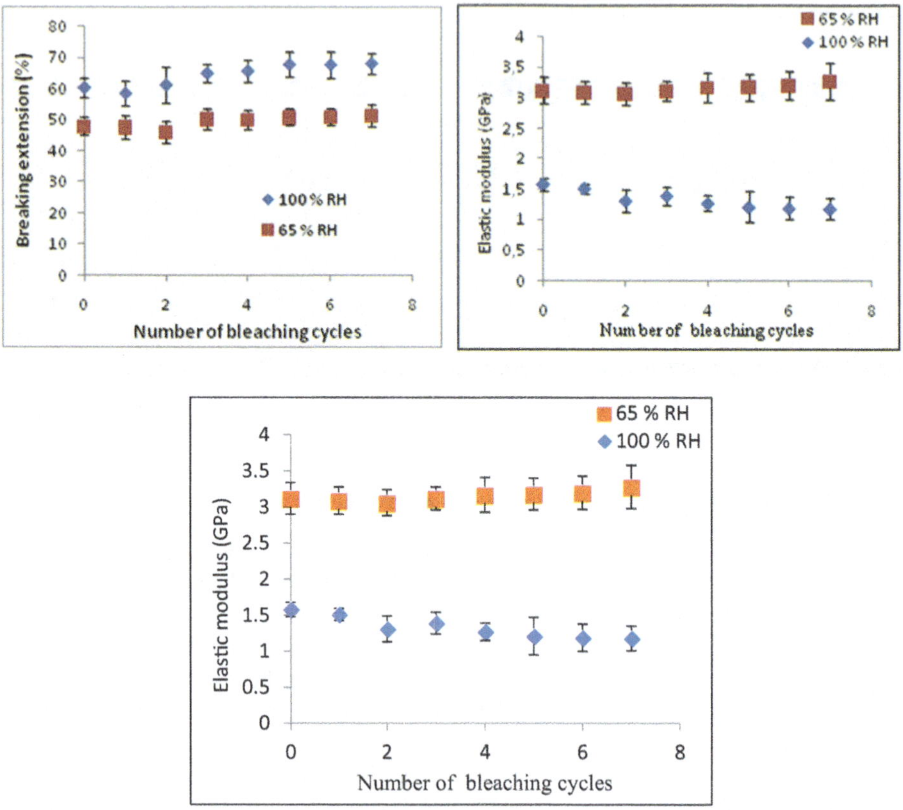

Fig. 2. Mechanical properties of bleached hair

3.3 Thermal Analysis of Bleached Hair

Figure 3 summarizes the results for denaturation enthalpies and peak temperatures, respectively, for the bleached samples. We observed that the first four bleaches lead to a roughly linear decrease of the denaturation enthalpie. It has decreased by 39% from 5,73 to 3,5 J/g after seven bleaching cycles.

The result showed also an increasing evolution of the denaturation temperature as a function of the bleaching cycles. Indeed, this evolution is rapid until the 4th cycle and slow beyond. The temperature of the untreated hair is 236.2 °C, after the fourth bleaching, it rapidly increases to 243.57 °C and reaches 245.22 °C after the seventh bleaching.

In agreement with our results, (Leroy et al. 1992) observed a decrease in denaturation peak area and an increase of the enthalpie through bleaching.

The variation of both temperature and denaturation enthalpie demonstrated that bleaching leads to largely homogeneous damage in IFs and IFAPs.

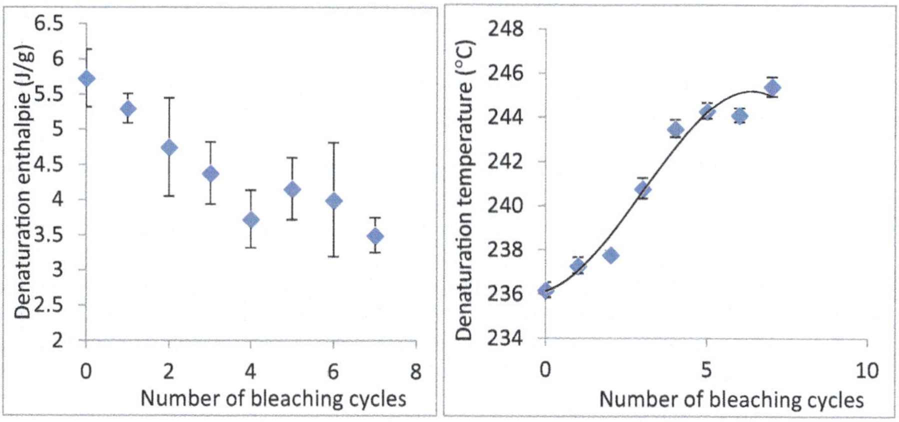

Fig. 3. Effect of bleaching on the denaturation enthalpie and temperature

3.4 Amino Acid Analysis

The Table 1 presented the variation in cystine and cysteic acid levels following the bleaching treatment. This result showed a highly significant increase in cysteic acid and a significant decrease in cystine after the bleaching treatment. Indeed, after three cycles of treatment, the hair lost almost 12% of cystine, on the other hand the level of cysteic acid increased from 27 μmol/g to 214 μmol/g and 297 mol/g respectively after 3 and 7 treatment cycles. No significant variation in the others amino acids levels was observed after the bleaching treatment.

Table 1. Cystine and cysteic acid variation of bleached hair.

	Non treated hair	Bleached hair			
	μmol/g	3 bleaching cycles	P_{value}	7 bleaching cycles	P_{value}
Half-cystine	1388	1083	0,027	963	0,016
Cysteic acid	27	214	0,004	297	0,001

4 Conclusion

In this study, the effect of bleaching treatment on hair properties was examined. The hair cuticles were lifted and removed. Importantly, we have demonstrated that bleaching treatment induced a large change on wet tensile strength of the hair. This change depended on the bleaching time. It was due to the breakdown of the disulfide bonds which contribute largely to the wet strength of keratin fibers. The damage due to the bleaching treatment was also observed on the variation of the thermal properties and the cystine and cysteic acid levels of the hair.

References

Alexander, P., Fox, M., Hudson, R.F.: The reaction of oxidizing agents with wool. 5. The oxidation products of the disulphide bond and the formation of a sulphonamide in the peptide chain. Biochem. J. **49**, 129–138 (1951)

Beyak, R., Meyer, C.F., Kass, G.S.: Elasticity and tensile properties of human hair i. single fiber test method. J. Soc. Cosmet. Chem. **20**(10), 615–620 (1969)

David, W., et al.: Permanent waving: utilization of the post-yields lope as a formulation parameter. J. Soc. Cosmet. Chem **29**, 685–701 (1978)

Fernandes, M., Cavaco-Paulo, A.: Protein disulphide isomerase-mediated grafting of cysteinecontaining peptides onto over-bleached hair. Biocatal. Biotransform. **30**(1), 10–19 (2012)

Inoue, T., et al.: Structural analysis of the cell membrane complex in the human hair cuticle using microbeam X-ray diffraction: relationship with the effects of hair dyeing. J. Cosmet. Sci **58**, 11–17 (2007)

Leroy, F., et al.: Thermoanalytical investigations of reduced hair, In: 8th International Hair Science Symposium of the German Wool. Research Institute, Kiel (1992)

Masukawa, Y., et al.: Damage to human hair caused by repeated bleaching combined with daily weathering during daily life activities. Exog. Dermatol. **3**, 273–281 (2004)

Nogueira, A.C., et al.: Impairment of hair mechanical properties by sun exposure and bleaching treatments. J. Cosmet. Sci. **55**, 533–537 (2004)

Robbins, C.R., Kelly, C.: Amino acid analysis of cosmetically altered hair. J. Soc. Cosmet. Chem. **20**, 555–564 (1969)

Robbins, C., Kamath, Y.: Hair breakage during combing. III. The effects of bleaching and conditioning on short and long segment breakage by wet and dry combing of tresses. J. Cosmet. Sci. **58**, 477–484 (2007)

Wortman, F.J., Souren, I.: Extensional properties of human hair and permanent waving. J. Soc. Cosmet. Chem **38**, 125–140 (1987)

Comparative Dielectric Analyses of Hybrid Frayed Cotton Fibers Reinforced Unsaturated Polyester

Asma Triki[1(✉)], Bechir Wanassi[2], Mohamed Ben Hassen[2,3], and Ali Kallel[1]

[1] Laboratory of Composites Materials, Ceramics and Polymers, University of Sfax, Sfax, Tunisia
trikilamacop@yahoo.fr
[2] Laboratory of Textile Engineering, University of Monastir, Monastir, Tunisia
[3] College of Engineering, Industrial Engineering Department, Taibah University, Medina, Saudi Arabia

Abstract. This study focuses on the use of waste hybrid cotton fibers, provided by the textile industry, as reinforcement in unsaturated polyester composites. Hence, to evaluate the performance of white yarns and indigo denim fabrics on the recycled cotton fibers adhesion in unsaturated polyester matrix, dielectric measurements were performed in the temperature range 0–150 °C and the frequency range 0.1 Hz–1 MHz on two UP composites (#1 and #2) whose white yarns/indigo denim fabrics relative volume fractions were 1:3 and 1:1, respectively. This dielectric study revealed a better reinforcement/matrix adhesion occurred for the relative volume fraction 1:3. This dielectric analysis was in accordance with thermal and vibration ones based on differential scanning calorimeter (DSC) and FT-IR technique, respectively.

Keywords: Waste cotton fibers · Unsaturated polyester composites · Dielectric relaxation · FTIR · Interfacial polarization effect

1 Introduction

Global awareness of environmental issues has led scientists to develop recycling processes to produce valuable products from textile industry cotton waste (Wanassi et al. 2016). These low cost fibers can be reused in many fields among them textile, automobile, biogas etc. They present similar advantages in comparison with those of natural raw fibers which make them a potential alternative to replace the most common synthetic fibers in textile reinforced composites (Romanzini et al. 2016; Baccouch et al. 2020). Indeed, waste cotton fibers have been used as reinforcement in thermoplastic polymers and have shown satisfactory mechanical properties (Borsoi et al. 2011). Further, cotton/unsaturated polyester composites have been studied and have revealed an improvement in structural integrity of the material for a specific mechanical test (Hashmi et al. 2007). Recent studies have found that waste cotton fiber can replace partially the traditional glass fiber in unsaturated polyester composites for optimized mechanical

S. Msahli and F. Debbabi (Eds.): CIRATM 2020, SPM 17, pp. 301–306, 2022.
https://doi.org/10.1007/978-3-031-08842-1_48

properties (Romanzini et al. 2016). However, despite these numerous advantages, natural fibers reinforced polymeric composites posses some limitations originating from hydrophilic nature of the natural fibers. As results, poor compatibility and poor interfacial adhesion may occur between the reinforcement and the polymeric matrix which leads to poor mechanical properties of the composites. To overcome such limitation different chemical treatments have been used (Li et al. 2007; Baccouch et al. 2020). Among them, alkaline treatment has shown an improvement of the adhesion between waste cotton fibers and epoxy matrix with varying either the concentration or the soaking times (Baccouch et al. 2020). In addition, hybridization treatment of natural fibers with other natural fibers or synthetic ones represents an ecological treatment of these so as to promote their adhesion in polymeric matrix (Omri et al. 2013).

The purpose of this study is to evidence the effect of indigo frayed cotton fibers on hybrid frayed ones adhesion in unsaturated polyester matrix. For this reason, dielectric study has been performed on two unsaturated polyester hybrid biocomposites #1 and #2 in which the relative volume fractions of white frayed cotton fibers/indigo frayed cotton fibers have been fixed to 1:3 and 1:1, respectively. This dielectric analysis has been correlated to thermal analysis based on differential scanning calorimeter and vibrational one based on FT-IR technique in order to probe the reinforcement/matrix interfacial region.

2 Materials and Method

2.1 Materials

The matrix material used in this study was based on a commercially available unsaturated polyester, whose Trade Name is 'ENYDYNE® H 68372 TA' supplied by Polynt/Composites. The matrix was mixed with initiator Methyl ethyl Ketone peroxide, MEKP, and Cobalt octanone at a concentration of 1.5% w/w before introducing the reinforcement. Textile wastes cotton made of white yarns and indigo denim fabrics have been used in the relative volume fractions 1:3 and 1:1 as reinforcements for composites #1 and #2, respectively. Textile wastes were chopped into small pieces with a length between 5 cm and 10 cm before using a fraying machine to transform these wastes into fibers. A needle punching machine in the laboratory of SITEX Company was then applied to the obtained frayed cotton fibers twice in order to transform them into nonwoven structure. The composites #1 and #2 were manufactured in the Solutions composites Company using the classical 'contact mould method' (Akgul 2001). The obtained composites had 3% fiber weight fraction.

2.2 Method

Differential scanning calorimeter (DSC) allowed evaluating the glass transition temperature (T_g) of the UP matrix and its composites #1 and #2. The DSC experiment was carried out on samples weighing between 3 and 10 mg by means of a Jade DSC instrument (PerkinElmer). This latter operated in the temperature range 0–150 °C according to a heating-cooling-heating cycle and in a nitrogen environment purged at 20 mW. The heating rate was fixed at 5 °C/min for each measurement step. The T_g value of each sample was determined according to construction of the lines done by Pyris software (Perkin Elmer) from the midpoint value of the jump in heat flow in the second heating run.

Dielectric measurements were carried out with an Alpha Dielectric/impedance Analyser (Novocontrol) in the temperature range from 0 °C to 150 °C and a frequency interval from 10^{-1} to 10^6 Hz. In the dielectric characterization, a circular gold electrode (2 cm in diameter) was sputtered on both surfaces of the sample to ensure good electrical contact with the gold-plated measuring electrodes. A sinusoidal voltage was applied creating an alternating electric field. This produced polarization in the sample, which oscillated at the same frequency as the electric field, but had a phase angle shift δ. This phase angle shift was measured by comparing the applied voltage to the measured current, which was separated into capacitive and conductive components. Measurements of capacitance and conductance were used to calculate: • Real part of the permittivity (apparent permittivity) ε' which is proportional to the capacitance and measures the alignment of dipoles. • Imaginary part of permittivity (loss factor) ε'' which is proportional to the conductance and represents the energy required to align dipoles and move ions. Accordingly, the complex permittivity was given by the following equation: $\varepsilon^* = \varepsilon' - j\,\varepsilon''$ (1). Dielectric experiments were conducted in isothermal runs with fixed temperatures and scanning frequencies from 10^{-1} to 10^6 Hz.

FT-IR technique was performed on the reinforcement NT1 and NT2 and its constituents of both composites #1 and #2, respectively. FT-IR measurements were investigated by means of Perkin Elmer FTIR, in ATR mode and in the wavenumber range 400–4000 cm^{-1} with a resolution of 2 cm^{-1}. Samples were grinded by using an agate mortar and a small quantity of each sample powder was analyzed.

3 Results and Discussion

Figure 1 shows the DSC thermograms of the UP matrix and its composites in the second heating runs. Obtained T_g values of each sample by Pyris software is mentioned in this figure. These values reveal an enhancement of T_g for both composites in comparison with that of UP matrix. This enhancement is slightly higher for the composite #1. The variation of T_g associated with the glass transition of the polymer from a glassy state to a rubbery and flexible one may originate from greater segment/filler interactions than segment/segment interactions (Triki et al. 2012).

Isothermal variations of the dielectric permittivity for both composites #1 and # 2 reveal a decreasing tendency with frequency and an overall increase with temperature at low frequencies.

A similar tendency is observed for the isothermal variations of loss factor. Hence, the decreasing tendency of the dielectric parameters at low frequencies and low temperatures represent a characteristic of the low frequency phenomenon. This latter overlaps the water dipoles polarization effect. However, at higher temperatures, −1 slope of loss factor at the same frequency region represents a characteristic of a dc conductivity effect.

The use of Kramers-Kronig relation allows eliminating the dc-conductivity effect from the measured loss spectra. This relation is as follows:

$$\varepsilon''_{d\acute{e}riv} = -\frac{\pi}{2}\frac{\partial\,\varepsilon'}{\partial\ln(\omega)} \approx \varepsilon'' \quad (1)$$

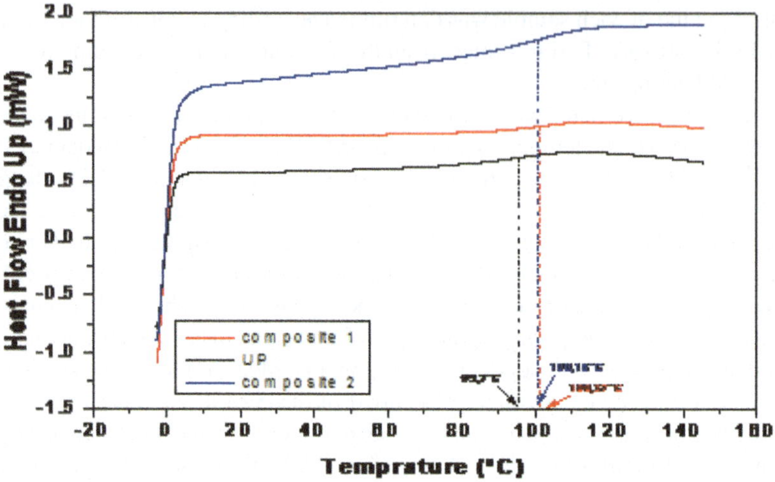

Fig. 1. DSC thermograms of the UP matrix and its composites #1 and #2 in the second heating run.

Figure 2 gives illustrations of $\varepsilon''_{d\acute{e}riv}$ frequency dependence plots for both composites at 125 °C. Two dielectric relaxations are detected and attributed to the interfacial polarization effect or Maxwell-Wagner-Sillars (MWS) relaxation and α relaxation. These relaxations are taken into account when analyzing dielectric parameters (ε', ε'') in Cole-Cole presentation and according to the Havriliak-Negami model (Havriliak and Negami 1996). This analysis reveals that MWS relaxation is in accordance with Havriliak-Negami response for both composites. However, its dielectric strength is lower for the composite #1. As result, the composite #1 exhibits a better reinforcement/matrix adhesion. Such result is consistent with the thermal analysis which reveals its higher glass transition temperature. Furthermore, vibrational analysis based on FTIR technique has shown a lower hydrophilic character of indigo cotton yarn in comparison with the white one. Accordingly, the use of the needle punching twice in elaborating the nonwoven frayed cotton gives rise to an additional stretching vibration originating from cellulose. Similar result has been obtained when hydrophilic character of Alfa fibers was decreased by the increase of the needle punching number in our previous research (Wanassi et al. 2016). Its intensity increases with the relative volume fraction of indigo frayed cotton fibers. Mechanism adhesion of both composites is based on chemical bonds formed by secondary bonding. However, the higher decrease of the OH stretching vibration located at 3448 cm^{-1} for the composite # 2 may be related to the change of the molecular dynamic of the polymer matrix as shown in thermal analysis section. Indeed, this decrease may be attributed to hydrogen bonds disruption occurred between polymer matrix molecular chains and caused by the reinforcement.

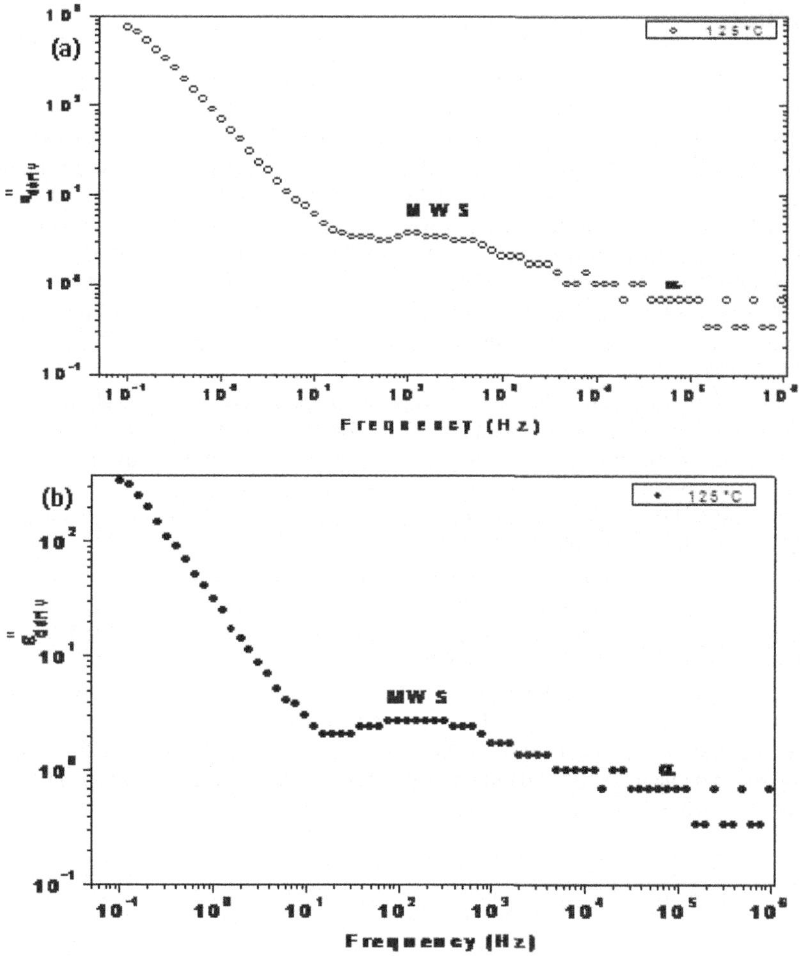

Fig. 2. Frequency dependence of the ohmic conduction free dielectric loss $\varepsilon''_{dériv}$, for the composite # 1 (a) and the composite # 2 (b).

4 Conclusion

Dielectric analyses have been investigated on hybrid frayed cotton fibers reinforced UP matrix. These analyses reveal a better reinforcement/matrix adhesion when indigo frayed cotton fiber volume fraction is higher than that of white frayed cotton fibers. This enhancement may be attributed to a lower hydrophilic character of frayed indigo cotton fibers according to vibrational analysis based on FTIR technique. Mechanism adhesion of both composites is based on chemical bonds formed by secondary bonding. Thermal analysis based on DSC is consistent with dielectric and vibrational analyses.

References

Wanassi, B., Azzouz, B., Ben Hassen, M.: Value-added waste cotton yarn: optimization of recycling process and spinning of reclaimed fibers. Ind. Crops Prod. **87**, 27–32 (2016)

Romanzini, D., Angrizani, C., Amico, S. C., Zattera, A.J.: Influence of stacking sequence on the mechanical and dynamic mechanical properties of cotton/glass fiber reinforced polyester composites, Mater. Res. **19** (2016). https://doi.org/10.1590/1980-5373-MR-2016-0058.

Baccouch, W., et al.: Enhancement of fiber-matrix interface of recycled cotton fibers reinforced epoxy composite for improved mechanical properties, Mater. Res. Exp. **7** (2020). https://doi.org/10.1088/2053-1591/ab6c04

Borsoi, C., Scienza, L.C., Zattera, A.J.: Obtenção e caracterização de compósitos utilizando poliestireno como matriz e resíduos de fibra de algodão da indústria têxtil como reforço. Polímeros. **21**(4), 271–279 (2011)

Hashmi, S.A., Dwivedi, U.K., Chand, N.: Graphite modified cotton fibre reinforced polyester composites under sliding wear conditions. Wear **262**(11), 1426–1432 (2007)

Li, X., Tabil, L.G., Panigrahi, S.: Chemical treatments of natural fiber for use in natural fiber-reinforced composites : a review. J. Polym. Environ. **15**, 25–33 (2007)

Omri, M.A., et al.: Effect of wool and thermo-binder fibers on adhesion of alfa fibers in polyester composite, J. Appl. Phys. **114 (2013)**

Kaynak, C., Akgul, T.: Handbook of Composite Fabrication. Polestar Scientifica Exeter (2001)

Triki, A., Guicha, M., Ben Hassen, M., Arous, M.: Comparative study of the dielectric properties of natural-fiber–matrix composites and e-glass–matrix composites, J. Appl. Polym. Sci. **129** (2012). https://doi.org/10.1002/APP.38499

Havriliak, S., Negami, S.: A complex plane analysis of α-dispersions in some polymer systems. J. Polym. Sci. Part C Polym. Symp. **14**(1), 99–117 (1966)

Triki, A., Ben Hassen, M.: Effect of mechanical consolidation on adhesion mechanism of hybrid non-woven Alfa fibers reinforced unsaturated polyester composite. J. Adv. Phys. **11**(8), 3940–3946 (2016). ISSN 2347-3487

Synthesis and Characterization of Carboxymethylcellulose from Posidonia Oceanica Pulp in Unconventional Media

Imen Landolsi[1(✉)], Ibtissem Moussa[2], Narjes Rjiba[1], and Mohamed Hamdaoui[1]

[1] Textile Materials and Processes Research Unit, Monastir University, Monastir, Tunisia
andolsi-imen@hotmail.com

[2] Research Unity of Applied Chemistry and Environment, Monastir University, Monastir, Tunisia

Abstract. Posidonia oceanica marine wastes, cellulosic biomass, which are widely available in Tunisia, were used as a starting raw material to prepare cellulose derivatives. To accomplish this, the synthesis of carboxymethyl-cellulose (CMC) from totally chlorine free-bleached Posidonia pulp was conducted in homogeneous reaction conditions using DMAc/LiCl as solvent system for cellulosic raw materials and monochloroacetic acid (MAC) as the etherifying reagent. The obtained CMC sample was characterized by estimating the degree of substitution (DS), then using several analytical tools such as FTIR to confirm the formation of the acetylic group, TGA and DSC to study the thermal stability, and SEM to study the morphological appearance of CMC powdered sample.

Keywords: Posidonia ocenica · Cellulose · Homogeneous reaction conditions · Carboxymethyl cellulose (CMC)

1 Introduction

Carboxymethyl cellulose (CMC) is the most important water soluble derivative. It is available in several grades (technical, semi-purified, purified and extra purified) and it is widely employed in many areas of industry and human life such as detergents, paper coating, textile sizing and printing, ceramic, food, toothpaste, pharmaceuticals, etc. (Heinze and Koschella 2005). Also, biopolymers, mainly those from renewable sources such as cellulose have received much attention in research because of their structural versatility, ready biodegradability, and relatively low cost (Srinivasa and Tharanathan 2007; Cao et al. 2000). And nowadays, attention is focused on the recovery of waste ("waste-to-value"). In this context, the valorization of *Posidonia oceanica*, a widely available marine residue in Tunisia, was carried out.

The CMC production in industrial scale is carried out by slurry process in a heterogeneous reaction conditions by swelling cellulose in a solvent without dissolving

This project is carried out under the MOBIDOC scheme, funded by the EU through the EMORI program and managed by the ANP.

S. Msahli and F. Debbabi (Eds.): CIRATM 2020, SPM 17, pp. 307–312, 2022.
https://doi.org/10.1007/978-3-031-08842-1_49

the polymer before functionalization (Heinze et al. 1994). However, Homogeneous processes offer more even distribution of substituents than heterogeneous processes (Heinze et al. 1994). The current study focuses preparation and characterization of homogeneous carboxymethyl cellulose from totally chlorine free-bleached posidonia pulp.

2 Material and Methods

2.1 Materials

P. oceanica leaves collected from the coast of Monastir (Tunisia) was used. This raw material was washed in order to eliminate sand and impurities. Then it was dried and milled to fine particle size.

2.2 Methods

Cleaned and crushed raw material was dignified using soda-anthraquinone pulping process. Then, the obtained pulp was bleached by the TCF (Totally Chlorine Free) bleaching processes, using only the peroxide.

Then, the carboxymethylation in solution under homogeneous reactions conditions of the obtained bleached pulp was carried out, using LiCl/DMAc (lithium chloride in N, N-dimethylacetamide) as solvent system. It is a non-derivatizing solvent, i.e., one that causes dissolution without forming covalent bonds. This step was followed by reaction with a derivatizing agent (monochloroacetic acid (MCA)) to get the desired product (CMC). Indeed, oven dried bleached pulp of cellulose (1 g) was kept at 130 °C under magnetic stirring in 60 ml of N,N-dimethylacetamide (DMAc) over a period of 2 h. After cooling to 100 °C, 3 g of anhydrous LiCl were added to the slurry. Then cooling down to room temperature under stirring and let standing overnight. Next day a suspension of monochloroacetic acid (5.8 g) in 20 ml of DMAc were added and stirred during 10 min at room temperature, followed by a suspension of NaOH40% (20 ml) in 20 ml DMAc. Afterwards, the suspension was vigorously stirred for 24 h at 70 °C. Finally, the reaction mixture was precipitated into 300 ml of ethanol. After filtration, the solid product obtained was washed extensively with ethanol to remove undesirable by-products and dried at 50 °C.

3 Results and Discussions

The prepared material was further characterized using the analytical tools such as degree of substitution titration, FTIR, TGA, DSC and SEM.

3.1 Degree of Substitution (DS)

The determination of the degree of substitution was carried out by the titration method. This technique was widely described in the literature (Salmi et al. 1994). Under these reaction conditions, CMC with degree of substitution of 1.25 was formed.

3.2 FTIR

The effective formation of the acetylic group was confirmed using FTIR studies. The FTIR spectra were performed using a FTIR-8400S SHIMADZU FOURIER TRANSFORM INFRARED SPECTROPHOTOMETER with a wavelength range from 600 to 4000 cm^{-1}. Indeed, Figure 1 shows the infrared spectra of CMC from Posidonia leaves. The appearance of an intense absorption band around 1620 cm^{-1} confirms the presence of the carboxylate group (-COO^- Na +). Thereby giving an evident that during the modification process hydroxyl groups of cellulose are replaced with carboxyl groups. The absorption bands at 1318 and 1420 cm^{-1} are attributed respectively to the COO stretching in CO_2H and binding in OCH_2 (Adinugraha et al. 2005).

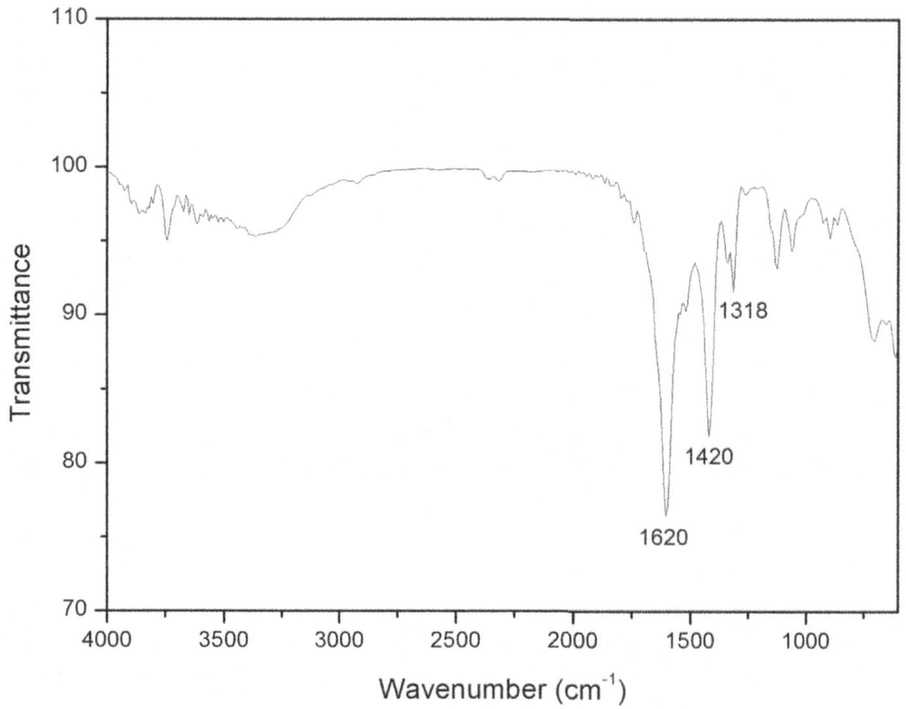

Fig. 1. FTIR spectra of a CMC sample prepared in solution with degree of substitution of 1.25

3.3 TGA

To study the sample's thermal stabilities, TGA Q500 V20.10 Build 36 instrument was used to measure the weight loss at different temperatures with the heating range 35–800 °C and heating rate of 10 °C per minute.

The initial decomposition temperature (IDT) was calculated from the end of the initial straight-line portion of the curve, where the actual decomposition was believed to have begun.

The IDT of CMC was 280 °C, and it showed a rapid weight loss between 280 °C and 380 °C (Fig. 2). At 380 °C, the residual mass was about 80%. Above 380 °C, the rate of weight loss decreased, indicating the occurrence of a different pyrolysis reaction mechanism. As the temperature continued to increase above 700 °C, weight loss continued with a slower rate. Complete decomposition of CMC did not take place, as the residue mass was about 60% when the temperature reached 800 °C.

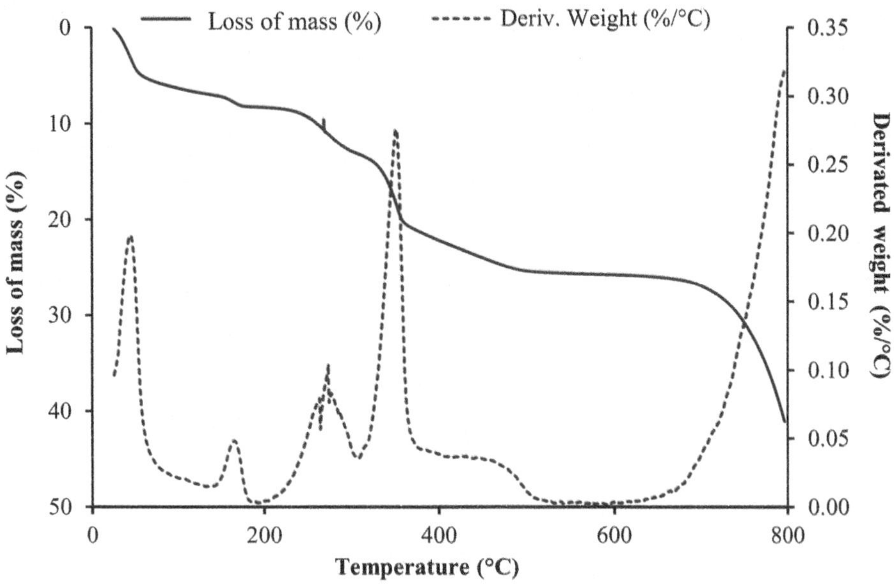

Fig. 2. TGA and DTA curves of the CMC sample prepared from Posidonia oceanica in solution

3.4 DSC

The thermal analyzer DSC Q200 V24.4 Build 116 was used to study the thermal behavior of CMC thus synthesized. Under nitrogen atmosphere at a flow rate of 50 mL, two DS scans were performed: a first scan was made from 25 to 180 °C with a heating rate of 10 min^{-1}, then was cooled to clear the thermal history of the sample. Then the second scan was carried out from 25 °C to 250 °C with a speed of 10 $°C.min^{-1}$.

CMC showed a sharp endothermic peak at 100 °C and an exothermic peak at 220 °C. The glass transition temperature Tg of CMC was around 100 °C. Differential scanning calorimeter (DSC) analysis showed that an exothermic CMC reaction began in the air when temperature exceeded 200 °C, and it had an exothermic peak at 220 °C (Fig. 3). For the thermal decomposition reaction of CMC, the loss of mass (Fig. 2) and the exothermic phenomena were simultaneous. The release of heat associated with CMC mainly occurred within the temperature range of 200 °C to 250 °C. Within this temperature range, hydrogen and oxygen of various functional groups split off the polymer and CMC carbonized. These splitting reactions not only induced weight losses, they also released a great deal of heat.

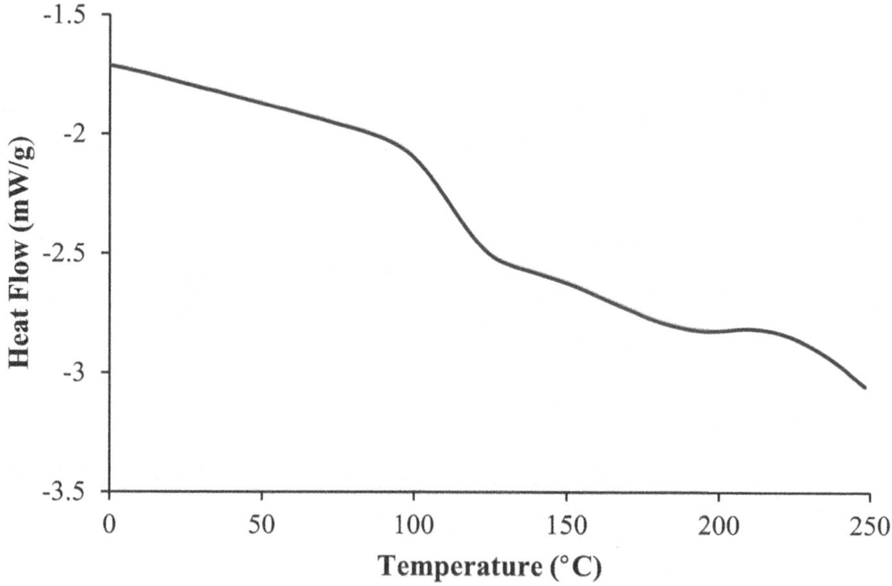

Fig. 3. DSC curve of the CMC sample prepared from Posidonia oceanica in solution

3.5 SEM

The morphological characterization of CMC powdered sample was studied using a high resolution SEM instrument. The sample was examined using a JEOL JSM-IT100 transmission electron microscope. SEM micrographs below show that the cellulose dissolution was carried out successfully (Fig. 4).

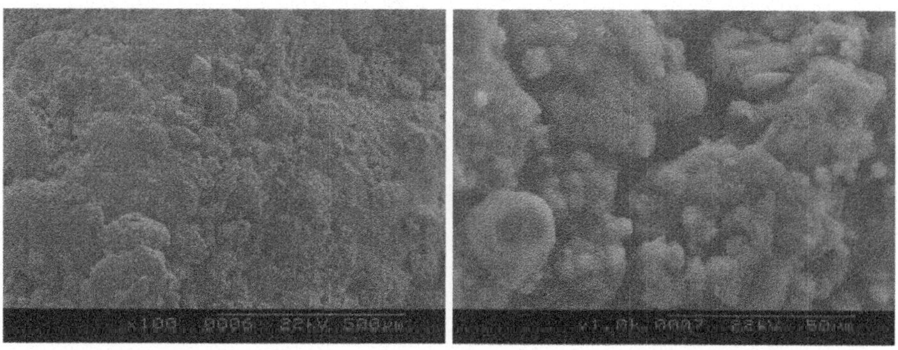

Fig. 4. SEM micrographs of CMC produced from Posidonia leaves by homogeneous reaction conditions

4 Conclusion

Carboxymethylation of cellulose from Posidonia Oceanica through homogeneous process using DMA/LiCl as cellulose solvent system has been successfully carried out with a DS of 1.25. The formation of acetylic group was confirmed by FT-IR Spectroscopy. Based on the DSC results, supported by ATG analyses, the loss of mass and the exothermic phenomena were simultaneous. Which means that thermal decomposition reaction of CMC, not only induced weight losses, they also released a great deal of heat. SEM images confirmed that the prepared CMC have a powdered aspect and the dissolution of cellulosic structure was done successfully, which can be used as a sizing yarn agent in textile industries.

References

Adinugraha, M.P., Marseno, D.W., Haryadi. Synthesis and characterization of sodium carboxymethylcellulose from Cavendish banana pseudo stem (Musa cavendishii LAMBERT). Carbohy. Polym. 62 (Novembre), 164–169 (2005)

Cao, X., Sessa, D.J., Wolf, W.J., Willett, J.L.: Static and dynamic solution properties of corn amylose in N, N-Dimethylacetamide with 3 LiCl. Macromolecules **33**(9), 3314–3323 (2000)

Heinze, T., Erler, U., Nehls, I., Klemm, D.: Determination of the substituent pattern of heterogeneously and homogeneously synthesized carboxymethyl cellulose by using high-performance liquid chromatography. Die Angewandte Makromolekulare Chemie **215**(1), 93–106 (1994)

Heinze, T., Koschella, A.: Carboxymethyl ethers of cellulose and starch – a review. Macromol. Symp. **223**(1), 13–40 (2005)

Salmi, T., Valtakari, D., Paatero, E., Holmbom, B., Sjoholm, R.: Kinetic study of the carboxymethylation of cellulose. Ind. Eng. Chem. Res. **33**(6), 1454–1459 (1994)

Srinivasa, P.C., Tharanathan, R.N.: Chitin/Chitosan — safe, ecofriendly packaging materials with multiple potential uses. Food Rev. Intl. **23**(1), 53–72 (2007)

Synthesis of Iron Oxide Pigments from Iron Waste (Mill Scale)

Nesrine Touzi[1,2](✉) and Karima Horchani-Naifer[1]

[1] LPCMMA, National Research Center in Materials Sciences CNRSM (Technopole Borj Cedria), Soliman, Tunisia
nesrine.touzi89@gmail.com

[2] Department of Physics, Bizerte Science Faculty FSB, Bizerte, Tunisia

Abstract. Mill scale is one of solid waste materials in steel. This work treat the valorization of Mill scale and transformed it into an iron oxide pigments with a high value (non-toxic chemically stable/has a durable and varied paint color). The process of synthesis of iron oxide pigments is carried out either thermally in laboratory batch kiln by calcination at a well-defined temperature (oxidation/reduction) either chemically (addition of precursor or reagent/red-ox reaction). Mill scale and iron oxide pigment (color) were characterized using X-ray diffraction (XRD) analysis, *Differential Scanning Calorimetry* DSC, scanning electron microscopy (SEM). The results revealed that mill scale waste can be successfully converted into grade iron oxide pigment. Chemical analysis has shown that Mill scale contains 72.36% total iron in the form of iron oxides (FeO (31.2%), Fe_3O_4 (33.52%), Fe_2O_3 (34.6%). Thermal analysis by DSC shows exothermic peak which corresponds to dehydration and de-oiling of Mill scale and endothermic peak indicating a possible recrystallization of iron oxides (FeO) which is stable above a temperature of 570 °C.

Keywords: Mill scale · Recycling · Oxidation · Reduction · Iron oxide pigments

1 Introduction

The steel industry is criticized worldwide for the generation of large quantities of solid waste often having a negative impact on the environment. Reusing them as an alternative raw material in other industries is a promising environmental solution. The objective of this work is to offer reliable and feasible solutions for the recycling and recovery of solid waste, mainly mill scale. Annually, several tons of carbon dioxides are produced during steel cooling operations in the order of 40 kg per ton of steel produced. This presents a threat to the environment and a loss for the industry by admitting that the scale is rich in iron oxide (about 70%) so it is better to recover it and reuse it in various sectors besides the cement plant (mill scale is used as a simple addition to the mixture at a low dose). The ever-increasing importance of iron oxide pigments is based on their non-toxicity,

S. Msahli and F. Debbabi (Eds.): CIRATM 2020, SPM 17, pp. 313–319, 2022.
https://doi.org/10.1007/978-3-031-08842-1_50

their chemical stability, their low price and a wide variety of colors ranging from yellow, orange, red, brown to black. Indeed, nowadays, various studies and patents have been filed to obtain iron oxide pigments. Synthetic red iron oxide pigments are conventionally prepared by the reaction of iron salts with alkalis, as described in Penniman *and al* and in the Laux process which use acids and alkalis or the nitro-benzene process or by copperas red process. The aim of all the processes is to synthesize the best synthetic iron oxide pigments made up of well-defined compounds with known crystal structures (Table 1); Goethite (α-FeOOH, yellow color, by precipitation), Hematite (α- Fe_2O_3 (red color, by copperas process, calcination between 600 and 900 °C for 3 h), Maghemite (γ- Fe_2O_3, brown color, by calcination) magnetite at 200 °C for 3 h), magnetite (Fe_3O_4, black color, by reduction).

2 Materials and Methods

2.1 Materials and Analyses

The characterization technique must be sensitive enough to detect as many components present as possible using tools and materials such as elementary analysis by X-ray diffraction (DRX), scanning electron microscopy (SEM), thermal analysis by differential scanning calorimetry (DSC) as well as Fourier-transform infrared spectroscopy (FTIR) of Mill Scale and finished products (pigments). Indeed, this work consists in synthesizing different pigments (colors) in the first step and the synthesis of pigments from a percentage mixture studying mill scale with the pigment already obtained (the pigment is thus treated as a catalyst for the reaction) in the second step. Thus, we proceed according to the chemical phase and the expected particle morphology sometimes by heat treatment and sometimes by precipitation of aqueous media. For the red color (hematite), the Copperas Red method was carried out by adding sulfiric acid (H_2SO_4) to the Mill scale to obtain iron sulfate ($FeSO_4$ $7H_2O$). This product contains a fine powder white ($FeSO_4$) and dark (impurities solid particles, has been calcined (heat in the oven at temperatures between 600 and 900 °C for 3 h. The color of the resulting product (red pigment) giving shades of red while going through brown colors, orange, reddish brown, bright red, brown as the temperature increases, these are the different pigments or structures.

2.2 Methods to Synthesis of Iron Oxide Pigments

The synthesis of a well-defined color pigment is done according to several processes. Thus, we proceed as follows (Table 1):

Table 1. Reaction equations for the production of different iron oxide pigments

Color	Reaction	Process
Red	$6\ FeSO_4.x.\ H_2O + ½\ O_2 \rightarrow Fe_2O_3 + 2Fe_2(SO_4)_3 + 6H_2O$ (1) $2\ Fe_3O_4 + ½\ O_2 \rightarrow 3\ Fe_2O_3$ (2) $2FeOOH \rightarrow Fe_2O_3 + H_2O$ (3)	Copperas process Calcination Calcination
Yellow	$2\ FeSO_4 + 4NaOH + ½\ O_2 \rightarrow 2\ \alpha - FeOOH + 2Na_2SO_4 + H_2O$(4)	Precipitation
Orange	$2\ FeSO4 + 4NaOH + ½\ O_2 \rightarrow 2\ \gamma\ -FeOOH + 2Na_2SO_4 + H_2O$(5)	Precipitation
Black	$3\ Fe_2O_3 + CO \rightarrow 2Fe_3O_4 + CO_2$ (6)	Reduction
Brown	$2\ Fe_3O_4 + ½\ O_2 \rightarrow 3\ \gamma\text{-}\ Fe_2O_3$ (7)	Calcination

*Hematite (red color):Mill scale was ground and filtered to 40 mesh using a sieve. It was then allowed to react with the requisite amount to ..M sulphiric acid (H2SO4) to convert it into sulphate salts(FeSO4 xH2O) following these equations :

$$[Fe, FeO, Fe_2O_3, Fe_3O_4] + H_2SO_4 \rightarrow FeSO_4xH_2O \quad (8)$$

$$6FeSO_4.H_2O \rightarrow 2\ Fe_2O_3 + Fe_2(SO_4)_3 + 3\ SO_2(deg) + 6H_2O(deg) \quad (9)$$

$$Fe_2(SO_4)_3 \rightarrow Fe_2O_3 + 3SO_3(deg) \quad (10)$$

The precursor H_2SO_4 analytical reagent (250 ml) was added to 400 g of Mill scale in a glass beaker. There is a release of heat resulting from an exothermic reaction. The mixture is left to dry in the sun for 6 days (from the 5th day, a green product has been appeared) with agitation from time to time. The acidic environment was created to facilitate the conversion of iron oxides to ferrous or ferric ions in an aqueous solution. The resulting muddy solid product, containing a fine white powder and dark solid particles, was then used as a starting material for the preparation of hematite (red pigment, calcination at temperature between 600 and 900 °C in an oxidizing medium) and according to the flow sheet for the preparation of red iron oxide from mill scale below (Fig. 1).

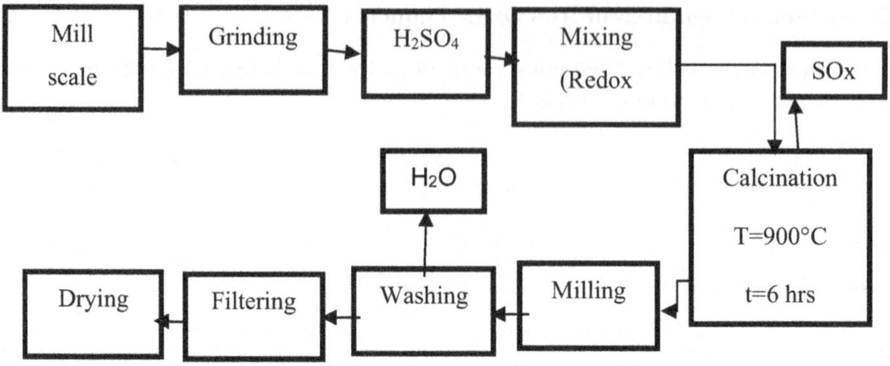

Fig. 1. Flow sheet for the preparation of red oxide from mill scale. *Magnétite (black color):The synthesis of magnetite (black color) takes place by reducing the hematite Fe2O3 at a temperature of 500 ° C and in the presence of the reducing gas CO.

3 Results and Discussions

At the beginning, it is necessary to know the chemical composition and iron oxide phase change diagram of our product. This is illustrated in the table below (Table 2):

Table 2. Analyzes requested and results

Analysis	Methods	Parameter: results	
SiO_2, MgO, Al_2O_3, CaO, Fe_2O_3, Fe, Fe_3O_4 ZnO, FeO	X-ray fluorescence spectrometric	Fe(%): 72.36 SiO_2(%): 0.51 Fe_2O_3 (%): 34.6 Al_2O_3(%): 0.012 MgO(%): <0.01	FeO(%): 31.2 Fe_3O_4 (%): 33.52 ZnO(%): <0.01 CaO(%): 0.005 Apparent: density (g/cm^3): 5.6
Apparent density	Measurement using a graduated cylinder		
Particles size	Sieving method		

The characterization of the raw material (Mill Scale) was done by X-ray fluorescence spectrometric method. This Mill Scale resulting from hot rolling of steel has a percentage of 72.36 of iron, 34.6% of Fe_2O_3 (Hematite), 33.52% of Fe_3O_4 (Magnetite), 31.2% of FeO (Wustite). This confirms that it is rich in iron and iron oxide, resulting in a rentable recovery. It has also been identified by scanning electron microscopy (SEM) and X-ray diffraction DRX which consolidates these results. The red pigment (Hematite) was obtained by the Copperas process by adding concentrated sulfuric acid (H_2SO_4) to the mill scale to obtain iron sulphate ($FeSO_4 . 7H_2O$). This product, containing a fine white powder and dark solid particles, has been calcined (heating in the oven at temperatures between 600 and 900 °C for at least 3 h. The color of the resulting product (red pigment) giving shades of red while passing through brown colors, orange, red brown bright red, brown as the temperature increases (these are the different structures/pigments). The determination of the phase was carried out by DRX which shows that the products are according to the color of the pigment; Hematite (red), Magnetite (black), Maghemite (brown), Goetite (yellow).

3.1 SEM

The observation of the crushed Mill scale with SEM showed a homogeneous structure composed of Iron oxide grains of different sizes and shapes ranging from 1 μm to 14 μm and has average diameters of around 4 μm (Fig. 2). The exploration by image of the red iron pigment with a scanning electron microscope (SEM) shows an aggregate of more or less rounded grains formed of iron oxides and gangue.

Fig. 2. SEM images of mill scale

3.2 FTIR

The FTIR spectrum of Mill scale shows peak in the range of 3860 – 3635 cm^{-1} corresponds to the stretching vibration intermolecular hydrogen bond (O-H) existing between the adsorbed water molecular and indicates the lower amount of hydroxyl group.

3.3 DSC

Thermal analysis by DSC shows exothermic peak which corresponds to dehydration and de-oiling of Mill scale and endothermic peak indicating a possible recrystallization of iron oxides (FeO) which is stable above a temperature of 570 °C.

3.4 DRX

3.4.1 DRX of Mill Scale

The interpretation of the results of the Mill scale XRD analyzes presented in the Fig. 3 and Table 2 shows that the crystalline iron phases are: wustite FeO (predominant) in large quantity, followed by magnetite (Fe_3O_4) and hematite (Fe_2O_3) which are in the form of iron oxides. The influence of non-ferrous compounds, in quantities of approximately 4% (identified in the X-ray fluorescence spectrometric analysis) was not taken into account in this work.

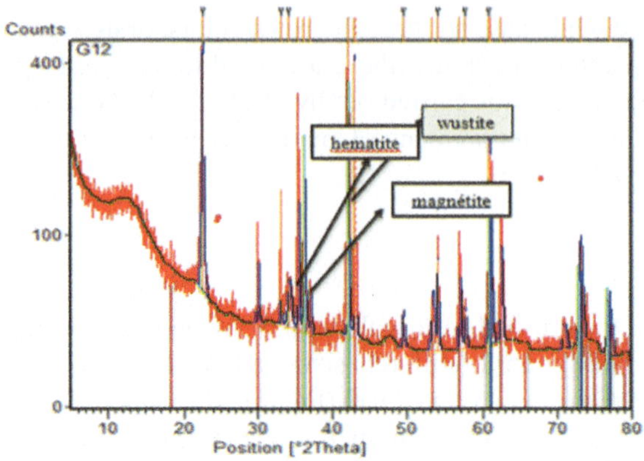

Fig. 3. Characterization of iron cristalline phase in the mill scale by XRD

3.5 DRX of Pigments

-Magnetite: The XRD analysis of the product showed largely the presence of magnetite phase (89%) The slight broadening of the XRD lines can be interpreted in terms of poor crystallinity of the precipitated magnetite and small size of crystallites. In fact, the stoichiometry of this product as determined using XRD technique is near than that of Fe_3O_4. This suggests that the crystal structure is slightly distorted (leading to broadened XRD lines due to the deficiency of Fe ions). The magnetite stoichiometry obtained in this study is acceptably close. The correct stoichiometry ensures that there is little inclusion of other phases, like hematite, into the synthetic magnetite to reduce the denseness of the black color achieved. -Hematite: The calcination of mill scale-derived iron sulphate resulted in hematite phase as detected by X-ray diffraction. From the XRD lines it can be said that the product is fairly crystalline. This product seems to be relatively pure.

4 Conclusion

This work has shown that it is possible to prepare magnetite (Fe_3O_4, color black), hematite (Fe_2O_3, color red), goethite (alpha-FeOOH, color yellow) and maghemite (gamma-Fe_2O_3, color brown) pigments with acceptable purity from mill scale iron waste using natural and low-cost effective methods. The formation of iron oxide precursors has facilitated the precipitation of goethite in an aqueous medium and the synthesis of hematite through calcination. The good color tones is thanks to the small particle sizes and purity of pigment products.

References

Bienvenu, Y., Rodrigues, S.: Fabrication des poudres métalliques à partir de déchets pulvérulents, ENSMP, Centre des matériaux, CNRS UMR 7633, France (2007)

Penniman, Jr., R.S., Zoph, N.M.: Process of Manufacturing İron Compound, West Coast Kalsomine Co,1920, Calıfornıa (1917)

Westerhaus, A., et al.: Color İntensive İron Oxide Balck Pigments Obtained by the Nitrobenzene Reduction Process and a Process for İts Preparation, ktiengesellschaft, Bayer (1990)

Schwertmann, U., Cornell, R.M.: Iron oxides in the laboratory, Preparation and Characterization, VCH Verlagsgesellschaft, Weinheim, Germany (1996)

Low, K.O., Darul Ehsan, S.: (MY): Iron oxide pigments from Mill Scale, IOP Specialists Sdll-Bhd-, selangor Dal'ul Ehsan (MY),4,Birch Stewart Kolasch &Birch Po Box 747 Falls Church, VA 22040-0747 (2006)

Oulsnam, B.T., Stoke-on-Trent : Coloring Pigment and Method Of Manufacture,Corveagh Limited, Dublin. Ireland, United Kingdom,4, Disederius Erasmus, Edelweiss Springs. South Africa

Keglevich de Buzin, P.J.W., et al.: Prodution of ferrous sulfate from steelmaking mill scale. Int. J. Sci. Eng. Res. **5**(4), 353 (2014). ISSN:2229-5518

Saiful Quddus, Md., et al.: Synthesis and Characterization of Pigment Grade Red Iron Oxide from Mill Scal*e*, Institute of Glass and Ceramic Research and Testing (IGCRT), Bangladesh Council of Scientific and Industrial Research (BCSIR), Dr. Qudrat-i-Khuda Road, Dhanmondi, Dhaka-1205, Bangladesh,9, International Research Journal of Pure & Applied Chemistry,Dr. Surendra Reddy Punganuru (2018)

Comparative Life Cycle Assessment of Dental Scrubs Made of Polyester/Cotton and 100% Cotton

Thi Anh Dao Tran[1(✉)], E. Thauvin[2], F. Ferraz[2], Emilie Drean[3], Laurence Schacher[3], and Adolphe Dominique[3]

[1] Ho Chi Minh City University of Food Industry, Ho Chi Minh City, Vietnam
daotta@hufi.edu.vn
[2] ENSISA, 11 rue Alfred Werner, 68093 Mulhouse Cedex, France
[3] École Nationale Supérieure d'Ingénieurs Sud Alsace, Mulhouse, France

Abstract. Due to increasing public awareness and the growing perception of social cognizance about the environment, the textile industry has focused on producing environmentally friendly products, with the environmental impact of a textile product being evaluated as a key point/factor/area for sustainable development.

The aim of this study was to obtain empirical data concerning the environmental impact associated with the life cycle of dental scrubs. The dental scrubs used in this study were the most commonly available products on the French market. They were made of different compositions of fabric (100% cotton and two blends of cotton/polyester) and assumed to be manufactured in two different countries (Tunisia and China). The environmental impact was analyzed to determine which of these scenarios has the best ecological assessment in terms of material and manufacturing site by applying the main principles of the life cycle assessment (LCA) methodology and by using Simapro® software in order to quantify the impacts of the different stages of the life cycle. The results show that the 35% cotton and 65% polyester scrub has less environmental impacts than the other two, especially for aquatic and terrestrial pollution. The majority of the environmental impacts are found to be much more limited when the scrub is made in Tunisia than when it is from China. The results obtained can be used as reference information when considering the ecological aspect of medical clothing or for the development of new dental workwear that will provide the best possible ecological balance.

Keywords: Life cycle · Dental scrubs · Environmental impact · Polyester · Cotton

1 Introduction

Life cycle analysis is becoming not only a scientific methodology but also a marketing tool to communicate the commitments that companies make in order to convey a good image to their customers (Toprak and Anis 2017). The aim of this study was to compare the environmental impact associated with the life cycle of dental scrubs made from

S. Msahli and F. Debbabi (Eds.): CIRATM 2020, SPM 17, pp. 320–327, 2022.
https://doi.org/10.1007/978-3-031-08842-1_51

different materials (polyester/cotton and 100% cotton), and from two manufacturing sites (China and Tunisia) (Thauvin et al. 2013). The results give us an overview of the ecological aspect of dentist's workwear and help us to determine which of these cases has the best ecological balance sheet in terms of material and place of manufacture. The life cycle analysis of these traditional dental scrubs was carried out in France.

2 Materials and Methods

2.1 Description of the Dental Scrubs

We carried out life cycle analyses on three dental scrubs currently available on the French market (Tran et al. 2013). Each scrub has the same design: waist length, V-neck, short sleeves and 2 pockets (Fig. 1), but they are made from different materials: a scrub in 100% cotton, a scrub in 50% cotton and 50% polyester, and a scrub in 35% cotton and 65% polyester. All fabrics weigh 200 g/m^2 and have a 2/1 twill weave structure.

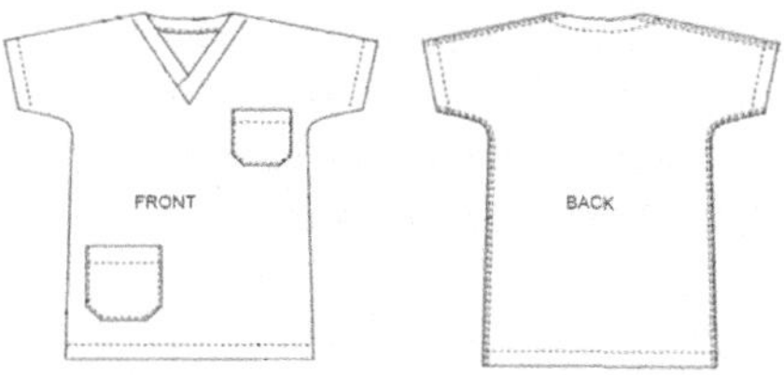

Fig. 1. Design of dental scrub

Regarding the mode of use of dental scrubs, we made assumptions based on the results of the survey of French dentists (Tran et al. 2013): lifetime 1 year, use twice a week, washing after each use at 60 °C and ironing, 104 washing loads during their lifespan.

2.2 Functional Unit and Reference Flow

To carry out the comparative study of the different scenarios, a Functional Unit (or unit of comparison) which is the fact of «wearing a dental scrub for one day» was determined to be applied for the calculations for a product lifespan of one year.

2.3 Study Boundary and Process Flows

The analysis was carried out over the entire life cycle of the scrubs: from the extraction of raw materials to its end of life. The LCA comparison of the scrubs was undertaken using the data from a literature review (Tran et al. 2013; Thauvin et al. 2013) and the Sima Pro® software to create life cycle models of each scrub type which could then be analysed to determine various environmental impacts. Figure 2 shows the process flow for dental scrubs.

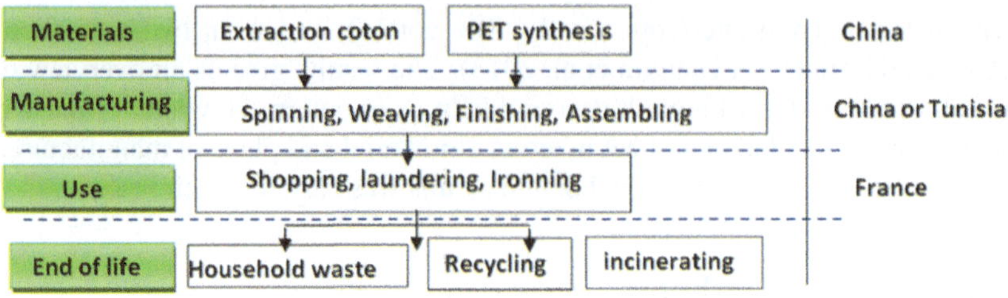

Fig. 2. Process flow for dental scrubs

2.4 Life Cycle Analysis of the Products

The information used to define the data relating to the life cycle of the dental scrubs was entered into the flow inventory in order to enter it in the Simapro Software, this is where we brought the data back to the reference flow. The data corresponding to the logistics (transport) was processed in the next step.

- **Raw materials**: the origin of the dental scrubs' materials were also studied as it is important to integrate this geographical criteria when assessing a product's life cycle.
- **Manufacturing:** It is considered that all stages of manufacturing (from spinning to clothing) are fully integrated in the same company.

  - *Energies:* The energy mix of different manufacturing sites (Tunisia and China) is presented in Table 1 (the energy mix information of China was taken from the Ecoinvent database).
  - *Assembling*: Based on the measurements of the referenced dental scrubs, the assembling operations process of the scrub (totalling 20 operations), we determined the data concerning the weight of a dental scrub's fabric (247,78 g), the time used to manufacture (0.57 h) and cut a scrub (0.07 h). The data of the energy consumed for 1kg of fabric used for sewing and cutting processes is calculated and presented in Table 2. They were also entered into the software.

Table 1. Energy mix from the different manufacturing countries

Technology	Thermal steam	Cogeneration	Gas turbine	Hydraulic	Wind turbin	Natural gas	Nuclear	Petroleum
Tunisia	40,2%	28,8%	27,9%	1,3%	1,8%			
China	78,6%	0,1%		15,9%	0,06%	0,3%	2,13%	2,87%

Table 2. Energy consumption for the production of a dentist scrub.

	Power (W)	Cutting time 1 scrub (h)	Energy consumption for a scrub (247,78 g) (Wh)	Energy consumption for 1 kg of fabric (Wh)
Sewing machine	85	0,57	85 × 0,57 = 48,45	195
Cutting table	930	0,07	930 × 0,07 = 65,1	262

- **Sale:** The distribution channels for selling dental scrubs have been assumed according to the results of a survey of dentists in France (Tran et al. 2013) and are presented in Table 3.

Table 3. Types of distribution channel for selling dental scrubs

Online shopping				Store	Total
Catalogs	Exhibition	Internet	Representant		
40,4%	5,4%	24,6%	5,8%	23,7%	100%

- **Laundering:** We assumed that most French dentists have a washing machine class C with a capacity of 5 kg. The datas for the laundering process is listed in Table 4.

Table 4. Consumption of energy and materials for laundering process

	Capacity	Energy consumption at 60 °C (kWh)	Water consumption (liter)	Laundry detergents consumption (g)
Machine class C	5 kg	0,74 kWh/wash	125L/wash	108 g/wash
Dentist scrub	5 kg/255,77 g = 20 scrubs	0,74 kWh/20 = 0,037 kWh/scrub	125L/20 = 6,25 L/scrub	108 g/20 = 5,4 g/scrub

- End of Life:

We have considered three ways to end the life of waste: household waste, incineration and recycling. After the clothes are collected, the next step is carried out in the following ratio [8]: 40% is reused, 40% is recycled and about 20% is incinerated. Knowing that 15% of textiles are collected, the rest will be thrown away as household waste and thus we can calculate the proportion of total waste: 6% reuse (0.4 * 15%), 3% incineration and 85% domestic waste.

2.5 Logistics (Transport)

The logistics takes into account all routes of transportation during the manufacture of dental scrubs made of different materials. The distance of different trips and means of transport are presented in Table 5.

Table 5. Transport during the manufacture of dental scrubs made of different materials

Activity		Country of manufacture		Mode of transport	Distance (km)
		Tunesia	China		
Cotton harvest	Tunesia	Chengdu-Gangzhou		Truck	1 679
		Gangzhou-Port of Tunis		Ship	15 173
		Port of Tunis-Industrial zone (IZ) of Tunis		Truck	16
			Chengdu-Nanping	Truck	1 910
	USA	Texas-Port of Houston		Truck	578
		Port of Houston-Port of Tunis		Ship	9 473
			Port of Houston-Gangzhou	Ship	20 313
Polyester synthesis		Wilmington-Port of Wilmington		Truck	45
		Port of Wilmington-Port of Tunis		Bateau	9 473
			Port of Wilmington-Gangzhou	Bateau	20 198
Manufacturing and logistics		Nanping-Port of Gangzhou		Truck	781
		Port of Gangzhou-Port of Marseille		Ship	6 935
		Port of Marseille-Paris		Paris	777
			IZ of Tunis-Port of Tunis	Truck	16
			Port of Tunis-Port of Marseille	Ship	866
			Marseille-Paris	Truck	777
Sale (in France)		Warehouse (Paris)-Big cities of France		Van	395
		City of France-Home (round trip)		Car	40

Table 6. Environmental indicators of the CML 2 baseline method

Environmental indicators	Unit	Equivalence
Depletion of natural resources	kg Sb	Antimony
Acidification	kg SO_2	Sulfur
Eutrophication	kg PO_4	Phosphate
Global warming	kg CO_2	Carbon dioxide
The destruction of the ozone layer	kg CFC-11	Trichloromethane
Human toxicity	kg 1,4 - DB	Number of deaths
The eco-toxicity of drinking water	kg 1,4 - DB	Number of deaths
Aquatic eco-toxicity	kg 1,4 - DB	Number of deaths
Terrestrial eco-toxicity	kg 1,4 - DB	Number of deaths
Photochemical oxidation	kg C_2H_4	Ethylene

3 Results and Discussion

3.1 Assessment Method

In our study, we are interested in the impacts related to human health and the environment. We have therefore chosen the calculation method "The CML 2 baseline 2000," which gives values to the different types of pollution and corresponds to the impacts we want to analyze. It also takes into account the indicators mentioned in Table 6.

These units are expressed according to equivalences of polluting substances as each type of pollution is generated by different indicators or chemical substances.

3.2 Comparisons

3.2.1 Influence of the Manufacturing Sites

The results showed that between the two 'classic' manufacturing places considered (China and Tunisia) within the framework of this study, Tunisia generates less impact than China in all indicators except two environmental impact indicators: the destruction of the ozone layer and the depletion of natural resources (Fig. 3).

3.2.2 Influence of Fabrics

From the collected results shown in Fig. 4, we found that the environmental impact on each composition of fabrics differs for each category.

Therefore, depending on the preferred criteria of environmental effects, we can choose the corresponding materials. For example, the results showed that 100% cotton material caused less impact on resources and climate change, while 35% cotton and 65% polyester material caused less environmental pollution. This result is consistent with the fact that cotton cultivation generally has a larger impact on the environment (due to the consumption of a very large amount of water and pesticides) than polyester synthesis (Eija and Pertti 1999).

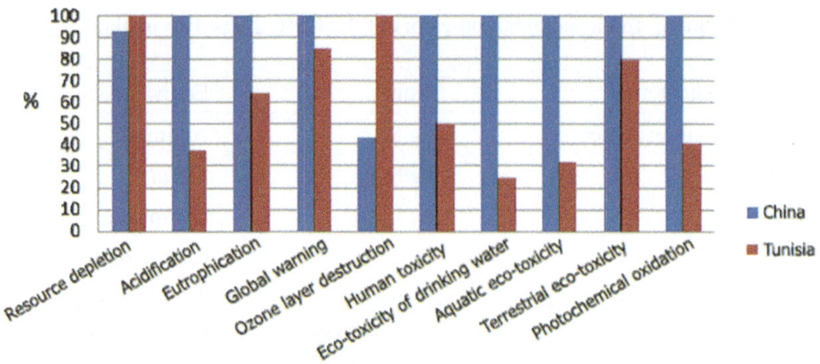

Fig. 3. Influence of manufacturing site on the environmental impact

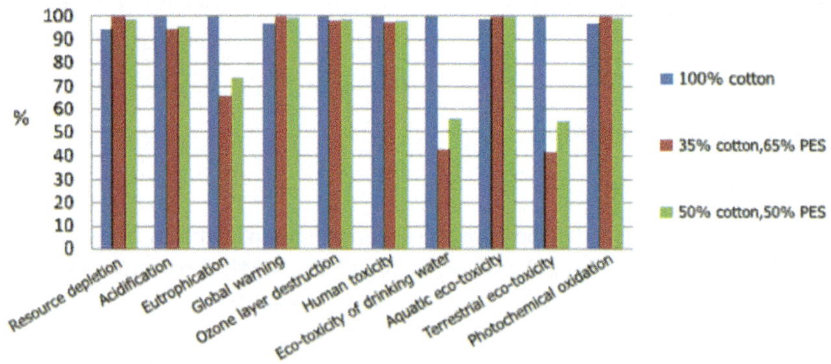

Fig. 4. Comparison of life cycle impacts of 100% cotton versus cotton/polyester scrubs

4 Conclusion

An analysis of the life cycle of dental scrubs made from different compositions of fabric (100% cotton and two blends of cotton/polyester) that were manufactured at two different sites (Tunisia and China) was made. The dental scrubs used in the study are the most commonly available products on the market.

Interpretation of the results revealed that the 35% cotton and 65% polyester scrub has less environmental impacts than the other two. The majority of the environmental impacts are found to be much more limited when the scrub is made in Tunisia than when it is from China. It is noted that this study is based on several hypotheses obtained from existing research which were then cross-referenced during an in-depth literature review. However, we did not compare the impacts of different end-of-life scenarios (recycling, reuse or disposal).

The results of this study can be used as reference information when considering the ecological aspect of medical clothing. This information can also be useful for developing new more efficient and environmentally friendly dental workwear. The study can be continued on other new materials when considering a new type of dental scrub that can provide comfort and protection to dental surgeons while respecting the environment.

References

Toprak, T., Anis, P.: Textile industry's environmental effects and approaching cleaner production and sustainability, an overview. J. Text. Eng. Fashion Technol. **2**(4), 429–442 (2017). https://doi.org/10.15406/jteft.2017.02.00066

Eija, M.K., Pertti, N.: Life cycle assessment environmental profile of cotton and polyester-cotton fabrics. AUTEX Res. J. **1**(1), 8 (1999)

Tran, D., Schacher, L., Adolphe, D.C., Arnold, M., Reys, G.: AUTEX Conference, Book of Proceedings, Dresden, Germany, p. 67 (2013)

Thauvin, E., Ferraz Ferreira A.F., Tran, D.: Analyse et Développement d'EPI Eco-conçus pour le milieu médial, University project, École Nationale Supérieure d'Ingénieurs Sud-Alsace (ENSISA), France (2016)

SimaPro I World's Leading LCA Software, Information sur. http://www.pre-sustainability.com/simapro

Treatment of Tannery Effluent Based on Electrochemical Process Combined to UV Photolysis

Sameh Jallouli[1(✉)], Imen Ben Chabchoubi[1,2], Olfa Hentati[1,3], and Mohamed Ksibi[1,3]

[1] Laboratoire Génie de l'Environnement et Ecotechnologie (GEET), Université de Sfax, Ecole Nationale d'Ingénieurs de Sfax (ENIS), Route de Soukra, Km 3,5; B.P. 1173, 3038 Sfax, Tunisie
sameh.jallouli.etud@fss.usf.tn

[2] Institut Supérieur de Biotechnologie de Monastir, Université de Monastir, Rue Taher Haddad, 5000 Monastir, Tunisie

[3] Institut Supérieur de Biotechnologie de Sfax, Université de Sfax, Route de Soukra, Km 4,5; B.P 1175, 3038 Sfax, Tunisie

Abstract. The use of electrochemical process in tannery wastewater treatment is a widely employed technique. Its main advantages includes environmental compatibility, amenability to automation and cost effectiveness. In the present study, an attempt has been made to characterize physicochemical parameters of tannery effluent and evaluate the electrocoagulation (EC) treatment alone or in combination with UV-photolysis (EC-UV) to remediate tannery effluent. The results revealed that wastewater contained high chloride ($Cl^- = 1573$ mg/L), high sulfide ($S^{2-} = 204$ mg/L) and to have a high chemical oxygen demand (COD = 7376 mg O_2/L). Removal of COD performances was well below the prescribed permissible limits for effluent discharged when combined EC-UV process was used compared to EC process alone. To confirm the efficacy efficacy of the physicochemical treatments, avoidance assay was carried out with terrestrial organisms, *Eisenia andrei*.

Keywords: Tannery effluent · COD efficiency removal · EC-UV treatment process · Earthworms · Avoidance test

1 Introduction

Tanning industry is one of the oldest industries in the world and plays a significant role in the economy of a country through employment and export earnings (Paul et al. 2013). But the tannery industry wastewater typically includes hazardous pollutants such as acids, chromium salts, tannins, solvents, sulfides and dyes (Elabed et al. 2019) that have severe impacts once discharged in the environment. So, a number of research works have been carried to remove organic pollutants and suspended solids from tannery effluents using different technologies, such as adsorption (Hossain et al. 2019), chemical coagulation (Song et al. 2004), ozonation (Mella et al. 2018), and biological treatment (Hasegawa et al. 2011). In this context, the method of treatment does not always achieve

S. Msahli and F. Debbabi (Eds.): CIRATM 2020, SPM 17, pp. 328–334, 2022.
https://doi.org/10.1007/978-3-031-08842-1_52

satisfactory performance due to the toxicity of the tannery wastewater. Thus, it is necessary to investigate powerful process for the treatment of tannery wastewater. In fact, the electrochemical process has been actually described as a clean and an environmentally sustainable wastewater treatment technology. For instance, the combinations of EC with other methods are gaining interest and have been reported by several researchers (GilPavas et al. 2019; Thakur and Chauhan 2007). The aim of the study was to evaluate the treatment efficacy by electrocoagulation (EC) process and EC combined with UV photolysis of discharged polluted water collected from a tannery located in Moknine in center-east of Tunisia. Physicochemical properties of the effluent during the electrochemical process were monitored and spiked soils with raw, diluted and treated effluents were tested for acute toxicity to earthworms.

2 Materials and Methods

2.1 Physicochemical Parameters of the Tannery Effluent

The effluent used in this study was supplied from a Tunisian tannery. To achieve the aforementioned objectives, an exhaustive characterization of the effluent was performed based on the laboratory studies and standard methods. For this purpose, pH, electric conductivity (mS/cm), the COD (mg/L), total suspended solids (TSS mg/L) and chloride (Cl^-) and sulfide (S^{2-}) concentrations (mg/L), were determined.

2.2 Electrocoagulation of Tannery Effluent

Two pairs of aluminum plates of rectangular shape 75 mm × 40 mm were used as electrodes. In the upper part of the EC reactor, the electrode plates were firmly assembled in an upright position and arranged parallel to each other with a gap between the anode and cathode plates, using a non-conducting support to avoid any short circuits. The electrodes were positioned in the EC reactor at a distance of 3 cm and were connected to terminals of direct current power supplier (TM 501-2) during 5 h.

2.3 Photolysis of the Tannery Effluent

The UV photolysis was conducted in a 120 μL cylindrical laboratory-scale photoreactor, operating in a closed recirculating circuit driven by a centrifugal pump. The tannery effluent was irradiated out using an 11 w low-pressure mercury lamp (Philips, Holland) with a wavelength in the range of 200–280 nm. The UV photolysis process was run during 3 h. Samples collected at different time intervals (5 h after EC treatment and 5 h EC + 3 h UV treatments) were analyzed for the reduction of COD.

2.4 COD Measurement

Chemical oxygen demand (COD) was determined according to Knechtel method (Knechtel 1978) according to the following formula:

$$\%\ \text{Removal COD} = \frac{C_0 - C_F}{C_0} \times 100$$

where C_0 and C_F are the initial and final oxygen concentrations (mg/L) before and after treatment, respectively.

The absorption spectra were recorded at different stages of photolysis in the range of 200–450 nm.

2.5 Soil and Solutions Collection and Preparation

The area selected for soil sampling is located in the northern Sfax at approximately 7 km far from the city of Sfax. These gardening lands belong to a private domain and are used for the production of fruit crops. 14 kg of soil were collected from a 0–10 cm soil depth. The physicochemical properties of the soil were determined on samples air dried and sieved to particle size <1 mm. The soil matrix was checked for its maximum water holding capacity (WHC_{max}). The tannery raw effluent (TWW) was diluted (DW) at 10, 15, 20 and 50% v/v in distilled water. Dilutions were chosen based on the study of Mlaik et al. (2011). The efficacy of single (EC) and combined treatment (EC-UV) of the crude waste was evaluated and compared to the control (Ctrl, soil mixed with distilled water), dilutions and to a laboratory prepared salty solution (SW) with the same conductivity value of the tannery raw effluent.

2.6 Avoidance Tests with Earthworms

The avoidance test is based on the natural behavior of the worms when they are given the choice between two different soils, one is clean (Ctrl) and the other is polluted (test soil). The test was conducted according the OECD 207 (1987) standard. Acute toxicity assay was evaluated in the natural soil thoroughly mixed to 50% of its WHC_{max} with distilled water, TWW, TW-EC, TW-EC-UV, SW and DW. Mature earthworms (*Eisenia andrei*) were bred in the laboratory and were selected for their weight and clitellum appearance. Four replicates were performed for each experiment, in plastic trays divided into two equal sections using a vertically introduced card. One side of the box was filled with 200 g Ctrl soil, while the other side contained the same amount of test soil. The dividers were then removed and ten mature worms were placed in the line separating both soils. The containers were covered by perforated lids to allow animal respiration and were kept under a photoperiod of 16:8 h/light: dark. The test period was 48 h at the end of which the plastic card was reinserted and the number (n°) of individuals at each section counted and the results were expressed in % of avoidance according the following formula:

$$\%\ \text{Avoidance} = [(C - T)] / N * 100$$

where C = n° of individuals in the control soil; T = n° of individuals in the test soil; N = total n° of individuals. Fisher exact test was applied for statistical significance at $p < 0.05$ level.

3 Results and Discussion

3.1 Determination of Physicochemical Parameters of the Tannery Effluent

The Table 1 showed that the analyzed parameters of the tannery effluent were higher than the permissible limits of discharge as indicated by the Tunisian norm (NT 2018-315).

Table 1. Characterization of raw tannery effluent (TWW) Compared to Tunisian norm for discharges in public canalizations

pH	Conductivity (mS/cm)	COD (mg/L)	TSS (mg/L)	$[Cl^-]$ (mg/L)	$[S^{2-}]$ (mg/L)	References
10.8	92	7376	2400	1573	204	**Present work**
6.5 < pH < 9	5	1000	400	700	3	**NT 2018-315**

Table 2. The COD (mg/L) and COD removal (%) before and after TWW single (EC) and combined (EC-UV) treatments

COD in raw TWW (mg/L)	Treatment	COD after treatment (mg/L)	COD removal (%)
7376	**EC**	1018.12	86
	EC-UV	429	94.12

The effluent is largely alkaline suggesting that it could negatively affected terrestrial and aquatic habitats (Gomes et al. 2016). The conductivity is more than 18 times higher than the indicated by the standard, suggesting the presence of large amounts of minerals or soluble salts (Diagne et al. 2017). The high COD (more than 7 times the norm) indicates the presence of large amounts of biologically resistant organic substances in the effluent (Saxena and Bharagava 2016). The tannery effluent is almost 5 times enriched in suspended solids suggesting its opacity to the sunlight and so the susceptibility to the eutrophication of receiving aquatic ecosystems (Kaur et al. 2010). The effluent is more than the double enriched in chloride and 68 times in sulfide than the tolerable limits. This reflects effectively the presence of soluble salts with acids, complexed metals and probably organic dyes residues used to color the animal skins (Amanial 2016). All these parameters suggest a possible endorsed toxicity in the effluent.

EC and UV photolysis processes were applied separately or combined in order to decrease COD and to reduce a supposed toxicity of the effluent. The first treatment allowed a decrease up to 86%, while the second allowed achieving 94% of COD removal (Table 2).

The high drawdown of COD means that the combined process EC-UV photolysis would be an interesting alternative for tannery wastewater treatment prior to be released in the environment or in the public canalizations. The smallest COD (429 mg/L) obtained after the EC-UV treatment is the half of the Tunisian standard for industrial effluent discharges to surface waters and publicly owned treatment works.

These results are in accordance with the previous work of Gutiérrez-Bouzán and Buscio (2018). Authors combined electrochemical and UV photolysis, enabling the reuse of water as part of an ecofriendly sustainable process and of recirculating economy. The Fig. 1 illustrates the absorption spectra of the tannery effluent before (TWW) and after EC (during 5 h) and EC-UV (during 5 and 3 h) treatments.

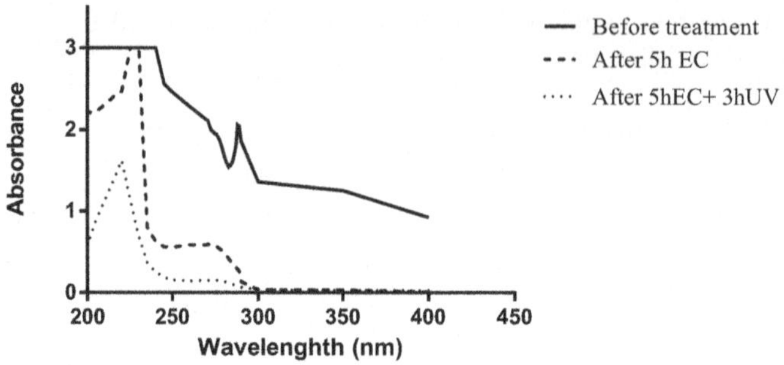

Fig. 1. Absorbance spectra of Moknine tannery effluent after 5h of EC and 5hEC+3hUV degradation

The results showed that the UV absorbance of dissolved organic matter in wastewater decreased drastically after the treatment with combined EC-UV process. The absorbance at 225 and 275 nm decreased, indicating the complex reaction process related to free substituents and aromatic rings from cyclic aromatic compounds present in leather industry wastewater. Moradi and Moussavi, (2019) confirmed the efficiency of the EC process followed by UVC/VUV since they obtained a COD removal up to 99.52% for the treatment of tannery wastewater.

3.2 Acute Toxicity Assay

The relative sensitivity of *E. andrei* earthworms towards untreated (TWW), treated (TW), salty water (SW) and diluted tannery wastewater (DW) was assessed by acute toxicity. The avoidance response (%) after 48h of exposure to voluntary spiked soils is shown in the Fig. 2.

During the test period no mortality was observed. Statistically, significant responses were obtained between control and modified soil. It can be noted that the mixed soil with the raw TWW was significantly (100%) unappreciated, reflecting and confirming its toxicity to earthworms. The single (TW-EC) and combined (TW-EC-UV) treatments, as well as the salty matrix (SW) and the less diluted effluent (DW10%) were similarly avoided ($95 \pm 5\%$). Avoidance sharply decreased to 15 ± 17, to -45 ± 13 and to $-40 \pm 18\%$, at dilutions of 15, 20 and 50%, respectively. This means that the mixed matrix with 15% DW is liked more than the previous combinations and could be an appropriate habitat, since the avoidance is less than 60%. But the habitat function is similarly improved when the dilution increases (20 to 50%). The avoidance percentage may decrease if the concentrations of toxic intermediates decrease during treatments and dilutions. This laboratory scale simulation, translates that the discharge into the soil of TWW can indirectly affect soil living organisms. Although the TWW was diluted, these results demonstrated the potential harmful effect of its compounds at short term on terrestrial biodiversity. The similarity between SW and single (EC) and combined treatment (EC-UV), might recall that after discarding all microorganisms and breaking down organic molecules, soluble minerals and salts in water could explain the toxicity to earthworms. These results lead to focus on more detailed dilution interval of the effluent

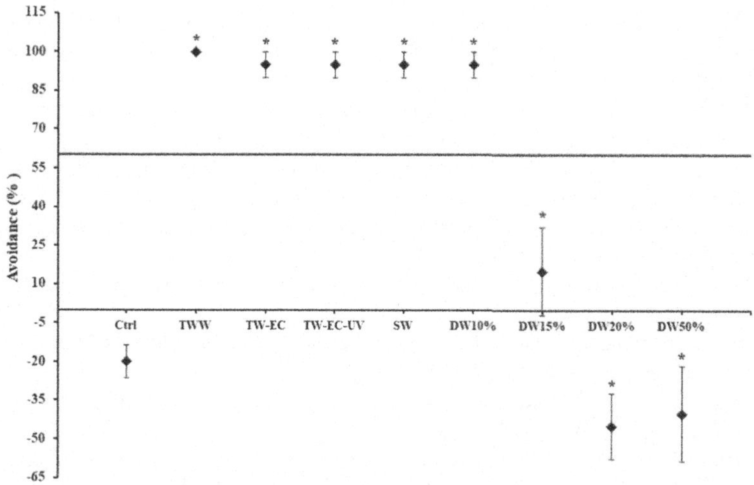

Fig. 2. Avoidance responses of *Eisenia andrei* to natural soil (Ctrl) contaminated with untreated (TWW), treated (TW-EC, TW-EC-UV), salty water (SW) and diluted tannery wastewater (DW). *: highly significant response at $p < 0.001$.

between 10 and 20%. Additionally, identification of the remaining micro-pollutants in electrocoagulation and UV photolysis treated water would provide rational explanation of the worms' behavior.

4 Conclusion

This study attempted to characterize the effluent released by a tannery industry in center east of Tunisia and to assess the efficiency of electrocoagulation alone or combined to UV photolysis for the removal of highly toxic components from samples. This effluent was demonstrated to be alkaline and contained organic and inorganic pollutants with high load values. The two-stage process of coagulation EC-UV is a proposal that requires further research, which should use a larger number of tannery wastewater samples, other advanced oxidation processes (AOPs), and involve detailed cost evaluation studies. Acute toxicity assay revealed significant effect of treated water by EC and EC-UV compared to the control and salty water.

References

Amanial, H.R.: Physico-chemical characterization of tannery effluent and its impact on the nearby river. J. Environ. Chem. Ecotoxicol. **8**(6), 44–50 (2016)

Diagne, I., et al.: Caractérisation physico-chimique et contamination métallique des eaux usées déversées au niveau de la baie de Hann (Dakar/Sénégal). Int. J. Biol. Chem. Sci. **11**(1), 462 (2017)

Elabed, A., et al.: Sustainable approach for tannery wastewater treatment: bioelectricity generation in bioelectrochemical systems. Arab. J. Sci. Eng. **44**(12), 10057–10066 (2019)

GilPavas, E., et al.: Optimization and toxicity assessment of a combined electrocoagulation, H2O2/Fe2+/UV and activated carbon adsorption for textile wastewater treatment. Sci. Total Environ. **651**, 551–560 (2019)

Gomes, H.I., et al.: Alkaline residues and the environment: a review of impacts, management practices and opportunities. J. Clean. Prod. **112**, 3571–3582 (2016)

Gutiérrez-Bouzán, C., Buscio, V.: Combining electrochemistry and UV for the simultaneous wastewater decolorization and reduction of salinity. AIMS Environ. Sci. **5**(2), 96–104 (2018)

Hasegawa, C., et al.: Biotreatment of industrial tannery wastewater using Botryosphaeria rhodina. J. Serb. Chem. Soc. **76**(3), 439–446 (2011)

Hossain, M.N., et al.: Treatment of tannery effluent using a bio-adsorbent. Int. J. Adv. Res. Innov. **5** (2019)

Kaur, A., et al.: Physico-chemical analysis of the industrial effluents and their impact on the soil microflora. Procedia Environ. Sci. **2**, 595–599 (2010)

Knechtel, J.R.: A more economical method in the determination of chemical oxygen demand. Water Pollut. Control **116**, 25–27 (1978)

Mella, B., et al.: Treatment of leather dyeing wastewater with associated process of coagulation-flocculation/adsorption/ozonation. Ozone: Sci. Eng. **40**(2), 133–140 (2018)

Mlaik, N., et al.: Treatment of unhairing effluents by activated sludge system. Environ. Prog. Sustain. Energy **30**, 337–346 (2011)

Moradi, M., Moussavi, G.: Enhanced treatment of tannery wastewater using the electrocoagulation process combined with UVC/VUV photoreactor: parametric and mechanistic evaluation. Chem. Eng. J. **358**, 1038–1046 (2019)

NT, 2018-315 (Norme Tunisienne) : Protection de l'environnement rejets d'effluents dans le milieu hydrique Norme homologuée par arrête du ministère des affaires locales et de l'environnement et le ministre de l'industrie et des petites et moyennes entreprises du JORT 26 (2018)

OECD 207: Guideine for Testing of Chemicals, Earthworm, Acute Toxicity Tests (1984)

Paul, H., Antunes, P., Covington, A., Evans, P., Phillips, P.S.: Bangladeshi leather industry: an overview of recent sustainable developments. J. Soc. Leather Technol. Chem. **97**, 25–32 (2013)

Saxena, G., Bharagava, R.N.: Organic pollutants in tannery wastewater and bioremediation. Appr. Environ. Saf. **34** (2016)

Song, Z., et al.: Treatment of tannery wastewater by chemical coagulation. Desalination **164**(3), 249–259 (2004)

Thakur, S., Chauhan, M.S.: Electro-coagulation integrated with advance oxidation processes : Tech. Rev. Treat. Ind. Wastewater **5**(6), 6 (2007)

Nonwoven from Cotton Waste

Mohamed Ben Hassen[1,3(✉)], Mohamed Taher Halimi[2,3], and Bechir Wannassi[3]

[1] Department of Industrial Engineering, College of Engineer, Taibah University, Medina, Saudi Arabia
m.benhassen@taihbahu.edu.sa, benrayen@yahoo.fr
[2] Financial and Administrative Sciences Department, Community College Dammam, Imam Abdulrahman, Faisal University, Medina, Saudi Arabia
mthalimi@iau.edu.sa
[3] Laboratory of Textile Engineering, University of Monastir, ISET Ksar Hellal, Monastir, Tunisia

Abstract. A quality assessment of recycled fibers obtained from two different kind of post-industrial wastes indicate that in spite of the repeated mechanical constraints; their physical and mechanical properties enable them to be exploited in dry nonwoven manufacturing. According to the level average analyses, nonwoven strength, elongation and Young modulus are significantly affected by number of layer and needle punching passage number. Results indicate that needle punching passage number and the layer number are more influent on recycled fiber nonwoven properties then recycled fiber type.

Keywords: Non woven · Cotton · Waste · Drylaid · Needle punched

1 Introduction

Recycling is a response to critical challenges related to environmental protection. This is why, developed and under developing countries give interest for recycling and reuse of industrial wastes (Rigamontia et al. 2016; Kiziloglu and Serinkan 2015; Baumann and Tillman 2004).

Textile recycling is the method of reusing or reprocessing waste fibers, waste yarns, waste fabrics or used clothing by an appropriate manufacturing process (Wang 2006; Bartl et al. 2005).

Researchers worked on the recycling of a wide range of waste from spinning fiber to carpets, composite, end of life vehicles wastes, and the possibilities to reuse them specially on blend with virgin fibers (Halimi et al. 2008a, b; Wanassi et al. 2016; Wang et al. 2016; Sakthivel et al. 2012).

In this study, we are interested to produce nonwoven from two different kind of textile wastes: yarns and textiles garments wastes collected from Blue Denim manufacturer. The main idea is to study the effect of fiber quality, product characteristics and needle-punched parameters on the quality of nonwoven produced.

S. Msahli and F. Debbabi (Eds.): CIRATM 2020, SPM 17, pp. 335–339, 2022.
https://doi.org/10.1007/978-3-031-08842-1_53

2 Material and Method

2.1 Fiber and Nonwoven Testing

Yarn wastes used in this work are collected at the end of an Open-end process. Garment wastes were collected from the end of the weaving process of Society of Textile Industries (SITEX). All fiber tests are conducted according to ISO 139 standard. The mechanical properties of obtained non wovens including breaking strength and elongation were determined using ISO 13934-1(1999) standard.

2.2 Wastes Recycling Process and Non woven Preparation

The overall process used for non-woven manufacturing is presented in the Fig. 1. The non-woven consolidation is performed by Needle-punching machine. This method consists of mechanically interlocking fibers by repeatedly punching through the fiber web with an array of barbed needles (Fig. 2).

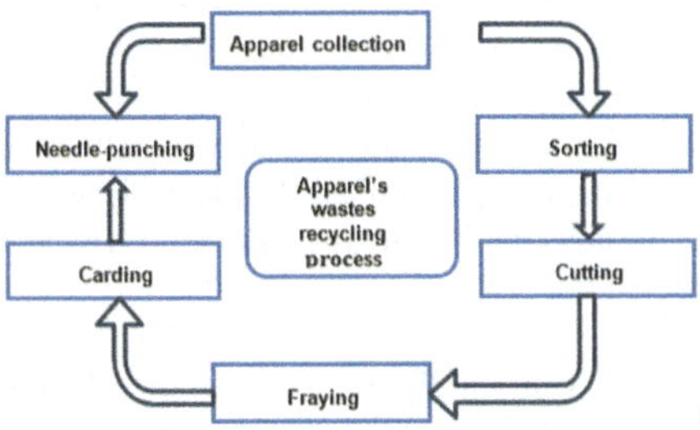

Fig. 1. Waste recycling process

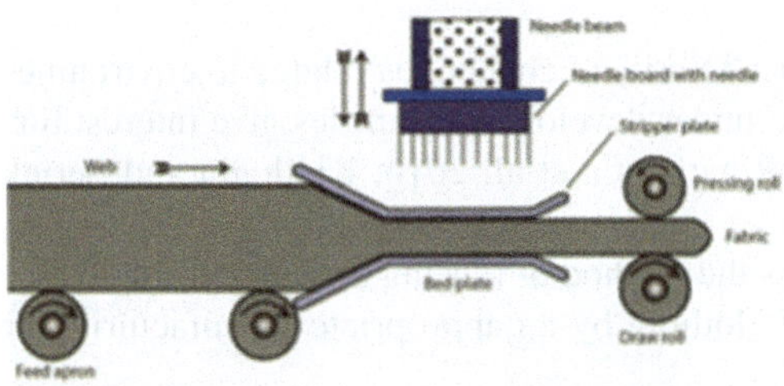

Fig. 2. Needle punching principle.

2.3 Design of Experiment

A literature study shows that the factors: fiber type (FT) needle punching passage (NPN) and layer number (LN), were influential on the quality of nonwoven. Table 1 show the levels of each factor considered in DoE. Therefore, L09 orthogonal array (Table 2) was selected for the experimentation and each experiment was performed three times and average in each experiment was considered as the response variable.

Table 1. Factors and levels used in fractional set.

Level	Fiber type (FT)		Needle punching passage number	Layer Number
	F	T	NPN	LN
1	75	25	1	2
2	50	50	2	4
3	25	75	3	8

Table 2. The design of trails for 9 experiments.

Sample	FT		NPN	LN
	Yw	Gw		n
1	25	75	2	2
2	25	75	3	4
3	25	75	1	8
4	50	50	2	4
5	50	50	3	8
6	50	50	1	2
7	75	25	2	8
8	75	25	3	2
9	75	25	1	4

3 Results and Discussions

Figure 3 presents a sample of nonwoven produced. Figure 4 illustrates effects of different parameters studies on non woven properties according to Minitab software.

Fig. 3. Non woven sample.

Figure 4 illustrates clearly that the layer number is the most influential factor on the nonwoven strength, then we note the needle punching passage number. Whereas the effect of fiber type is, relatively, limited. These results can be explained by the fact that with the increasing of layers gives more material to the structure and at the same time; the number of passages improve fibers overlapping with a deep penetration of barbed needles.

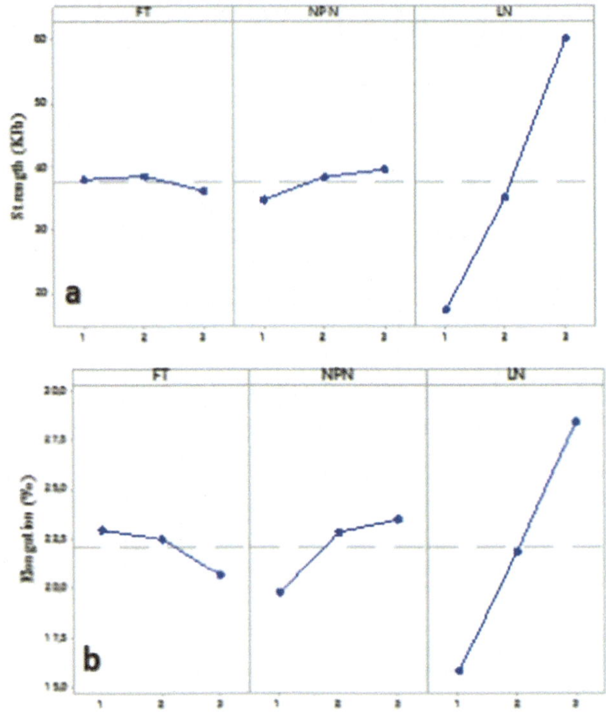

Fig. 4. (a) Main effect plot for nonwoven strength. (b) Main effect plot for nonwoven elongation.

It can be seen in the main effect plot (Fig. 4b) that fiber type has a negative effect on nonwoven elongation. Indeed, the adding of woven recycled fibers favors the elongation's property. In addition, results can be explained by the higher short fiber content in recycled woven that can be slipped inside the structure during breaking. Before being needle-punched action, the web was formed with parallel fibers. But during this process, the vertical action of needles causes an entanglement between fibers which forms the web and its orientation become more randomly. Therefore, if the passage numbers of fibers web on the needle-punching machine increase, the orientation of fibers become more and more randomly. For this reason, during dynamometric test the rupture of the specimen was carried out only when the fibers become parallel (under the action of stretching). Consequently, the nonwoven elongation increases when the passage number increase (fiber orientation more randomly). On the other hand, layer number and the needle punching passage numbers have an increasing effect on nonwoven elongation.

4 Conclusion

A quality assessment of recycled fibers obtained from two different kind of post-industrial wastes indicate that despite of the repeated mechanical constraints; their physical and mechanical properties enable them to be exploited in dry nonwoven manufacturing. According to the level average analyses, nonwoven strength, elongation and Young modulus are significantly affected by number of layer and needle punching passage number. Results indicate that needle punching passage number and the layer number are more influent on recycled fiber nonwoven properties then recycled fiber type.

References

Bartl, A., Hackl, A., Mihalyi, B., Wistuba, M., Marini, I.: Recycling of fiber materials. Process Saf. Environ. Prot. **83**(4), 351–358 (2005)

Baumann, H., Tillman A.M.: The Hitch Hiker´s Guide to LCA – An Orientation in Life Cycle Assessment Methodology and Application, Studentlitteratur, Lund (2004). ISBN: 9144023642

Halimi, M.T., Ben, H.M., Azzouz, B., Sakli, F.: Effect of cotton waste and spinning parameters on rotor yarn quality. J. Text. Inst. **98**(5), 437–442 (2008a)

Halimi, M.T., Ben, H.M., Azzouz, B., Sakli, F.: Cotton waste recycling: quantitative and qualitative assessment. Resour. Conserv. Recycl. **52**, 785–791 (2008b)

Kiziloglu, M., Serinkan, C.: Perception of strategical management in textile sector. Procedia. Soc. Behav. Sci. **207**, 306–314 (2015)

Rigamontia, L., Sterpib, I., Grossoa, M.: Integrated municipal waste management systems. Ecol. Ind. **60**, 1–7 (2016)

Sakthivel, S., Ramachandran, T., Archana, G., Ezhilanban, J.J., Sivajith Kumar, V.M.S.: Sustainable non woven fabric composites for automotive textiles using reclaimed fibers. Int. J. Eng. Res. Dev. **4**(7, 11–13 (2012)

Wanassi, B., Azzouz, B., Ben Hassen, M.: Value-added waste cotton yarn: optimization of recycling process and spinning of reclaimed fibers, Ind. Crops Prod. **87**, 27–32 (2016)

Wang, Y.: Recycling in Textiles. Georgia Institute of Technology, Woodhead Publishing (2008). ISBN: 9781845691424

Wang, S., Meixia, M., Weil, M., Xu, O., Jia, H.: Functional porous carbons from waste cotton fabrics for dyeing wastewater purification. Fibers Polym. **17**(2), 212–219

Tyrosol Valorization by Photocatalysis: Chemical Important Intermediate (Extended Abstract)

Hiba Khlifi[1(✉)], Leila Elsellami[1], and Francesco Parrino[2]

[1] Laboratoire de Recherche Catalyse et Matériaux Pour l'Environnement et les Procédés URCMEP (UR11ES85), Faculté des Sciences de Gabès, Université de Gabès, Campus Universitaire Cité Erriadh, 6072 Gabès, Tunisia
Khlifihiba194@gmail.com

[2] Department of Industrial Engineering, University of Trento, Via Sommarive 9, 38123 Trento, Italy

Keywords: Hydroxytyrosol · TiO2 · Selectivity enhancement · TiO2 fluorination

1 Introduction

Hydroxytyrosol is recently considered as the most powerful and expensive natural antioxidant due to its interesting properties which are fundamental for its well documented applications. Different Hydroxytyrosol synthetic approaches have been reported in the literature, which are unsafe operating conditions (chemical syntheses) in some cases and time demanding processes (biological syntheses) in the other cases. Making an alternative greener approach using heterogeneous photocatalysis is highly desirable. Photocatalytic partial oxidation of tyrosol, which is also a phenolic compound with a similar structure of Hydroxytyrosol, has been found to be a successful approach to synthesize Hydroxytyrosol. In this paper, we have investigated the photocatalytic partial oxidation of tyrosol to Hydroxytyrosol in aqueous suspension using TiO2 as photocatalyst. In order to enhance the selectivity towards this desired antioxidant, we have studied the effect of surface fluorination of TiO2 P25 on this selectivity. Even if the maximum selectivity value obtained (10%) is still relatively low, the obtained improvements open in perspective the route for a process optimization towards the practical application of this green reaction.

2 Materials and Methods

TiO2 P25 Evonik (ca. 80% anatase and 20% rutile, specific surface area, SSA, ca. 50 m^2g^{-1}), Fluorination of TiO2 carried out by dispersing TiO2 in a KF aqueous solution at pH 3.2. The suspension was stirred for 48 h in the dark. The powder was then recovered by centrifugation, washed several times with water and finally dried at 80 °C overnight. Photocatalytic runs have been performed in a cylindrical photoreactor containing tyrosol solution and 0.5 g/L of the photocatalyst. The suspension was kept at a

S. Msahli and F. Debbabi (Eds.): CIRATM 2020, SPM 17, pp. 340–342, 2022.
https://doi.org/10.1007/978-3-031-08842-1_54

constant temperature and irradiated with UV lamps. The reaction mixture was stirred for 30 min under dark before switching on the lamps. Samples were withdrawn at fixed times and filtered through a 0.2 μm filter. The quantitative determination of tyrosol and hydroxytyrosol was performed by means of HPLC at 280 nm, the eluent consisted of a solution containing 90% v/v of a 1% v/v CH3COOH aqueous solution and 10% v/v acetonitrile.

3 Results and Discussion

Figure 1 shows the normalized concentration of Tyrosol during irradiation in the presence of the bare and fluorinated P25. We conclude that surface fluorination slightly suppresses the degradation rate of Tyrosol.

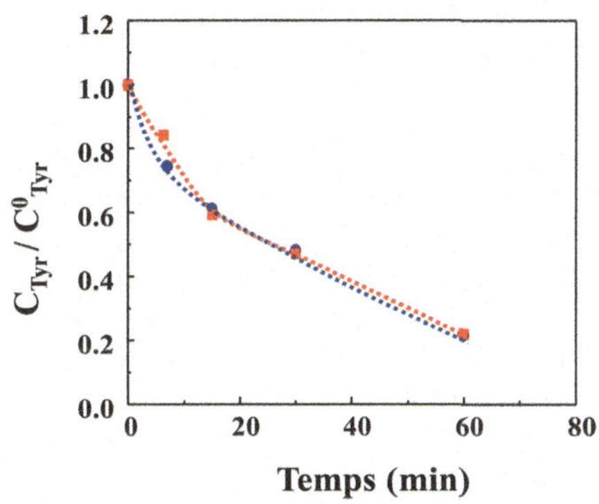

Fig. 1. Normalized concentration of tyrosol during irradiation in the presence of bare P25 (blue circles) and P25 fluorinated (red squares)

Figure 2 shows the Hydroxytyrosol concentration during irradiation in the presence of the bare and fluorinated P25. We concluded that fluorination increases the presence of Hydroxytyrosol in the reacting suspension.

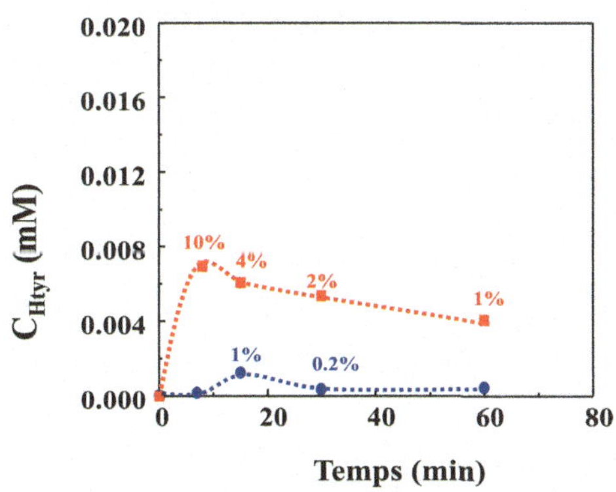

Fig. 2. Hydroxytyrosol concentration during irradiation in the presence of P25 (blue circles) and P25 fluorinated (red squares). The % values indicated the selectivity values calculated at the corresponding time

Recently, Mino et al. [1] demonstrated that surface fluorination reduces the hydrophilicity of the surface of TiO2. Therefore, in the present case, the lower density of surface hydroxyl groups favours desorption of Hydroxytyrosol whose surface chelation mechanism is hindered. Moreover, fluorination selectively addresses interfacial electron transfer, resulting in a more efficient generation of hydroxyl radicals [2]. The enhanced production of hydroxyl radicals justifies the higher production of Htyr in the presence of P25-F. Moreover, the above mentioned lower hydrophilicity of the surface results in reduced interaction with Hydroxytyrosol whose over oxidation is therefore limited.

4 Conclusion

Results show that partial oxidation of tyrosol to hydroxytyrosol could be improved by surface fluorination.

References

1. Mino, L., et al.: Beyond shape engineering of TiO2 nanoparticles: post-synthesis treatment dependence of surface hydration, hydroxylation, Lewis acidity and photocatalytic activity of TiO2 anatase nanoparticles with dominant 001 or 101 facets. ACS App. Nano Mater. **1**(9), 5355–5365 (2018)
2. Mrowetz, M., Selli, E.: Enhanced photocatalytic formation of hydroxyl radicals on fluorinated TiO2. Phys. Chem. Chem. Phys. **7**(6), 1100–1102 (2005)

Simulation and Modeling in Textile Materials and Process

Influence of Projectile Accelerator Parameters on the Weft Yarn Pulling Force Used in Electromagnetic Weft Insertion System

SeyedAbbas Mirjalili(✉) and Emad Owlia

Textile Engineering Faculty, Yazd University, Yazd, Iran
amirjalili@yazd.c.ir

Keywords: Electromagnetic weft insertion · Launcher parameters · Yarn pulling force

1 Introduction

There are different methods that the weft yarn is inserted in the warp shed such as shuttle, rapier, projectile and air/water jet [1–4]. Using electromagnetic field is another method for weft insertion. In electromagnetic weft insertion method, the electrical energy influences the projectile by magnetic field and then launches it. While the weft yarn that attached to the projectile is inserted in the warp shed. Therefore, this system seems very affordable in the terms of saving interface devices and preventing energy waste in the direction of deformation and achieving energy to final form. Also due to launching projectile without a hit, it could reduce a large amount of noise pollution in weaving machines especially projectile looms [5].

The higher weft insertion rate means the weft yarn should be accelerated by higher pulling force and endures higher tension. The greater weft yarn pulling force enhances probability of yarn rupture and consequently reduces the efficiency after an optimum value. Also, unevenness of the yarn pulling force influences fabric properties [6]. The measuring and influence of the yarn pulling force on the fabric properties have been investigated in numerous researches [6–11]. The pulling force can be considered as response function to the effect of other parameters in the weft insertion. In addition to yarn variables and the type of weft insertion system, the parameters of weft yarn accelerator strongly affect the weft yarn velocity, its pulling force and consequently the fabric quality.

During weft insertion, the puling force is applied to the weft yarn by accelerator. The level of this force correlates with the level of the residual stress in the yarn inserted as one of the quality indicators of the final product. The rate of applied force at specific time (during projectile acceleration) determines the weft yarn weave ability in these conditions, because the highest tension that applied to the weft yarn occurs during acceleration [12]. Each parameter of the weft yarn accelerator has an effect on the weft yarn velocity and its pulling force. Therefore, the knowledge about the impression of the parameters affecting on the weft yarn velocity helps us to control its pulling force and consequently the quality of the fabric.

S. Msahli and F. Debbabi (Eds.): CIRATM 2020, SPM 17, pp. 345–352, 2022.
https://doi.org/10.1007/978-3-031-08842-1_55

2 Materials and Methods

Two main kinds of electromagnetic launchers are called rail and coil guns [13, 14]. The rail gun is not suitable for weaving machine because the spark occurs between the rail and the projectile. Also, it needs a large power source that huge amount of the heat is produced at the contact between the rails and the projectile. This heat corrodes the electrical contact surface of the projectile and also increases friction by making the surface rough, which reduces the efficiency tremendously. Because there is no electrical contact and the spark between the coils and the projectile, the coil electromagnetic launcher seems to be suitable for using in weft insertion system.

Using electromagnetic force in weft insertion was proposed by Mirjalili at Yazd University for the first time [15]. An optical sensor is used to cut solenoid current when the projectile passes thoroughly in front of it. When the length of solenoid is more than two projectiles length, using this method to cut solenoid current is caused to not use all lunching capability in order to accelerate the projectile. An electromagnetic weft insertion system has been proposed in 2012 [16, 17] in which the velocity of projectile was determined by switching frequency of the electromagnetic coils via programmable logic controller. The stator of a linear asynchronous motor providing a running magnetic field can be used to move a small-sized weft-picking device through the warp shed. They have not become established in the industry. A basic problem with linear electric motors is the need for additional cooling of the stator [2, 18].

A coil launcher system was designed and built. The projectile of *Sulzer* weaving machine was used. Therefore, the results of this model can be used for setting of projectile weaving machines too. In this system the launcher thrusts a projectile and is stopped in the other side with mechanical obstacle. The main components of this system are solenoid, control board and capacitor bank. Figure 1 shows the system parts.

By the keyboard on the control board the voltage that we want to apply to the solenoid and the proper time to cut solenoid current can be directly set. The maximum voltage applied to the solenoid is determined by the arrangement of capacitors. It is possible to reach a specific voltage by setting capacitors appropriate connection and then charge them to the desired voltage. Capacitor bank consist of twenty rows that each one has five parallel capacitors. These rows can be connected with each other in series and parallel. After turning off the system the energy remained in capacitors loss in resistances that placed in the system box. The temperatures of capacitor bank, control board, and resistances are checked by temperature sensors. Three fans function to decrease their temperatures. They work faster while the temperature of each part increases. The control board is connected to other parts by three cables. Control Power Lines cable applies required voltage of other sections. Control Input Lines cable send data to control board (such as temperature sensor data, operation correctness of capacitors, and solenoid). Control Output Lines cable send data from control board (such as turn on/off LED lamps and capacitor power, launch message, charge and discharge of capacitors). Table 1 shows system specification (Figs. 2, 3, 4 and 5).

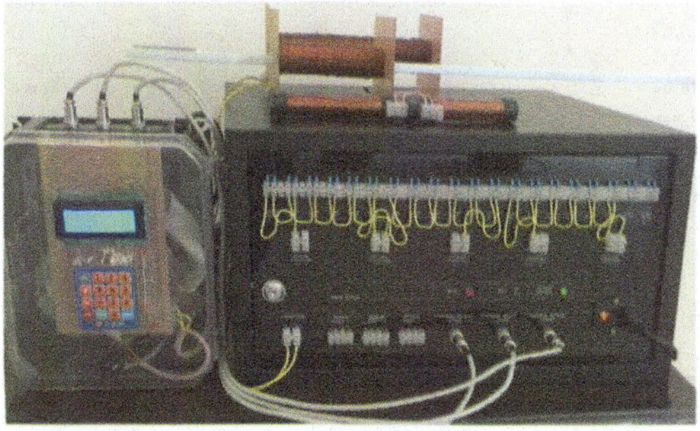

Fig. 1. The principle components of electromagnetic weft insertion

Table 1. System specification.

System part	Parameter	Value	
Projectile	Length	89×10^{-3} (m)	
	Width	14.3×10^{-3} (m)	
	Height	6.35×10^{-3} (m)	
	Weight	40×10^{-3} (kg)	
	Initial position (inside the solenoid)	3×10^{-3} (m)	
Solenoid		Number 1	Number 2
	Length	150×10^{-3} (m)	250×10^{-3} (m)
	Inner width	17.6×10^{-3} (m)	17.6×10^{-3} (m)
	Inner height	26.6×10^{-3} (m)	26.6×10^{-3} (m)
	Wire diameter	54×10^{-3} (m)	74×10^{-3} (m)
	Wire resistance	0.0894 (ohm/m)	0.0455 (ohm/m)
	Coil number	(1, 2 & 3) $\times 10^3$	(1, 2, 3 & 4) $\times 10^3$
Capacitor	Capacity	4700×10^{-6} (F)	
	Voltage	60 (V)	
	Number	100	

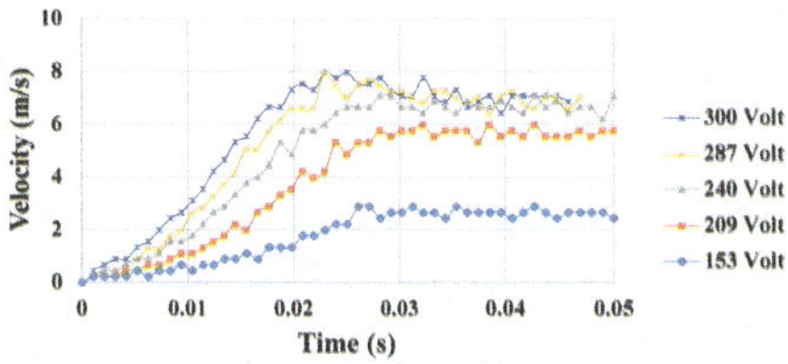

Fig. 2. The effect of voltage on projectile velocity. (Constant parameters: Solenoid number 1; Current cut time = 0.019 (s); Penetration = 3 (mm); Coil number = 200 (coils/cm)).

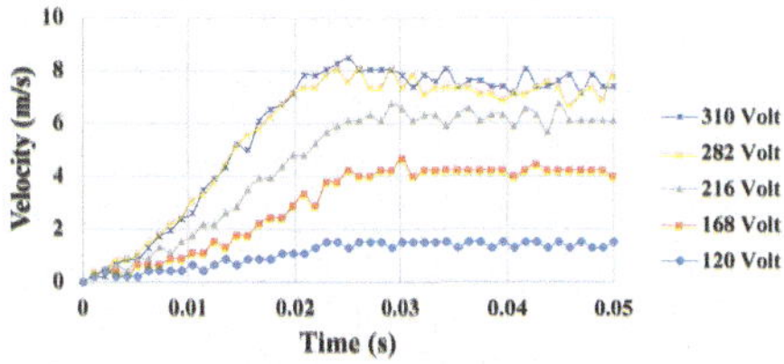

Fig. 3. The effect of voltage on projectile velocity. (Constant parameters: Solenoid number 2; Current cut time = 0.019 (s); Penetration = 3 (mm); Coil number = 160 (coils/cm)).

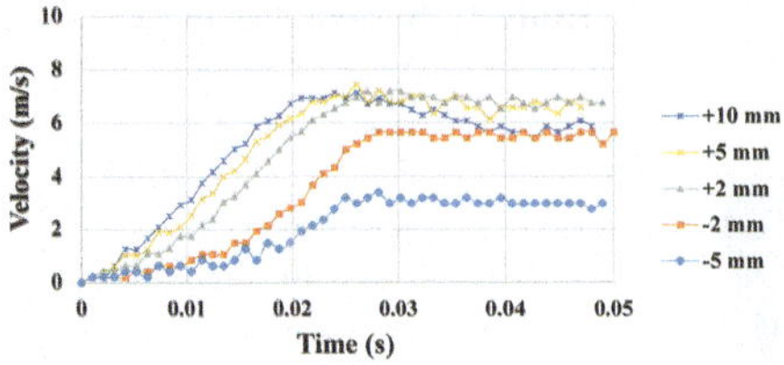

Fig. 4. The effect of projectile penetration on its velocity. (Constant parameters: Solenoid number 1; Current cut time = 0.019 (s); Voltage = 250 (V); Coil number = 200 (coils/cm)).

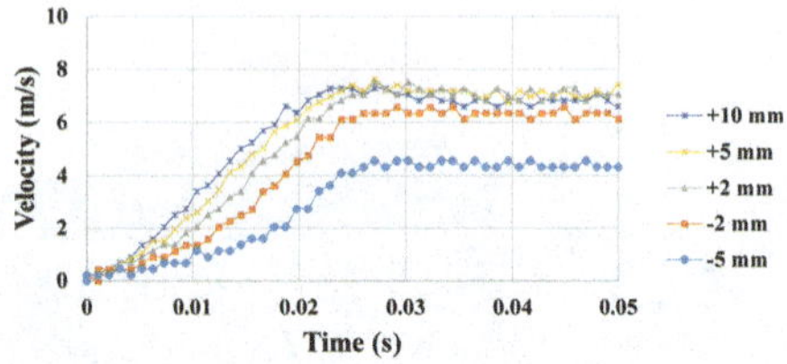

Fig. 5. The effect of projectile penetration on its velocity. (Constant parameters: Solenoid number 2; Current cut time = 0.019 (s); Voltage = 250 (V); Coil number = 160 (coils/cm))

3 Results and Discussion

The projectile velocity calculated by image processing. A guide mark is located on a very light paper caught by the projectile. This mark traced by a high frame rate camera (SONY DSC-RX10M2 with 960 frames per second) to record projectile movement throws the solenoid. Figures 2, 3, 4, 5, 6, 7 and 8 show the variations of projectile velocity in each case. The other solenoid parameters that were constant listed in the parenthesis below each fig. Figures 6, 7 and 8 show the initial projectile acceleration is approximately constant with increasing the current cut time and the coil number per centimeter. Thus, for achieving more productivity the changes of these parameters have no significant effect on the initial acceleration and the initial yarn pulling force. When the studied parameters are higher than its optimum value, the exit velocity is lower than the maximum velocity. In addition to maximum velocity of the projectile, the amount of exit velocity is very important, either.

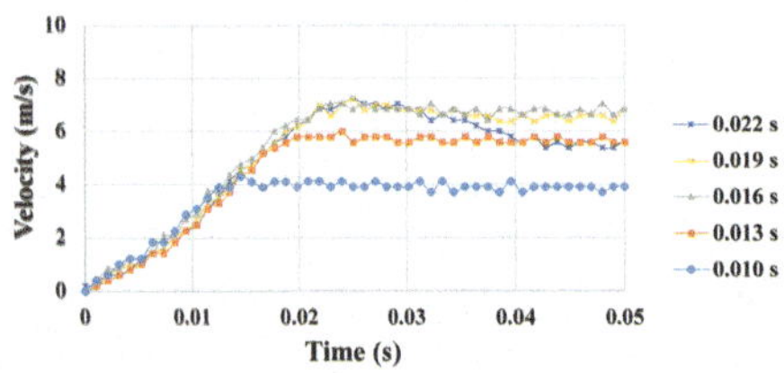

Fig. 6. The effect of solenoid current cut time on projectile velocity. (Constant parameters: Solenoid number 1; Voltage = 250 (V); Penetration = 5 (mm); Coil number = 200 (coils/cm))

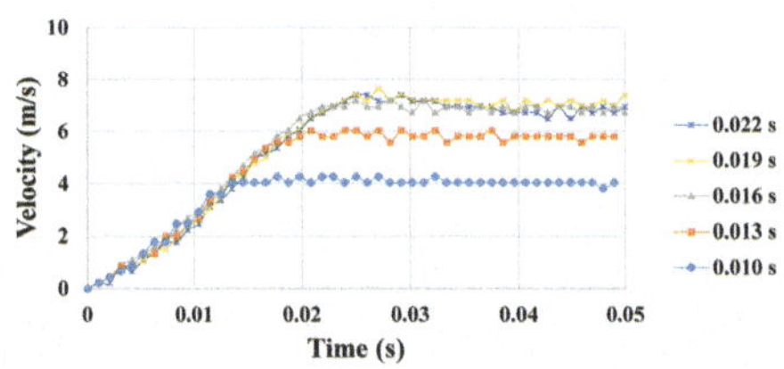

Fig. 7. The effect of solenoid current cut time on projectile velocity. (Constant parameters: Solenoid number 2; Voltage = 250 (V); Penetration = 5 (mm); Coil number = 160 (coils/cm))

The maximum velocity determines the rate of arising force that applied to the yarn. Reduction of velocity for any reason causes wasting energy; in addition, the weft yarn would endure higher tension. Figures 9 and 10 show the maximum and exit velocity of

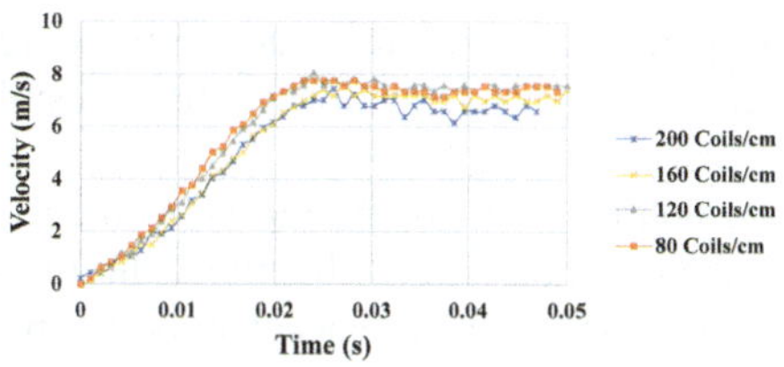

Fig. 8. The effect of coil number on projectile velocity. (Constant parameters: Current cut time = 0.019 (s); Voltage = 250 (V); Penetration = 5 (mm)).

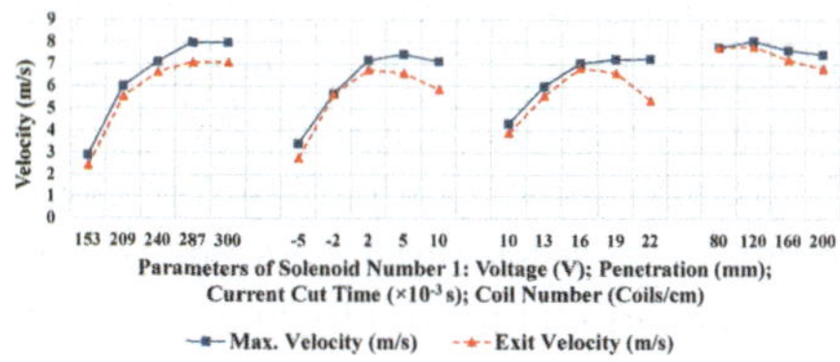

Fig. 9. Comparison the effect of solenoid number 1 parameters on the maximum and exit projectile velocity

projectile in each case. As seen the projectile velocity variation is significant in initial increases of voltage. But it is inconsiderable for higher voltages.

When the initial position of projectile is far from the solenoid (negative penetration values), its maximum and exit velocity are lower than when the projectile is positioned in the solenoid. This is due to a large amount of produced electromagnetic energy which is wasted while a little portion of it is applied to the projectile. Moreover, the solenoid current cut time (0.019 (s)) is equal in this experimental case. Therefore, the electromagnetic field remains and applies negative force to projectile when it passes the middle of the solenoid. Then the projectile velocity decreases after the optimum penetration such as the effect of current cut time. There is also an optimum value for coil number. The magnetomotive force in an electromagnet is related to coil number and the current. First when the coil number increases, the magnetomotive force increases and causes higher projectile velocity. Then in more coil number, the increase of inductance (L) is more than the increase of solenoid resistance (R). Because the inductance is related to the square of coil number, and the solenoid resistance is related to wire length. Therefore, the time constant (L/R) increases and according to exponential trend of solenoid current, it causes lower projectile velocity.

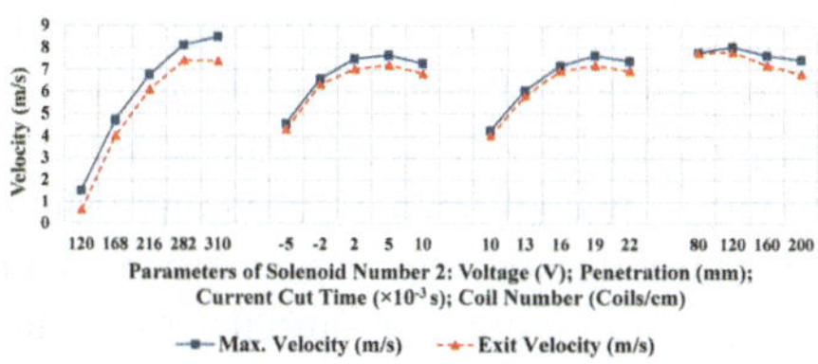

Fig. 10. Comparison the effect of solenoid number 2 parameters on the maximum and exit projectile velocity maximum and exit projectile velocity

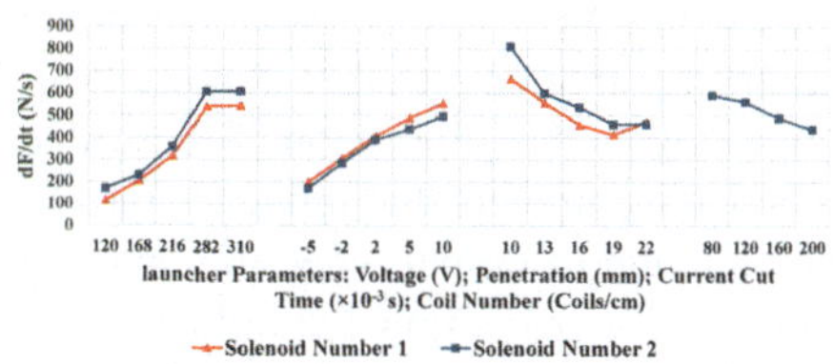

Fig. 11. Comparison the effect of launcher parameters on the maximum force variation in applied

Each yarn can withstand specific tensile force which is related to the yarn structure and conditions of tensile applied. The maximum force that applied to the weft yarn occurs during weft yarn acceleration. Therefore, the force variation is different in each weft insertion system. Also, the weft yarn accelerator has very important role in the weft yarn puling force. The effect of launcher parameters on maximum applied force during acceleration is shown in Fig. 11. Acceleration time is defined when the projectile velocity variation is less than 0.01 (m/s).

To control the yarn pulling force and fabric quality, it is essential to know the sensitivity of each accelerator parameters on the yarn pulling force. Therefore, the linear regression model is used to determine the impression coefficient of accelerator parameters (voltage, penetration, current cut time and coil number) on outputs (maximum and exit velocity of the projectile, maximum yarn pulling force and the rate of applied force). Table 2 shows coefficients of studied parameters in linear regression model.

Table 2. Impression coefficient of system parameters.

Parameters	Constant	Voltage (V)	Penetration (mm)	Current Cut time (s)	Coil number (Coils/cm)
Max Velocity (m/s)	−6.126	0.034	0.214	0.265	−0.007
Exit Velocity (m/s)	−2.482	0.027	0.176	0.205	−0.014
Max. Yarn Pulling Force (N)	−2.885	0.060	(Sig > 0.05)	0.474	−0.017
Rate of Applied Force (N/s)	245.506	2.721	24.262	−22.649	−0.852

Adjusted R-Squares are 82.7%, 68.2%, 89.6% and 87.7% for the maximum velocity, the exit velocity, the maximum yarn pulling force and the rate of applied force respectively. The results show that these accelerator parameters can predict the maximum velocity, the maximum yarn pulling force and the rate of applied force properly. In the other words, the assumed parameters of electromagnetic weft insertion system are main parameters that can predict outputs well. So, the linear regression model can be used to investigate the impression of these studied inputs on the outputs. But, there are other effective parameters that influence the exit velocity. The coefficient of each input shows the variation of output when the input changes one unit and other inputs are constant. Figures 12 and 13 show the influence of accelerator parameters on outputs. According to these figs, it is observed that the solenoid current cut time has the most influence on studied outputs. Its efficacy on maximum yarn pulling force is more than the other outputs. The penetration has no significant impression on the maximum yarn pulling force and the coil number has negative effect on the studied outputs. On the other hand, this force is applied to the yarn during solenoid current cut time and causes the impact on the rate of applied force to be more than other inputs. Therefore, the solenoid current cut time has a very important role to control the yarn pulling force and the fabric quality in this electromagnetic weft insertion system.

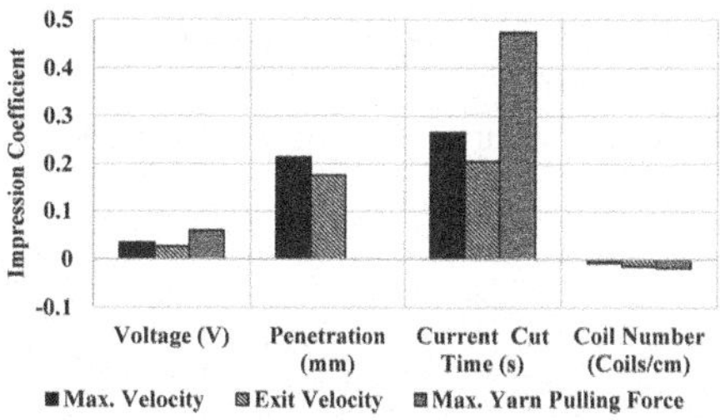

Fig. 12. Comparison the effect of voltage, penetration, solenoid current cut time and coil number on maximum velocity of projectile, exit velocity of projectile and maximum yarn pulling force

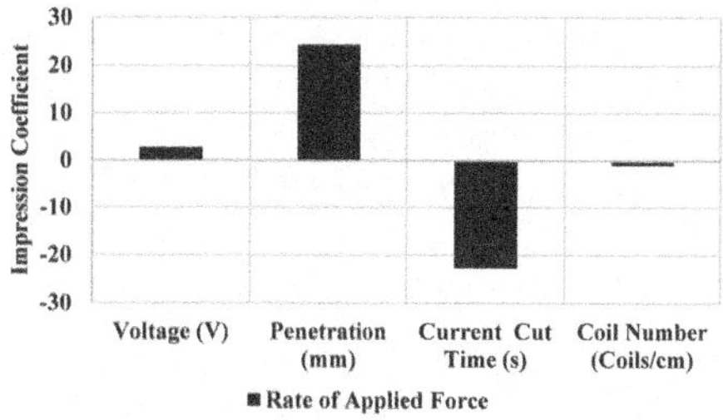

Fig. 13. Comparison the effect of voltage, penetration, solenoid current cut time and coil number on the rate of applied force

References

1. Gandhi, K.L.: The Fundamentals of Weaving Technology, pp. 117–160. Woodhead Publishing Limited, Cambridge (2012)
2. Choogin, V.V., Bandara, P., Chepelyuk, E.V.: Mechanisms of Flat Weaving Technology. pp. 83–108. Woodhead Publishing Limited, Cambridge (2013)
3. BEHERA, B.K.: Weaving Technology for Manufacturing High Performance Fabrics, pp. 253–278, Woodhead Publishing Limited., Cambridge (2008)
4. Hari, P.K., Behera, B.K.: Developments in weaving machines. Indian J. Fibre Text. Res. **19**, 172–176 (1994)
5. Talukdar, M.K.: Noise pollution and its control in textile industry. Indian J. Fibre Text. Res. **26**, 44–49 (2001)
6. Chahal, V., Mohamed, M.H.: Measuring filling yarn tension and its influence on fabrics woven on a projectile weaving machine. Text. Res. J. **56**, 324–333 (1986)
7. Haque, Md., M.L: Effect of weft parameters on weaving performance and fabric properties. J. Sci. Technol. **4**(2), 62–69 (2009)
8. Adanur, S., Gowayed, Y., Thomas, H., Ghosh, T.: On-line measurement of fabric mechanical properties for process control. pp. 196--A09, pp. 1–10, Auburn University (2000)
9. Nosraty, H., Jeddi, A.A.A.: Influence of controlled weft yarn tension of a single nozzle air-jet loom on the physical properties of the fabric. Text. Res. J. **76**(8), 637–645 (2007)
10. Adanur, S., Qi, J.: Property analysis of denim fabrics made on air-jet weaving machine Part I: experimental system and tension measurements. Text. Res. J. **78**(1), 3–9 (2008)
11. Adanur, S., Qi, J.: Property analysis of denim fabrics made on air-jet weaving machine Part II: effects of tension on fabric properties. Text. Res. J. **78**(1), 10–20 (2008)
12. Blanchonette, I.: Tension measurements in weaving of singles worsted wool yarns. Text. Res. J. **66**(5), 323–328 (1996)
13. Lee, S.J., Kim, J.H., Song, B.S., Kim, J.H.: Coil gun electromagnetic launcher (eml) system with multi-stage electromagnetic coils. J. Magnet. **18**(4), 481–486 (2013)
14. Harold, E., Bukiet, B., Peter, W.: Maximum projectile velocity in an augmented railgun. IEEE Trans. Magn. **30**(4), 1433–1436 (1994)
15. Mirjalili, S.A.: Using electromagnetic force in weft insertion of a loom. Fibre Text. East. Eur. **13**3(51), 67–70 (2005)

16. Sheng Mei1, L., Bo, Z.: Study of a new electromagnetic weft insertion mechanism based on the PLC. Appl. Mech. Mater. **130–134**, 1426–1429 (2012)
17. Sheng Mei1, L., Bo, Z.: Study of a electromagnetic weft insertion in textile machine. Adv. Mater. Res. **591–593**, 498–501 (2012)

Parametric Numerical Study of Mechanical Response of Agave Fibers Reinforced Composites

Yosra Glouia(✉), Asma El Oudiani, Imen Maatoug, and Slah Msahli

Textile Engineering Laboratory, University of Monastir, 5070 Ksar Hellal, Tunisia
gyosra1@yahoo.fr

Abstract. This paper deals with studying mechanical response of Agave fibers reinforced composites by means of finite element method. Simulation is performed thanks to ABAQUS software. The mechanical behavior of the composite at macro scale as well as meso-scale is simulated. In particular, the response to tensile and three-points bending test was studied. Comparison of Experimental and FEM tensile and three-point bending strength shows very good agreement. A parametric study is finally presented in order to analyze the effect of varying some parameters (aspect ratio, fiber's diameter, fiber's orientation, etc.) on the mechanical response of the simulated material thanks to DIGIMAT software. The originality in this work lies in the fact of studying the influence of the significant variability of the properties of natural fibers on the response of the effective properties of the simulated composites.

Keywords: Finite element analysis · Agave fibers composites · Mechanical behavior · ABAQUS · DIGIMAT

1 Introduction

Composites are the multi-phase and wonder materials becoming an essential part of today's structure due to their advantages such as low weight, corrosion resistance and high fatigue strength (Wilson 2003). Despite their interesting performance, it was necessary to think of ecofriendly by using natural fibers (Wambua et al. 2003). Composite microstructures are determined by the physical and mechanical properties of the individual materials. Analytical methods provide reasonable prediction for relatively simple configurations. Finite element analysis is a solution for these limitations. Literature reveals that researchers have not yet considered in their models the irregularity in the properties of natural fibers, especially in terms of diameter and length. In this context, the present work aims to establish a detailed parametric numerical study of mechanical response of Agave fibers reinforced composites by approaching as much as possible the real geometric properties of this category of fibers.

S. Msahli and F. Debbabi (Eds.): CIRATM 2020, SPM 17, pp. 353–359, 2022.
https://doi.org/10.1007/978-3-031-08842-1_56

2 Materials and Methods

2.1 Agave Fibers Reinforced Composites

According to the studies of Chaabouni (2005), Msahli (2007) and El Oudiani (2003) Agave fibers are very interesting and have attractive properties for composite applications thanks to their lightness. Agave Americana L. fibers density is equal to 1.36 which is considered as a low weight compared to other hard vegetable fibers. In this work, two categories of bio composites are investigated, one fabricated with chemically treated Agave Americana L. fibers and epoxy resin. While in the second one polyester matrix is added to these fibers. These composites are made using hand layup technique as shown in the following Fig. 1.

Fig. 1. Mold for the manufacture of composites reinforced with Agave fibers

To be able to model and simulate the mechanical behavior of any material with finite element software, several essential parameters must be available especially material geometry and dimensions as well as behavior law. The prepared composite plates (Fig. 2) are geometrically as well as mechanically characterized in order to deduce their properties. Their elastic parameters are summarized in Table 1.

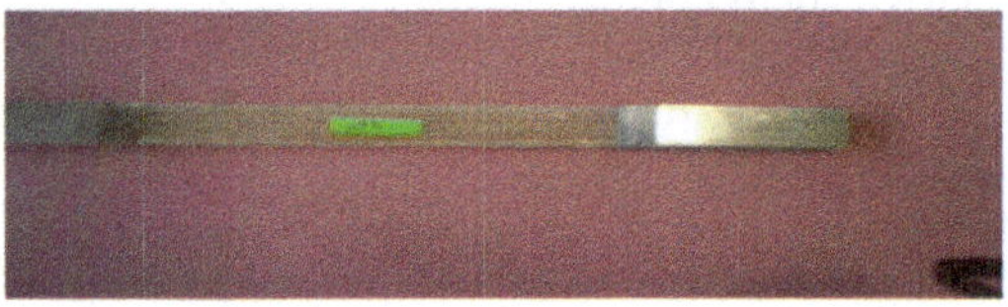

Fig. 2. Polyester agave fiber composite sample

Table 1. Elastic parameters of Agave based composites

Matrix	Agave fraction (%)	E (GPa)	ν	Stress (MPa)	Strain (%)
Polyester	15	1.48	0.32	28	4.8
Polyester	2	1.07	0.32	31	9
Epoxy	10	2.34	0.333	48	3.2
Epoxy	20	2.317	0.333	24	1.1

These values are very important because they will be useful either to validate finite element models proposed in this study or to be used as inputs for our FE models.

2.2 Finite Element Modelling

Two different techniques of modelling have been employed in the literature, one technique does not reproduce all the fibers and the matrix but considers the layer as a unique homogenous mean. This approach most likely does not allow the detailed investigation of complex phenomena. The Meso heterogeneous model (Da Selva et al. 2015; Bavan and Kumar 2013; Gehring 2013; Sun and Vaidya 1995; Kumar et al. 2016; Prasad et al. 2014; Rajesh et al. 2016; Houshyar et al. 2009) instead reproduces every single fiber and the matrix. The modeling and simulation at small scale will allow to perform a parametric study in which we will analyze and conclude the effect of some parameters and a special attention is paid to the enormous variability in the properties of natural cellulosic fibers, in fact, this point will be treated with details in this paper.

3 Results and Discussions

3.1 Macro Scale Modelling: Tensile and 3-point Bending Test Simulation

Agave reinforced composites are considered as a three dimensional plate (rectangular shape) when analyzing them at macro scale. Referring to Chaabouni's study (Chaabouni 2005), these composites have an elastoplastic behavior. Elastic parameters introduced in material's modulus in ABAQUS are those of Table 1. Concerning plastic behavior, we used experimental data and we calculate true stress and true strain thanks to the following formulas:

$$\sigma_{vraie} = \sigma_{nom}(1 + \varepsilon_{nom}) \tag{1}$$

$$\varepsilon_{vraie} = Ln(1 + \varepsilon_{nom}) \tag{2}$$

We illustrate in the following Fig. 3 the results of simulation of the tensile test of 15% Agave polyester composite. We draw for this material the numerical and the experimental curve. Then, we conclude from the two curves the error generated by the model.

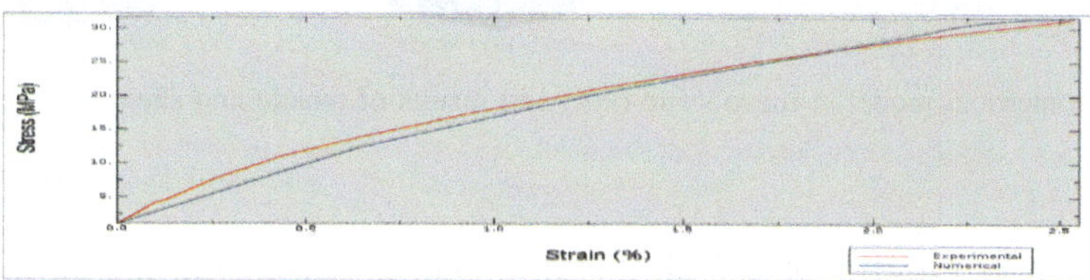

Fig. 3. Experimental vs numerical curve of Agave (15%) polyester reinforced composite under tensile test

For these composite, experimental and numerical curves present the same shape with an error in the range 7%. This proves the convergence of the proposed model.

To simulate the three-point bending test, we consider the same constitutive law for the model used in the simulation of the tensile test. Geometrical model as well as simulated material are illustrated in Fig. 4.

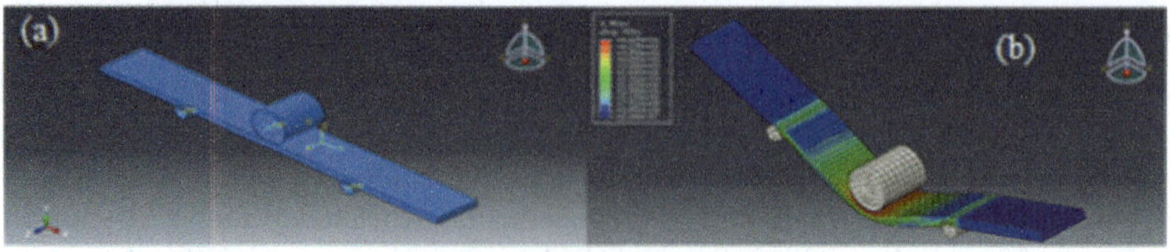

Fig. 4. Geometrical model (a) and simulated material under three-point bending test

Flexural modulus by means of FE modeling are calculated and compared to experimental ones. The error is between 3% and 12% for Agave fibers reinforced polyester and epoxide composites.

It is necessary to improve the models by further minimizing the deviation or error on the one hand and on the other hand by taking into account the two phases in the suggested model. For this purpose, we perform modeling and simulation of some mechanical tests at the scale of fibers.

3.2 Meso Scale Tensile and Shear Test Simulation

Concerning meso scale modelling, the model must contain two phases: matrix and fibers. In this situation we work with representative volume element and parameters are deduced by homogenisation technique thanks to Periodic boundary condition command implemented in ABAQUS (Omaiery Omairey 2018). In the present work, we study the composite structure under the orthotropic hypothesis (Chandana et al. 2015). Therefore, 9 engineering constants of the rigidity matrix are determined. The following Fig. 5 resumes the geometrical model and results of tensile as well as shear tests simulation.

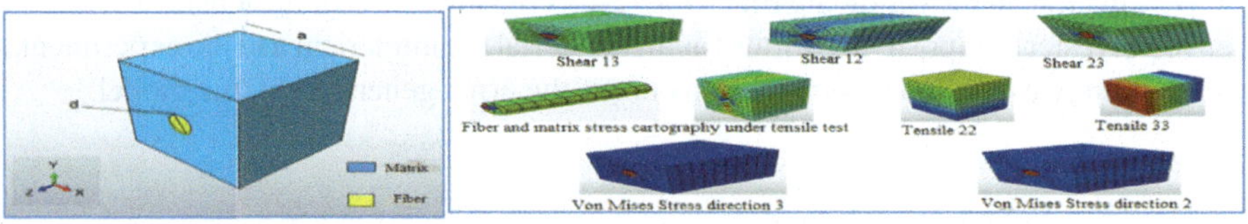

Fig. 5. Geometrical model at meso scale (left) and results of tensile and shear tests simulation (right)

We summarize also in Table 2 values of experimental versus numerical Young's modulus and the error between them for Agave fibers reinforced composites with different fiber mass ratio.

The proposed method analysis shows a good agreement between model and experimental results, it also have the advantage to permit determination of shear as well as young's modulus in the three different directions.

Table 2. Experimental versus numerical Young's modulus and error between them

Composite	Experimental E (MPa)	Numerical E (MPa)	Error (%)
PES 15% agave	960	1000.03	4.2
PES 24% agave	950	1034.55	8.9
EPX 10% agave	2200	1970.34	10.4
EPX 20% agave	2200	2047.643	10.9

3.3 Parametric study

3.3.1 Aspect Ratio Effect

Aspect ratio is a very determining parameter and influencing the mechanical properties of composite materials, the longitudinal modulus E11 among others (Yao and Chen 2012). It is defined by the ratio between the length's fibers and their diameter. The following Fig. 6 illustrates different representative volume element with various aspect ratios (1, 5, 10 and 20).

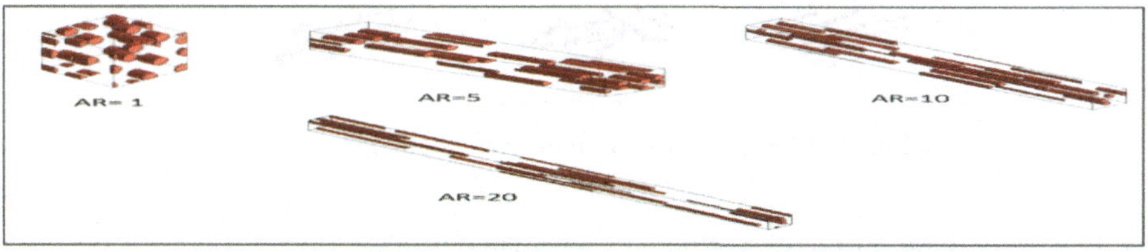

Fig. 6. RVEs with various values of aspect ratio

Obtained results reflect that the evolution of the longitudinal modulus E11 is proportional to the increase in the value of the aspect ratio.

3.3.2 Reinforcement Young Modulus's Effect

Results of simulation prove that young modulus's of overall composite depends on that of the reinforcement. Their evolution is proportional.

3.3.3 Fiber Orientation Effect's

We draw in Fig. 7 diverse RVEs with different fiber orientation angles (0°, 30°, 45, 60°, 75° and 90°). Results of this variation are shown in Fig. 8.

The previous figure proves that young modulus's of the simulated composite is inversely proportional to fiber orientation angle's.

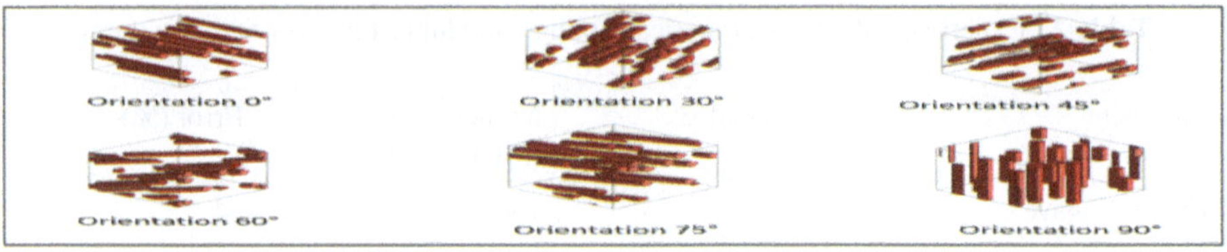

Fig. 7. RVEs with different fiber orientation angles

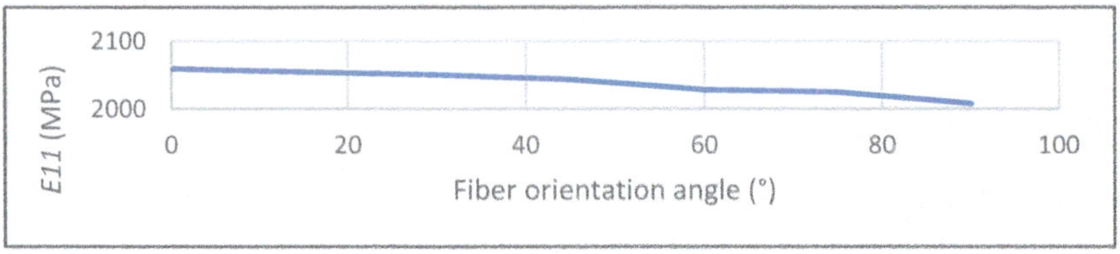

Fig. 8. E11 evolution function of fiber orientation angles

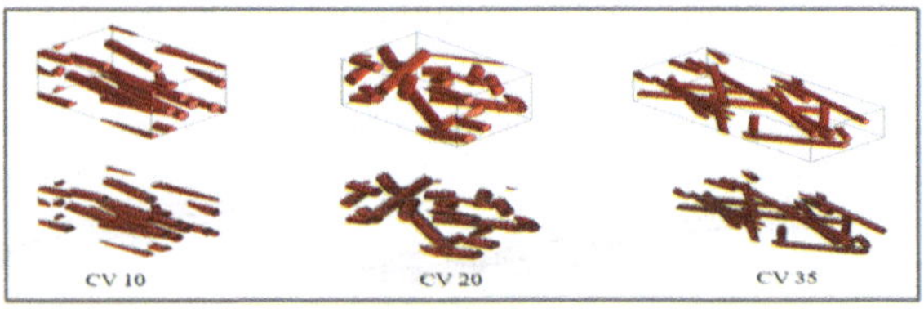

Fig. 9. RVEs with variation in fiber length distribution

3.3.4 Effect of Irregularity in Fiber Length Distribution

The results summarized in Fig. 8 show that the E11 modulus is characterized by an insignificant decrease of about 0.03% for coefficient of variation values less than 20%. Beyond this value, we observe a drop in the longitudinal modulus of about 1.07% (Fig. 9).

3.3.5 Effect of Irregularity in Fiber Diameter Distribution

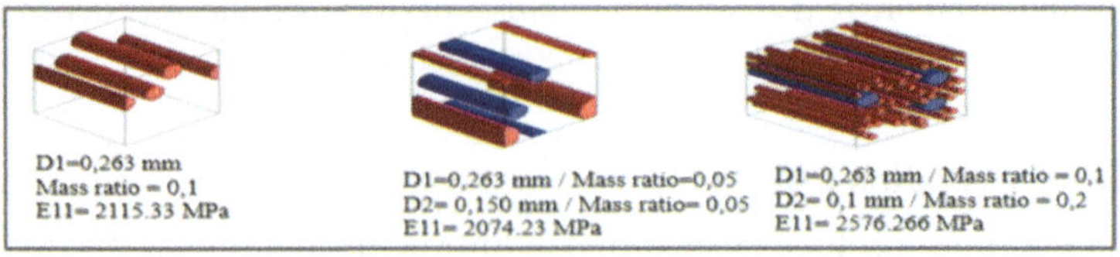

Fig. 10. RVEs with variation in fiber diameter distribution, fiber's mass ratio and their relative E11 modulus

Materials with a larger mass fraction of fibers having a finer diameter have the highest stress levels (Fig. 10).

4 Conclusions

Modeling and simulation of fiber reinforced composites at the level of reinforcement permit to ensure more convergence between FE model values and experimental ones in the one hand and in the other hand it allows to establish a parametric study.

References

Wilson, M.J.: Finite element analysis of glass fiber reinforced thermoplastic composites for structural automotive components, Nottingham (2003)

Chaabouni, Y.: Microstructure characterization of Agave Americanan L. fiber; study of Agave fibre reinforced composites, France (2005)

El Oudiani, A.: Study of the mechanical behavior of Agave fibers Americana L, Tunisia (2003)

Wambua, P.: Natural fibers: can they replace glass in fiber reinforced plastic? Compos. Sci. Technol. **63**(9), 1259–1264 (2003)

Msahli, S.: Mechanical behavior of Agave Americana L. fibers: correlation between fine structure and mechanical properties. J. Appl. Sci. **7**(24), 3951–3957 (2007)

Omairey, S.: (2019): Development of an ABAQUS plugin tool for periodic RVE homogenization. Eng. Comput. Springer **35**, 567–577 (2018)

Sri chandana, B.: Mechanical properties of fiber reinforced composites using finite element method. Int. J. Mech. Eng. Robot. Res. **4**(1), 1–11 (2015). ISSN 2278–0149

Yao, Y.: The effects of fiber's surface roughness on the mechanical properties of fiber reinforced polymer composites. J. Compos. Mater. **47**(23), 2909–2923 (2012)

Sewing Thread Consumption for 602 and 605 Cover Stitches Using Geometrical and Multi-linear Regression Models

Sarah Malek[1(✉)], Adolphe Dominique[2], and Boubaker Jaouachi[1]

[1] Textile Engineering Labóratory, University of Monastir, Monastir, Tunisia
maleksarah88@gmail.com
[2] LPMT, ENSISA, Mulhouse, France

Abstract. This work aims to determine the amount of sewing thread required to stitch a specific length of woven fabric using two cover stitch types of class 600. In fact, two different methods, geometrical method and a multi-linear regression method, are proposed, to select the best one for the prediction of thread consumption in order to help industrialists to evaluate accurately the consumed amount of sewing threads.

Keywords: Thread consumption · Cover stitches of the class 600 · Geometrical models · Multi-linear regression models

1 Introduction

To product a garment, the essential components are the fabric and the seam thread. The consumption of these two materials must be studied to predict the consumption quality in order to avoid the over quantity and the waste of money. Regarding the importance of this topic, studies does not stopped until now. Indeed, recently, Midha has predicted sewing thread consumption for lockstitch based on regression model (Midha et al. 2016). Abher has studied a geometrical model to determine the consumption of sewing thread for 504 over-edge stitch (Abher et al. 2018) Furthermore, Gazzah et al. has developed a model, based on 602 geometrical cover-stitches, to calculate the relative consumed sewing thread length (Gazzah et al. 2017).

In this study, we worked on two cover stitches of class 600, which are used in industries to sew clothing (namely for denim garments, for the laying of mesh fabric and lace underwear and lingerie…).

2 Material and Methods Used for Thread Consumption Prediction

Six commercial denim fabrics were chosen in our study. They have different characteristics namely thicknesses and blend compositions (Table 1).

S. Msahli and F. Debbabi (Eds.): CIRATM 2020, SPM 17, pp. 360–366, 2022.
https://doi.org/10.1007/978-3-031-08842-1_57

Table 1. Fabric properties.

Parameter	Warp yarn density (ends/cm)	Weft yarn density (picks/cm)	Linear density (tex)		Mass per square meter		Thickness		Tensile properties		
Fabric Composition n°			Warp	Weft	Mass (g/m²)	CV (%)	Thickness (mm)	CV (%)	Breaking force (N)	Elongation at break (%)	CV (%)
1	26	17	11.9	20	323	1.9	1.2	1.4	413.63	12.66	1.4
	30	22	12.5	16.67	387	2.1	1.04	1	1071.90	17.89	2.1
2	32	17	11.36	14.92	445	1.5	0.96	1.4	894.26	17.40	0.8
	29	22	12.5	27.78	342	1.2	0.90	1.6	624.21	30.45	1.2
3	31	21	12.5	27.78	334	1.6	0.85	1.2	557.24	33.64	0.6
	28	22	14.29	27.78	328	0.9	0.75	2.1	528.13	25.12	2.4

With:

- Fabric composition n°1: 100% cotton warp and weft threads;
- Fabric composition 2: 100% cotton warp yarn and weft yarn in 95% cotton and 5% elastane;
- Fabric Composition 3: 100% cotton warp yarn and weft yarn in 71% cotton, 5% elastane and 24% polyester.

Two commercial sewing threads, commonly used for sewing denim fabrics, were chosen (Table 2). The selection of these threads was based on their linear densities, which can probably affect the used amount of sewing thread.

Table 2. Sewing threads properties.

Properties	Thread N°1	Thread N°2
Linear density (tex)	63.50	95.00
Number of strands	3	3
Composition	100% PES	100% PES
Twist direction	S	S
Breaking force (N)	24.23	26.43
Breaking elongation (%)	21.12	16.90
Rigidity (N/m)	389.12	536.59
Twist/m	353.00	294.00
Tenacity (cN/tex)	38.16	27.82

With PES: Means Polyester material.

In this work, a geometrical method and a multi-linear regression method are considered. We tested the most useful cover stitch class 600 in sewn product industry. Two types of stitch usually used are chosen.

These stitches have double recovery (lower and upper):

(1) Cover stitch 602, Two needles, four threads;
(2) Chain stitch 605, Three needles, five threads;

Nevertheless, for the multi-linear regression, a statistical method based on experimental design using MINITAB-17 software was applied to conduct objectively this modeling technique.

3 Results and Discussions

3.1 Geometrical Models

Mathematical calculation based on geometrical shapes of cover stitch geometries was used to realize the stitch modeling. In our case, the cross –section of threads is circular and incompressible in nature. In addition, the same thread was used in all cases (needle, upper and lower loopers). Figure 1 presents geometrical model of cover stitch type 602 and 605.

The amount of sewing thread C needed for the 602 and 605 cover stitches, whose geometries are represented previously in Fig. 1, is estimated by the following formula (1, 2) and (3, 4) respectively.

$$C_{ul602} = B_1 + \sqrt{B_1{}^2 + SL^2} + 2 \times (\frac{1}{2}\left(\pi \times (d + \frac{d}{2})\right)) \quad (1)$$

$$C_{ll602} = 3 \times \frac{1}{n} + SL + \left(\left(\frac{1}{n}\right)^2 + B_1{}^2\right)^{1/2} + B_1 + 4 \times \frac{1}{2} \times \left(\pi \times (d + \frac{d}{2})\right) + 2 \times \frac{d}{2} \quad (2)$$

$$C_{ul605} = B_2 + \sqrt{(B_1 + B_2)^2 + SL^2} + \sqrt{B_1{}^2 + (3d)^2} + 2 \times \left(\pi \times (d + \frac{d}{2})\right) \quad (3)$$

$$C_{ll605} = 4 \times \frac{1}{n} + SL + \left(\left(\frac{1}{n}\right)^2 + B_1{}^2\right)^{1/2} + \left(\left(\frac{1}{n}\right)^2 + B_2{}^2\right)^{1/2} B_1 + B_2 + 6 \times \frac{1}{2} \times \left(\pi \times (d + \frac{d}{2})\right) + 2 \times \frac{d}{2} \quad (4)$$

where:

ul: Upper looper thread;
ll: Lower looper thread;
B1and B2: Distance between two needles expressed in cm (needle 1 and 2, 2 and 3);
SL: Stitch length;
d: Sewing thread diameter.

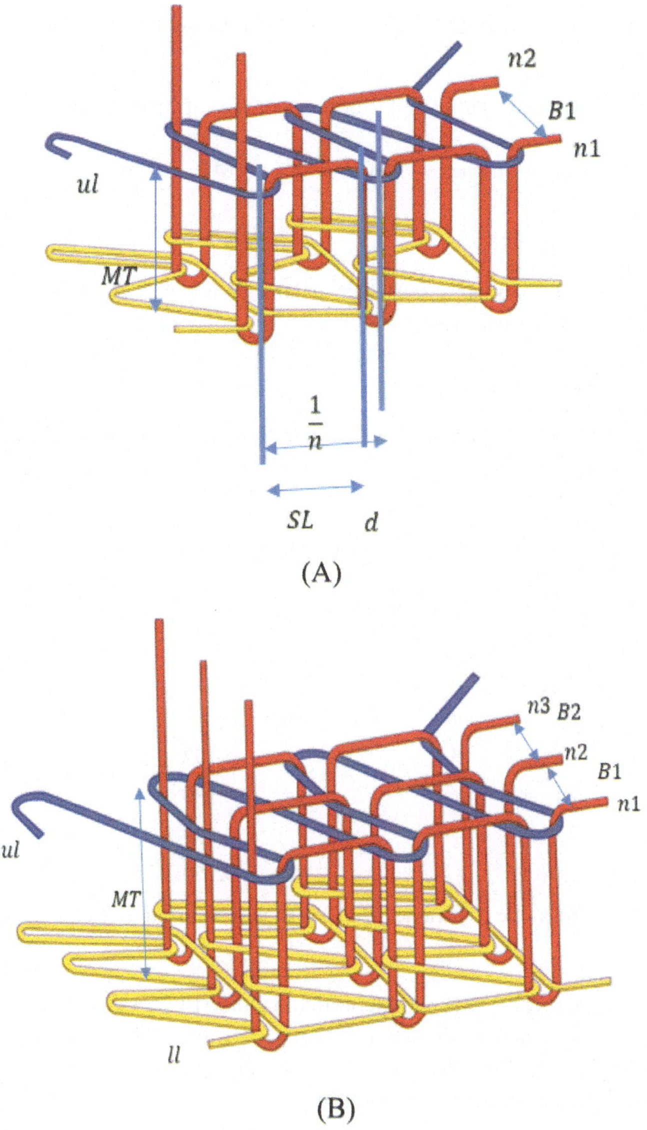

Fig. 1. Geometrical model of cover stitch type (A) 602 and (B) 605 presented in 3 Dimensions respectively

In order to verify these models, 72 samples were prepared by varying the different factors for cover stitch 602 and 36 samples were prepared for cover stitch 605.

For both cover stitch types 602 and 605, the industrial sewing machine type SIRUBA F007 (WL22-356) (Speed: 2000 rp/min, 3 needles) was used.

Table 3 presents the different factors and levels for each type of cover stitch point.

Table 3. Different factor for each type of Chain stitch.

Chain stitches	n	d	MT	B_1
602	*	*	*	*
605	*	*	*	

* Represents the existent parameter for each studied chain stitch type.

Where $B_2 = 0.3$ cm.

Table 4 presents the levels of each factor for preparing the samples of each type of cover stitch class 600.

Table 4. Levels of each factor.

Input parameter	n(stitches/cm)	d(cm)	MT(cm)	B_1(cm)
Level 1	3	0.024	0.240	0.3
Level 2	4	0.036	0.208	0.6
Level 3	5	-	0.192	
Level 4	-	-	0.180	
Level 5	-	-	0.170	
Level 6	-	-	0.150	

No value of parameter, which should be mentioned.

Comparing the experimental and theoretical values (Fig. 2) showed a good agreement between geometrical and experimental consumption values. The coefficient of regression ranged from 98.78% to 99.38% which reflects significant and efficient relationships between experimental and theoretical findings.

3.2 Multi-linear Regression Models

In order to help industrialists to use the best method to predict the consumed thread to sew denim garments, the multi-linear regression method was applied.

Thanks to MINITAB-17 software, the multi-linear regression models were determined (Table 5).

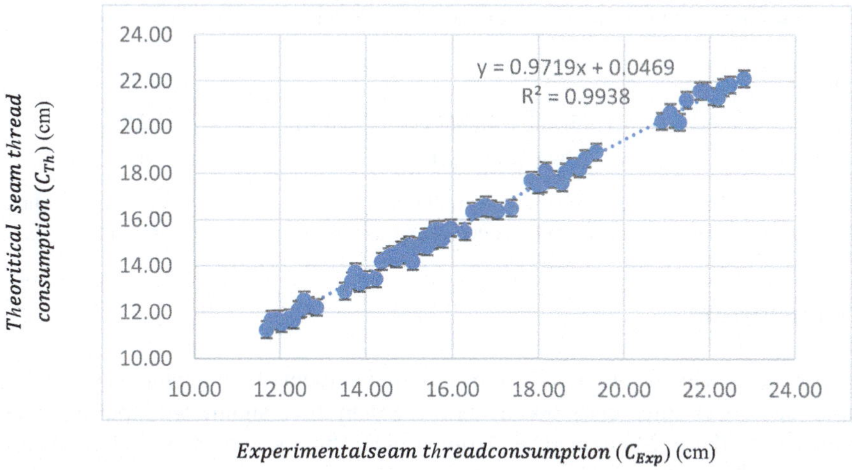

(a)

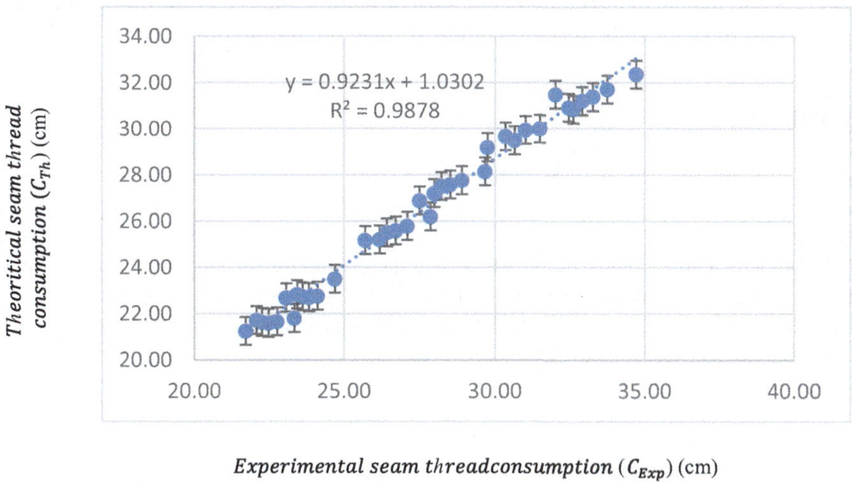

(b)

Fig. 2. Theoretical consumption evolution as a function of experimental results for cover stitch type 602 (a) and 605 (b)

The value of regression coefficient R^2 reflects the significance and relevance of the developed models. In our case, regarding the R^2 values, the multi linear regression models (97.05% and 99.63%) and the geometrical models (99.38% and 98.87%) can both be considered to determine the seam thread consumption.

Table 5. Multi-linear regression models for different cover stitch class 600.

Equation	R^2 (%)
$C_{602} = -3.414 + 2.5611n + 65.38d + 4.88MT + 14.020\ B_1$	97.05
$C_{605} = 3.376 + 4.2217n + 136.28d + 11.54MT$	99.63

Where: n: Stitch density expressed by stitch number/cm, d: Sewing thread diameter, MT: Material thickness and B1: Distance between two needles expressed in cm (needle 1 and 2).

4 Conclusion

This work deals with the prediction of the required amount of sewing thread using denim samples and cover stitch class 600 using more than three significant input parameters through two methods: Mathematical (geometrical technique) and statistical (multi-linear regressive technique) methods.

The proposed models can accurately predict the thread consumption based on chain stitch class 600 thanks to developed geometricalrelationships and multilinear regression models. Therefore, it can be used to better estimate the required thread.

References

Abher, R., Sheraz, A., Ateequr, N., Muhammad, B.R.: Geometrical model to calculate the consumption of sewing thread for 504 over-edge stitch. J. Text. Inst. **109**(11), 1418–1423 (2018)

Gazzah, M., Khedher, F., Jaouachi, B.: Modelling the sewing thread consumption of 602 cover-stitch based on its geometrical shape. Int. J. Appl. Res. Text. **5**(2), 1–15 (2017)

Midha, V.K., Sharma, S., Gupta, V.: Predicting sewing thread consumption for lockstitch using regression model. Res. J. Text. Appar. **20**(3), 155–163 (2016)

Application of Box-Behnken Design and Response Surface Methodology for Modeling of Cold Sizing Efficiency of Cotton Warp Yarns

Asma Rahmouni(✉), Sameh Maatoug, and Neji Ladhari

Textile Engineering Laboratory, University of Monastir, B.P. 68, 5070 Ksar Hellal, Tunisia
okasmus@hotmail.com

Abstract. Recently unconventional methods have been studied as an alternative to hot sizing method to prepare warp yarns to weaving process. The present study aims at developing and optimizing of cold sizing recipe ingredients in order to size 100% cotton warp yarns. Polyvinyl alcohol (PVA) was used as sizing agent and glycerol was used as plasticizer agent. In addition, a lubricant agent, which is a mixture of fatty acids, fatty alcohols, and emulsifiers, was used. Effects of PVA concentration (40 g/L–60 g/L), plasticizer content (0–10%), and lubricant content (0–8%) as independent variables on cold sizing efficiency was determined by response surface methodology based on three-level three-factor Box-Behnken design. Optimized values of cold sizing recipe composition was attributed.

Keywords: Box-Behnken design · Response surface methodology · Modeling · Cold sizing efficiency

1 Introduction

Warp yarns are subjected to various tensile strains and friction during weaving. The tensile strains mainly derive from the warp tension and the motions of shedding and beating up. The frictional effects mainly result from the relative movement between warps and back beam. Consequently, warp sizing is done to provide a protective coating and to lubricate the surface of the yarn to enable it to withstand the abrasive action, which occurs in the weaving process. In addition, the important reasons for applying sizes to warp yarns are to protect them from breaking during weaving and to decrease their hairiness so that the tendency for adjacent warp yarns to entangle will be reduced (Maatoug et al. 2007). However, a large amount of steam energy is consumed for maintaining high temperature throughout sizing operation (Tong 2014) and there is no doubt that the energy consumed in cotton warp sizing will be saved if paste temperature can be decreased. Therefore, there is an urgent need to reduce the temperature in cotton warp sizing for saving energy. Nowadays, there has been a growing interest in the use of economical methods in textile applications. Thus, cold sizing process, which is carried

S. Msahli and F. Debbabi (Eds.): CIRATM 2020, SPM 17, pp. 367–372, 2022.
https://doi.org/10.1007/978-3-031-08842-1_58

at room temperature, has certain advantages over conventional sizing in terms of saving in energy and time (Parekh and Harane 2012; Chavan and Patil 2014).

There is no doubt that quantity of different ingredient in the size recipe is recognized as an important effect on sizing efficiency. Thus, the optimization of sizing agent concentration, lubricant and plasticizer content in the sizing formulation is of the utmost importance for the weaving performance study. Maatoug et al. analyzed the mechanical properties of films based on size formulations contained maize starch, plasticizer i.e., glycerol, and lubricants viz., Avirol, glycerin, and tween 20 with different concentration and observed that tensile strength and extension of films were affected by hydrophilicity of plasticizer and its concentration (Maatoug et al. 2009). On another hand, Behera and Gupta conducted a comparative analysis of different varieties of natural starch, modified starch and synthetic size materials in terms of cohesion power, adhesion power, abrasion resistance and bending rigidity (Behera and Gupta 2009). The effect of lubricant agent on properties of size film is also examined and the results revealed that addition of lubricant improves some essential film properties. According Behera et al. (2008b), the addition of lubricant in the size paste causes the size film to get more pliable leading to improvement in weaving performance. Furthermore, no studies have been found so far on the effect of polyvinyl alcohol (PVA) size on cold sizing efficiency. This study is an attempt to investigate the PVA size recipes on cold sizing process of carded cotton warp yarns. A numerical method is presented by using the box-behnken design and the response surface methodology (RSM) for modeling the relationships between size recipe variables (PVA concentration g/L, lubricant content %, and plasticizer content %) and selected responses such as viscosity, cohesion power, adhesion power, add-on %, tensile properties, and hairiness of cold sized yarns.

2 Material and Methods

The following size materials and agents were used:

- PVA 2488 (high viscosity polyvinyl alcohol) with hydrolysis degree of 88% and molecular weight of 24.000 g/mol. AVIROL NW 94-PI was used as lubricant agent, which is the mixture of fatty acids, fatty alcohols, and emulsifiers. Glycerol GLY was used as plasticizer agents.

A single yarn sizing method was carried out under laboratory conditions in an apparatus simulating the sizing machine. 100% raw cotton carded yarns (with Nm11.5 and twist of 473 tr/m) and cotton roving (with a linear density of 1,1 Ktex) were cold sized with above prepared sizes under sizing speed of 3 m/min and squeezing pressure of 2 bar. Yarn and roving samples were dried and conditioned for 24 h at 65% RH, and then the properties of sized yarn and roving are tested. Parameters, which are responsible to characterize cold sizing efficiency, were carried according the following methods:

- Apparent viscosity of the size solutions was measured according to ISO6388 on a Rheotec RC 30 rotary viscometer.

- The cohesion power of size materials was tested by measuring the tensile properties of casted size film. The adhesion capacity of the size was quantitatively evaluated in term of maximum strength of slightly sized cotton roving (Zhifeng et al. 2017).
- The cold sized yarns were conditioned and tested on Lloyd tensile tester LR5K according to ISO 2062. The size add-on was calculated by taking the weights of the oven-dried sized and oven-dried unsized yarn samples of around 10 m, with three readings per sample.
- The hairiness of yarns was tested on Shirley Electronic Yarn Hairiness Tester according to ASTM D5647 standard test method.

3 Experimental Design

In this study, three independent experimental variables, each having three levels, were selected, namely the concentration of polyvinyl alcohol (Cpva), the lubricant content (lub%), and the plasticizer content (plas%). The three levels for these variables were coded as $-1, 0$, and 1, representing its lower limit, center point, and upper limit, respectively. All the coded and actual values of independent variables, along with their levels are shown in Table 1. The response surface methodology (RSM) using the Box-Behnken experimental model was employed to design the cold sizing process efficiency. A total of 15 runs were carried out with combination of different levels of variables, which included 3 replications of center points and 12 unique runs. For RSM, the most commonly used second-order polynomial equation developed to fit the experimental data and determine the relevant model terms can be written as:

$$Y = \beta 0 + \Sigma\ \beta i\ xi + \Sigma\ \beta ij\ xixj \quad (1)$$

where Y is the predicted response, βo is the interception coefficient, βi are the linear terms, βij are the interaction terms, xi and xj are the coded levels of the independent variables.

Table 1. Main specifications of the factors and their relevant coded levels

No.	Factor	Unit	Nominated variable	Coded levels		
				−1	0	1
1	Cpva	g/L	X1	40	50	60
2	%lub	%	X2	0	4	8
3	%plas	%	X3	0	5	10

4 Results and Discussions

According to the experimental data obtained and shown in Table 2, a statistical analysis including fitting of responses equation and analysis of variance was carried and presented in Table 3.

Table 2. Box-Behnken design and response results for the study of the three experimental variables in real values

R	Processing variables			Output variables						
	X1 (Cpva)	X2 (%lub)	X3 (%plas)	V	Adh	Coh	Add-on	St	E	H
1	40	0	5	45,5	76,32	12,6	4,22	17,72	9,9	37
2	60	0	5	175,6	110,2	13,8	7,72	18,9	8,9	31
3	40	8	5	45,2	77,0	10,2	3,6	17,02	9,8	44
4	60	8	5	210,1	101,2	15,5	5,48	18,61	9,6	47
5	40	4	0	49,5	75,7	13,6	3,93	17,48	8,7	38
6	60	4	0	175,1	92,9	14,7	5,54	18,00	8,3	39
7	40	4	10	41,7	74,2	8,06	4,13	16,88	9,8	37
8	60	4	10	168,2	83,2	16,9	7,52	18,31	9,9	40
9	50	0	0	86,5	91,6	15,5	4,14	17,79	8,0	47
10	50	8	0	92,9	71,8	13,1	4,22	17,26	8,2	57
11	50	0	10	84,9	78,9	14,9	6,72	17,21	9,0	46
12	50	8	10	88,2	78,35	12,12	5,95	17,42	8,7	54
13	50	4	5	99,0	79,67	10,86	4,23	18,05	9,1	52
14	50	4	5	103,0	75,46	10,53	4,29	18,25	9,4	49
15	50	4	5	99,5	74,97	11,96	4,27	18,02	9,5	47

According to results in Table 3, the calculated model for the size viscosity was statistically significant ($P < 0.05$; $R^2 = 99.2$). from the equation, neither the lubricant content nor the plasticizer content affected directly the viscosity of size solution, affecting only by the size concentration and by means of the interaction terms Cpva*Cpva and %plas*%plas.

There is no doubt that strong adhesion of the size to fibers is recognized as an extremely important behavior. The adhesion is capable of enhancing the strength of sized yarns by bonding the fibers in warps together. The weavability of sized yarns was greatly influenced by adhesion capacity of the size materials, which was selected by Behera et al. (2008a) as an important parameter to select the size material for any application. Based on the analysis of variance results (Table 3), the model for size adhesion power was statistically significant ($P < 0.05$; $R^2 = 93.8$). According to regression equation, only the size concentration parameter affect directly and by means of the interaction term the adhesion capacity of the size. The plasticizer and lubricant incorporation in the size recipe did not affect the size adhesion. Concerning the cohesion power of the size, the results of the analysis of variance lead to the model presented in Table 3. This model was statistically significant ($P < 0.05$; $R^2 = 90.7$). from the fitting equation, only the size concentration affect linearly the size cohesion (Coh), and the plasticizer content affected the (Coh) by means of interaction terms (%plas*%plas; and Cpva*%plas). This behavior was a consequence of the plasticizing effect of the glycerol used in this work and has been observed by other authors (Maatoug et al. 2009). Changing size add-on and chemical nature of size material can vary the size penetration and encapsulation of yarn, which in turn influence the size-fiber interactions. According to the results in Table 3, it was observed that the linear model of add-on percent could be fitted to data. It can also be observed (Table 3) that the parameters of the linear model (Add-on) in terms Cpva, %lub, and %plas were statistically significant ($P < 0.05$; $R^2 = 95.39$) and affected only directly the add-on percent of the size.

Table 3. Statistical results and response surface regression of viscosity, cohesion, adhesion, add-on, strength, elongation, and hairiness versus Cpva; %lub; %plas

Parameter	P-value	R^2	Response surface model in coded units
Viscosity (cP)	0.000	99.20	V = 100,52 - 13,62 Cpva + 19,57 Cpva*Cpva - 11,42%plas*%plas
Adhesion (N)	0.039	93.80	Adh = 76,70 + 10,53 Cpva + 7,92 Cpva*Cpva
Cohesion (MPa)	0.020	90.71	Coh = 11,117 + 2,069 Cpva + 1,529%plas*%plas + 1,947 Cpva*%plas
Add-on (%)	0.008	95.39	Add-on = 4,263 + 1,297 Cpva - 0,444%lub + 0,811%plas
Strength (N)	0.004	96.50	St = 18,108 + 0,590 Cpva - 0,541%plas*%plas
Elongation (%)	0.029	91.83	E = 9,383 + 0,523%plas - 0, 253 Cpva*Cpva - 0,618 Cpva*%plas
Hairiness (hair/m)	0.004	96.36	H = 49,57 + 5,125%lub - 10,78 Cpva*Cpva

To predict the weavability of warp sized yarns, Sweich and Sweich discussed methods for the quality control testing of yarns in relation to their weavability. Warp yarn breakages in weaving were considered as a function of the breaking strength, breaking elongation and irregularity of the warp yarns. The breakages were also examined as a function of yarn defects; neps; and thick and thin places (Swiech and Swiech 2000). Based on the analysis of variance results (Table 3), the calculated model for the strength (St) of cold sized yarn was statistically significant ($P < 0.05$; $R^2 = 96.5$). Neither the lubricant content, nor the plasticizer content affected directly the (St), only affecting directly by the size concentration Cpva and by mean of the interaction term (%plas*%plas). Moreover, the linear model (E) calculated using the sized yarn elongation at break data was statistical significant ($P < 0.05$; $R^2 = 91.83$). Certainly due to the plasticizing effect of glycerol on size cohesion (Coh), as explained earlier, the elongation at break of cold sized yarn was affected directly by the plasticizer content. Adding plasticizer agent to the size recipe improve yarn elasticity after sizing. Size concentration affect sized yarn elongation by means of interaction terms (Cpva*%plas; Cpva*Cpva). The increase of PVA concentration produce sized yarns more compact, improve the yarn strength, and reduce yarn elongation. As seen from Table 3, the linear model (H) was chose to fit the hairiness data of cold sized yarns, and was statistical significant ($P < 0.05$; $R^2 = 96.36$). According to that equation, only the lubricant content affect linearly the hairiness parameter, and the size concentration Cpva affected (H) by means of interaction term. Increasing size concentration decrease the yarn hairiness because PVA sizes gives very smooth and fine covering on the yarn.

5 Conclusion

In this study, the application of response surface methodology and Box–Behnken design from the point of view cold sizing tests was discussed. The three-level three-factorial Box–Behnken experimental design was applied in the study. Variables of model investigated in this study were PVA concentration (Cpva), Lubricant content (%lub), and plasticizer content (%plas) for cold sizing recipe. The mathematical model equations were derived for modeling the efficiency parameters of warp cold sizing recipe ingredients using Minitab software Ver. 17.0. In general, the most effect observed in the properties, which characterized the cold sizing efficiency, was that of the PVA concentration and glycerol content. The plasticizers agents (i.e. Glycerol) are low molecular and high polar component with high interaction with the hydroxyl groups of the PVA.

References

Behera, B.K., Mishra, R., Nakum, S.: Comparative analysis of size materials and their weaving potential. Indian J. Fibers Text. Res. **33**(2), 132–138 (2008a). ISSN 0975-1025

Behera, B.K., Mishra, R., Gupta, R.: Comparative analysis of mechanical properties of size film. II. Effect of blend composition and lubricants. Fibers Polym. **9**(4), 489–494 (2008b). ISSN 1229-9197

Behera, B.K., Gupta, R.: Comparative analysis of mechanical properties of size films. J. Text. Inst. **100**(4), 305–313 (2009). ISSN 1754-2340

Chavan, H.V., Patil, S.S.: Yarn performance with cold sizing. Melliand Int. **20**(2), 84–86 (2014). ISSN 0947-9163

Maatoug, S., Ladhari, N., Sakli, F.: Evaluation of the weavability of sized cotton warps. Autex Res. J. **8**(4), 239–244 (2007). ISSN 1470-9589

Maatoug, S., Ladhari, N., Sakli, F.: Mechanical behavior of starch films. J. Text. Inst. **100**(8), 731–737 (2009). ISSN 1754-2340

Parekh, H., Harane, R.: Innovation in sizing – starch free cold sizing. J. Text. Assoc. **72**(6), 400–401 (2012). ISSN 2347-2537

Swiech, T., Swiech, F.: Classical methods for the quality control of yarns and their suitability for preventing faults in processing. Przeglad wlokienniczy **54**(1), 13–16 (2000). ISSN 1731-8645

Tong, Y.: Research and application of energy saving technology in textile warp slashing process. Appl. Mech. Mater. **508**(1), 223–226 (2014). ISSN 1662-7482

Zhu, Z., Li, W., Liu, Y.: Electroneutral cornstarch by quaternization and sulfosuccinylation to improve the adhesion of cold starch paste to raw cotton for low temperature sizing. Fibers Polym. **18**(6), 1106–1114 (2017). ISSN 1229-9197

Effect of Enzyme Washing Parameters on the Residual Bagging Height of Denim Fabrics

Abir Ben Fraj[1,2](✉), Mouna Gazzah[2], and Boubaker Jaouachi[1,2]

[1] Textile Engineering Laboratory, National School of Engineers of Monastir, University of Monastir, Monastir, Tunisia
abir1991_benfraj@yahoo.fr

[2] Textile Engineering Laboratory, University of Monastir, Monastir, Tunisia

Abstract. This research focuses on the effect of process parameters on the residual bagging height of some denim fabrics. Denim garment was treated with neutral cellulase enzyme by varying the process parameters, namely concentration of enzyme, washing temperature, bath ratio, and time. The results show, that the properties of bagging are significantly affected by the quantity of enzyme, they are correlated positively. While the time, the bath ratio and temperature are not significant variables.

Keywords: Residual bagging height · Enzyme washing · Denim fabric · Principal component analysis PCA

1 Introduction

Denim is becoming the most popular clothing material. This popularity is due to the finishing treatment that has undergone on jeans such as stone washing, enzymatic washing, bleaching…These finishing treatments give the denim fabric comfort and softness.

During wear, the mechanical properties that establish aesthetic appearance and durability are also very important. Therefore, some researchers are focused on the effects of finishing treatments on mechanical properties. Patra et al. were analyzed the effect of neutral and acid enzymatic in terms of decrease in color depth and weight loss and results showed that the effect in weight loss was more pronounced in the case of acid cellulase due to their stronger action than neutral cellulase (Patra et al. 2018). Mondal and Khan studied the effect of enzymatic washing parameters namely concentration of enzyme, time and washing temperature on the physical and mechanical properties of denim fabric (Mondal and Khan 2014). Indeed, the elongation at break, weight loss, tensile strength, stiffness, shrinkage, water absorption, color fading and morphological values by SEM were discussed. They optimized the washing condition for the best value at 2% enzyme concentration at 55 °C for 40 min.

Until now, overall studies still focus on the effects of treatment in the mechanical, physical and sensorial properties but there are no studies on the effect of finishing treatment in the bagging properties. Hence, the aim of this work is to investigate the effect of enzymatic wash on the residual bagging height property of denim fabrics.

S. Msahli and F. Debbabi (Eds.): CIRATM 2020, SPM 17, pp. 373–379, 2022.
https://doi.org/10.1007/978-3-031-08842-1_59

2 Materials and Methods

2.1 Materials

Six denim fabrics having different fiber blend ratios were prepared into 50 × 100 cm^2 and sewed as tubes. Although these samples have different characteristics, they present the same weave (twill 3/1). These characteristics are shown in Table 1. The applied enzyme ULTRA REDEP (a nonionic agent), is very active for the dispersion of indigo molecules in the treatment bath. It has a strong anti-redeposition power, which keeps the indigo in suspension and prevents its redeposition on the fibers. The NEOZYME 166, cellulase neuter, was then used for washing a wide range of fabrics, producing a clean finish. It is used to hydrolyze cellulose chain molecules of the fabric to achieve a surface pattern effect within a color contrast and a softer touch.

Table 1. Characteristic of fabrics

Fabrics	Blend raw materials	Weight (g/m^2)	Warp density (Ends/cm)	Weft density (Picks/cm)
1	2% elasthanne 98% coton	297	43	23
2	1% elasthanne 99% coton	338	32	20
3	2% elasthanne 98% coton	370	33	19
4	3% elasthanne 7%PES 90% coton	421	39	20
5	2% elasthanne 8% PES 90%coton	420	34	22
6	1,5% elasthanne 98,5% coton	429	32	21

3 Methods

3.1 Enzyme Washing

The enzyme was applied in the bath ratio of 1/5, 1/8 and 1/10 at concentrations of 0.3, 2 and 4% and treatment time was 15, 45 and 75 min at 30, 45 and 65 °C. The process was carried out in a liquor containing 1g/l anti-redeposition agent (ULTRA REDEP) at pH = 7 with the fabrics in an industrial drum washing machine. After washing, the enzyme was deactivated by using sodium carbonate at the temperature 40 °C for 5 min. Then, the fabric was rinsed with warm water followed by cold water and finally, it was dewatered and dried at 90 °C.

3.2 Test of Bagging

The bagging deformation test of fabric samples was investigated using dynamometer type Lloyd (Fig. 1). During a bagging test, a removable hemisphere in the upper jaw applied a load on the sample up to a certain distance with a defined speed 60 mm/min. The number of the applied cycles was fixed at 5 cycles. Hereafter, the residual bagging height was measured after a relaxation time equals 30 min. This test was repeated for 5 times and the average is considered as the value of residual bagging height.

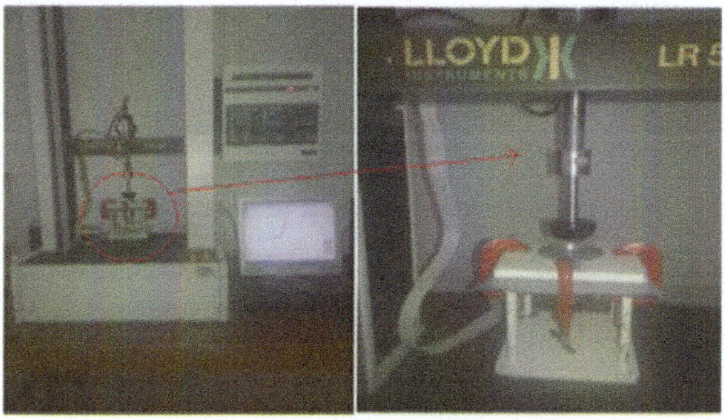

Fig. 1. Fabric bagging machine.

3.3 Principal Component Analysis (PCA)

The Principal Component Analysis (PCA) is a statistical method of description and reduction of studied parameters. Its goal is to find correlations between the data (inputs and outputs) and to evaluate the influence of the given i inputs on j outputs. Indeed, the PCA method amplifies, subjectively, the weight of some parameters while it dampens others that are less relevant to output values (Jaouachi 2012). The correlated parameters are grouped inside the trigonometrically circle. When the parameters are close to 1 and −1, they are considered as important. While they are considered insignificant, when they are close to the centre of this circle. The global variables defined in one of the circles are so-called positively correlated variables because they evolve in the same way. However, according to the rules of the PCA method, the groups of variables encircled together, which are opposed by the vertical axe (Y-axis) of this circle, are correlated negatively. Furthermore, the groups of variables which have symmetrically opposite direction relative to the horizontal axis are positively correlated. It means that the increase of one parameter of the group entails, as a consequence, the increase in the other group (Gazzah et al. 2015a).

4 Results and Discussion

The objective of this study is to investigate the effect of the enzyme washing parameters such as a temperature, a time of process, a quantity of enzyme applied, and the bath ratio on the residual bagging height of samples. Thus, the residual bagging height for the treated and untreated samples relative to the 18 experiments of the used Taguchi design was measured (Table 2). The obtained results are analyzed using MINITAB software (MINITAB® Release 14).

Table 2. Results of mean residual bagging height

Fabrics	Ms (g/m^2)	T (°C)	Time (min)	Qen (%)	Bath Ratio	MIUwp	MIUwf	Rbh (mm)
1	297	0	0	0	0	0.18	0.18	7.61
11n	365	30	15	0.3	1/3	0.23	0.21	8.7
12n	364	45	45	2	1/5	0.21	0.21	8.59
13n	370	65	75	4	1/10	0.22	0.22	9.89
2	338	0	0	0	0	0.17	0.19	8.52
21n	385	30	15	2	1/5	0.22	0.21	9.67
22n	393	45	45	4	1/10	0.26	0.22	9.06
23n	396	65	75	0.3	1/3	0.2	0.21	8.56
3	370	0	0	0	0	0.18	0.19	8.12
31n	434	30	45	0.3	1/10	0.23	0.22	9.13
32n	448	45	75	2	1/3	0.22	0.21	9.46
33n	440	65	15	4	1/5	0.22	0.2	10.34
4	421	0	0	0	0	0.18	0.18	9.24
41n	470	30	75	4	1/5	0.19	0.2	8.61
42n	463	45	15	0.3	1/10	0.2	0.21	8.74
43n	483	65	45	2	1/3	0.21	0.21	8.98
5	420	0	0	0	0	0.19	0.18	6.05
51n	458	30	45	4	1/3	0.2	0.21	9.5
52n	463	45	75	0.3	1/5	0.23	0.22	7.5
53n	454	65	15	2	1/10	0.21	0.23	9.33
6	429	0	0	0	0	0.17	0.17	11.02
61n	469	30	75	2	1/10	0.22	0.22	9.98
62n	470	45	15	4	1/5	0.2	0.21	10.31
63n	473	65	45	0.3	1/3	0.21	0.22	9.73

Figure 2 illustrates the Main Effects Plot for residual bagging height for fabrics. Results show that the temperature, time and bath ratio have no significant effect on the residual bagging height of fabrics (P-value (Fisher coefficient) = 0.529 >> 0.05 for temperature, 0.196 for time and 0.975 for bath ratio). However, the quantity of enzyme affects significantly the residual bagging height with P-values of 0.038. As well as the quantity of enzyme increases, the residual bagging height increases. In fact, the enzyme damages the cellulose, on the link 1,4b glucosique of the cellulose molecule. The hydrolysis of this link breaks the molecule in several pieces, which can be themselves divided (Khedher et al. 2009). Hence, a decrease in the degree of polymerization (DP) with a remarkable drops in mechanical strength, which justifies the significant decrease in the resistance to bagging (Gazzah et al. 2015b). This can limits the ability of fabric to return rapidly to its first state. We can also add that the enzymes specifically acting on the cellulose part, mainly on the surface of the fabric. This gives the desired look and at the same time, removes hairiness from surface (Patra et al. 2018).The removal of hairiness will cause an increase in abrasion resistance. During washing, the fibers on the surface of the fabric become more attached to each other and their movement in the yarn will be limited. For these reasons it can be concluded that enzyme washing promotes the increase of abrasion resistance. This conclusion is in a good agreement with Bhattacharjee's results (Bhattacharjee et al. 2019). In addition, the increase of abrasion resistance generates the increase of residual bagging height. This finding is in an ultimate accordance with Gazzah's study ones (Gazzah et al. 2014).

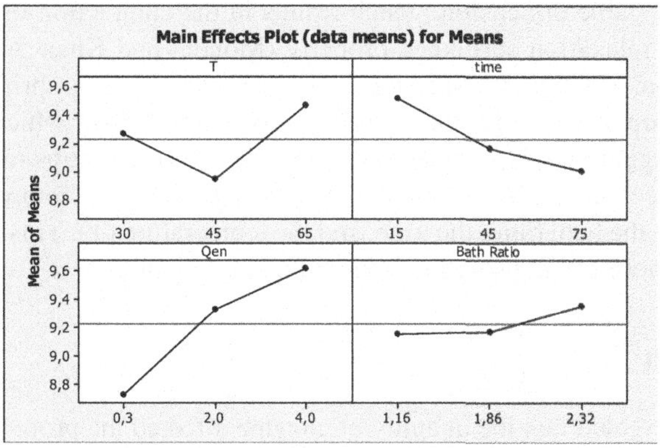

Fig. 2. Main effect plot for residual bagging height

Thus, using Principal Component Analysis (PCA) method Fig. 3 shows the different correlations between inputs and outputs. In fact, four groups A, B, C and D are found describing widely the relationships obtained.

The group A is composed by two outputs, the bagging recovery and the recovery speed. Therefore, group B is composed by four outputs, coefficient of friction in warp and weft direction (MIUwp, MIUwf), bagging resistance (Bresistance) and bagging fatigue (Bfatigue). However, concerning the group C, there composed by one output, residual bagging height (Rbh), and one input the quantity of enzyme (Qenz). Based on the ACP rules, it can be concluded that the quantity of enzyme and the residual bagging height are correlated positively. The increase or decrease of the input generated the increase or the decrease of the whole appearance and the shape of the bagged zone. This result can be explained by the shrinkage in the warp and weft fabric direction. During weaving, cotton fabrics are subjected to considerable tensions, particularly in the warp direction. In subsequent finishing processes, such as calendaring this stretch is increased

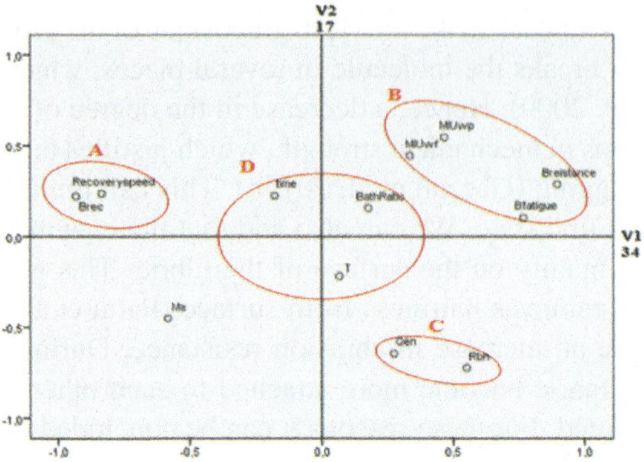

Fig. 3. Results with principal component analysis (PCA).

and temporarily set in the fabric. Hence, the fabric is in a state of dimensional instability. Subsequently, when the denim garment was thoroughly wetted in washing, it tended to revert its more stable dimension, which results in the contraction of the yarns. This effect is called a relaxation shrinkage property (Mondal and Khan 2014). Where the quantity of enzyme increases the shrinkage will be increased. This shrinkage will cause the increase of warp density, thus the friction coefficient will grow which causes the rise of the residual bagging height (Gazzah et al. 2015a). All the groups represent positive correlations between the inputs except the last group D, which is composed of three no significant inputs, the bath ratio, the time, and the temperature (T). This result is in good agreement with those found using the Main effect plot method.

5 Conclusion

This study was revealed that the quantity of enzyme affected the properties of bagging. While, washing temperature, bath ratio, and time are no significant inputs. The ideal condition for this treatment to minimize the residual bagging height it will be studied in our future research work.

References

Bhattacharjee, M., Dhar, K.A., Islam, Md.M., Rashid, Md.A.: Development of washing effects on reactive dyed denim fabrics: a value added approach of denim wash. Int. J. Text. Sci. **8**(2), 41–48 (2019). ISSN 2325-0119

Gazzah, M., Jaouachi, B., Sakli, F.: Modeling residual bagging behavior of denim fabrics as function of friction parameters. Milliand Int. **19**(2), 95–96 (2014). ISSN 0947-9163

Gazzah, M., Jaouachi, B., Sakli, F.: Effect of friction parameters on the residual bagging behavior of denim fabrics. Int. J. Appl. Res. Text. **3**(1), 32–43 (2015a). ISSN 2356-5594

Gazzah, M., Jaouachi, B., Sakli, F.: Recovery velocity evaluation of bagged denim fabrics as function of frictional parameters. Int. J. Clothing Sci. Technol. **27**(2), 302–314 (2015b). ISSN 0955-6222

Jaouachi, B.: Evaluation of spliced open end yarn performances using fuzzy method. J. Nat. Fibers **9**(4), 290–309 (2012). ISSN 1544-0478

Khedher, F., Dhouib, S., Msahli, S., Sakli, F.: The influence of industrial finishing treatments and their and their succession on the mechanical properties of Denim garment. AUTEX Res. J. **9**(3), 93–100 (2009). ISSN 1470-9589

Mondal, M.I.H., Khan, M.M.R.: Characterization and process optimization of indigo dyed cotton denim garments by enzymatic wash. Fashion Text. **1**(1), 1–19 (2014). ISSN 2198-0802

Patra, A.K., Madhu, A., Bala, N.: Enzyme washing of indigo and sulphur dyed denim. Fashion Text. **5**(1), 3–18 (2018). ISSN 2198-0802

Preference Mapping of Commercial Baby Diapers: Based on Hedonic Consumers Ratings and Expert Perception

Rahma Tilouche[1(✉)], Soumaya Essayeb[1], and Mohamed Ben Hassen[1,2]

[1] Textile Engineering Laboratory, Monastir University, Monastir, Tunisia
rahma.tilouche@enim.u-monastir.tn

[2] College of Engineering: Industrial Engineering Department, Taibah University, Medina, Saudi Arabia

Abstract. Baby diaper industry didn't offer a performant tool to identify the product attributes expected by consumers and essentially adapting the product performance to these expectation. Preference mapping were widely used in the food sector in order to draw the consumer liking profile. Comparing to other methods, firstly this technique is considered as fast and easy to interpret. Secondly, it helps to identify product attributes having the highest impact on overall consumer liking. In this paper, this technique was opted to be used for baby diapers samples available in the Tunisian market. Five samples were conceded for both sensory profiling by quantitative descriptive analysis by expert judges, as well sensory evaluation using the liking scale.

Keywords: Baby diapers · Preference mapping · Sensory profiling · Consumer ratings · Expert perception

1 Introduction

Obtained data from consumer and expert judges is considered as very complex to analyses. This is the reason why multivariate statistical techniques are intended to be used (Gaze et al. 2015). Preference mapping constitutes an easy method obtained through linking attribute intensity assessment of expert judges and hedonic consumer ranking (Gaze et al. 2015; Schlich et J.A.Mc Ewan 1992) (Gentner et al. 2012; Varela et al. 2014). Over the last decade, disposable baby diapers became a very popular product all over the world. This leads to a distinguish growth in term of use accompanied by important improvements in their properties and design (Environment Agency 2010). Previous studies of (Schlich et J.A.Mc Ewan 1992) and (Gentner et al. 2012) consider that experiment data issued from consumer's choices there are a complex interactions between the product and the consumer. In fact, they found that there are an out coming of consumer's personality and previous experience in the final results (Varela et al. 2014; Danzart 2009). In this paper, five samples of baby diapers were investigated: descriptive sensory profiles were determined and consumer test were accomplished. The obtained

S. Msahli and F. Debbabi (Eds.): CIRATM 2020, SPM 17, pp. 380–385, 2022.
https://doi.org/10.1007/978-3-031-08842-1_60

results may provide useful information for hygienic products manufacturer. In fact the results utility can be revealed in both development and production level, moreover when launching diapers quality control specifications.

2 Materials and Methods

2.1 Baby Diapers Tested Samples

The baby diapers examined in this study are from five brands available on the Tunisian hygienic product market (local and international brands). They were chosen according to their positions on the baby diaper consumer market, as well as they were the most popular.

2.2 Subjects

Industrial Experts: Seven experts (two women and five men) are requested to judge the surface quality of the five different diapers. Expert are members of an industrial tactile panel trained in a previous work (Tilouche et al. 2018). All judges maintained in the company a position where hand touch was necessary and was daily practiced.

Hedonic Consumer: In order to generate consumer preferences, the study involved a total of 200 women. All consumers were aged from 23 to 45 years, half of the consumers used at least 2,5 years of baby diapers for one child, and the other half used them for more than 5 years (2 children and more).

2.3 Procedure: External Preference Mapping (PREFMAPP)

The study suggests a simple approach, gathering data through 200 consumer's preference ranking (using a 10-point hedonic scale) and seven expert's description of six prefixed attributes: this is called "external preference mapping". Actually, this approach allows to the manufacturer adapting his product to his consumer tastes in a very rapid way.

Step1: Data gathering: The experiment was performed as follows:

Hedonic rating: Women were asked to formulate their preferences for the five different baby diapers brands (on a scale from 0 to 10, what rating do you assign to each baby diapers brand?

Description test: Experts are invited to judge the surface quality of the five different diapers according to six attributes respectively: wet-dry, soft, flexibility, thin-thick, elasticity, and fatty

Step2: Diaper sensory profiling: The sensory profiling was established using quantitative descriptive analysis.

Step3: Consumer clustering: This was through the hierarchical ascending classification (HAC) which is a simple iterative classification method. The principle is based on searching if individuals grouped within one class (intra-class homogeneity) are possibly similar while classes are more dissimilar (interclass heterogeneity).

Consumers distance proximity measurement: In this step, dissimilarity will be measured in order to assess the proximity between individuals, which means that two points have a dissimilarity equal to zero if they are coincided. The "HAC" upward classification uses distance measurements between products to form classes. The most direct method for calculating distances between n individuals in a multidimensional space containing p quantitative variables is to calculate Euclidean distance between x and y points position in space (Husson 2010) (since all consumers use the same rating scale) according to Eq. 1:

$$\textbf{Euclidean distance}(x_i, y_i) = \sqrt{\left(\left(\sum_{i=1}^{n} (x_i - y_i)^2\right)\right)} \quad (1)$$

Aggregation: Consumers who make the least difference in intraclass inertia will be aggregated. As a result, the disparity of all consumers is defined on the basis of the dissimilarity index, Ward criterion (Ward 1963) is used as an aggregation criterion. The dissimilarity "D Ward" between 2 individuals must be equal to half the square of the Euclidean distance d (Eq. 2).

$$D_{\textbf{Ward}}(C_1, C_2) = \frac{nC1nC2}{nC1 + nC2} d^2(gc_1, gc_2) \quad (2)$$

Step4: Preference mapping: In order to draw external preference mapping, two steps are needed to be done. First, products must be mapped according to their characteristics. This is through principal component analysis to obtain the products sensory profile. Subsequently, the scores assigned by each consumer group to the different baby diapers will be modeled according to the diaper's physical characteristics (scores assigned by experts on specific attributes). Finally, a space representation of consumers on the sensory profile according to the PREFMAP technique is gotten.

3 Results and Discussions

The creation of the external preference mapping of baby diapers requires the passage through three steps that will be detailed as follows.

STEP 1: Data gathering: Consumer and expert data are collected in order to be uses as entry database.
STEP 2: Baby diaper Sensory profiling: In this step, baby diapers will be mapped based on their sensory characteristics. The projection of expert's scores attributed to the diapers permit sensory profile. Results illustrates that 70, 37% of the information gathered in dimension two are on both axis F1 and F2. In order to describe 84, 28% of the total information (Fig. 1). This confirms that the diapers have been perceived by the experts as quite different (Fig. 2).

As it can be seen from Fig. 3, a closed angle from the origin, can demonstrate that "soft" and "wet-dry» attributes are quite correlated with each other. These results suggest strongly that the feeling of softness is a growing function of dry perception, in deed it is confirmed that this feeling is attacked in wet environment. Baby diapers corresponding

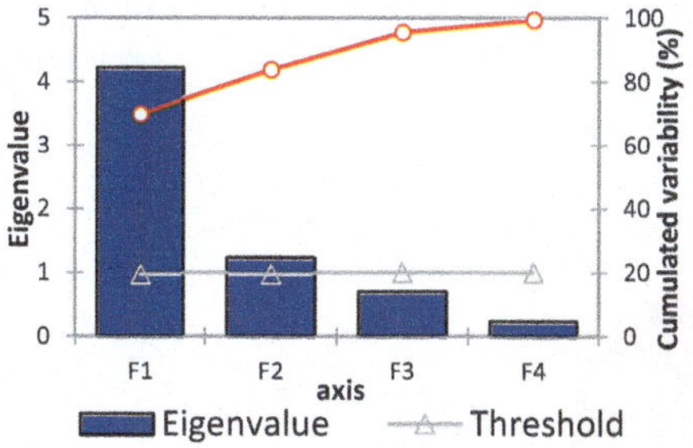

Fig. 1. Information variability

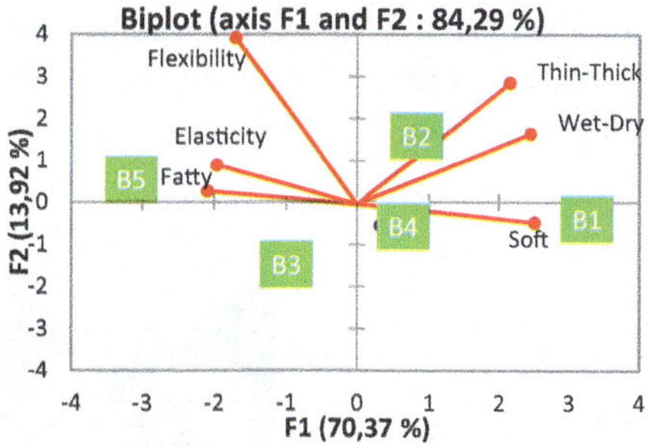

Fig. 2. Principal component analysis of the baby diapers (experts judgment).

to B1 and B2 brands are considered as the softest and driest diapers while presenting the thickest structure. However brand B5 diaper is considered as the fattest and the most elastic.

STEP 3: Consumer clustering.

Consumer profiling and gathering: Firstly, consumer profiling was drown as shown in Fig. 4. According to the correlation circles it can be noticed that on the first circle 50.18% of the gotten information are gathered on the first two axes F1 and F2 which can describe 78.14% of the total information. Next, Hierarchic Ascendant Classification (HAC) will be used in order to group consumers Fig. 5 below details a dendogram to visualize the progressive aggregation of consumers.

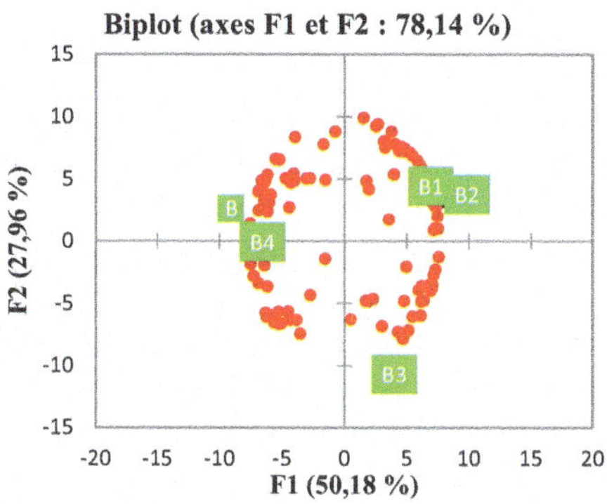

Fig. 3. Consumer profiling.

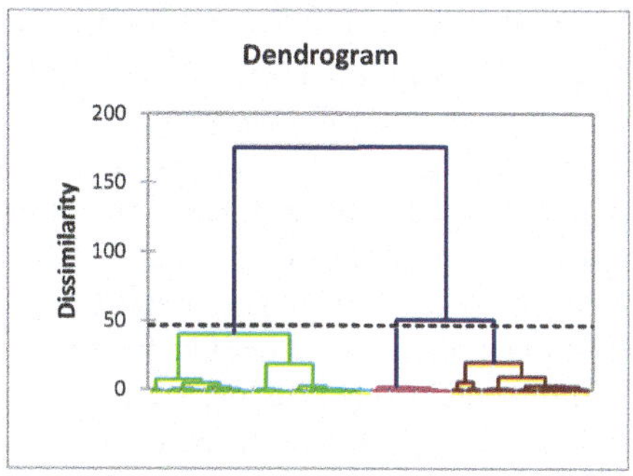

Fig. 4. Results after hierarchic ascendant classification (HAC).

At first the algorithm initially groups individuals and after that subgroups to finally, it gradually groups all consumers together. Three homogeneous groups that have been identified by Xlstat software. Presented class profile for each brand are shown in Fig. 6. We also proceed to the determination of the barycenter of each class which values will be used as input data when constructing the preference map.

STEP4: External preference mapping of baby diaper: In this level, baby diapers preference, will be modelled. From PrefMapp representation (Fig. 5), it can be noted that

Class 3 consumers have a maximum of preference terms (ideal point). In addition, it can be visualized that for each consumer, the map is separated into zones of: preferences and/or rejection.

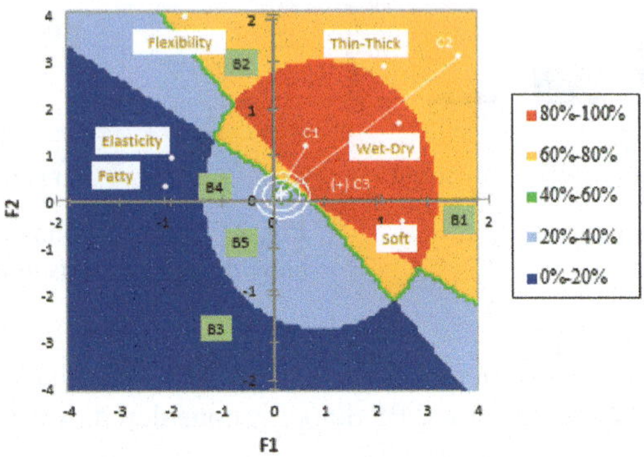

Fig. 5. Baby diapers brands Preference Mapping.

By synthesizing the information from the five diapers brands. Mapping showed a different dimension perceptual spaces that the three consumer groups have a global preference for B1 and B2 with 67% of satisfaction level. As presented those latter were positioned in an area that involved acceptance values above the overall value for 80% to 100% of consumers. Actually, this area include the ideal point besides brand B1 and brand B2 are considered the less fatty and the less rough. In addition, they are distinguished by their excellent softness and keeping dry properties. Otherwise, B4 and B5 brands have a very low preference with a 33% satisfaction level which is located in the limit zone on the preference mapping (in which the acceptance values were 20 to 40% above the overall acceptance). Definitely, this can be explained by the fact that: the diaper properties of these brands are fluctuating. Therefore, the consumer's preference cannot always be linear. Finally, the brand 3 is located in a rejection zone. Consumers have no appreciation for this brand in which the acceptance values were 0% to 20% above the overall acceptance. The brand B1 was considered as the first competitor of brand B2. In fact, the diapers brands positioning by consumers is comparable to the sensory profiling obtained from experts. The benchmarking shows that consumer words can give a relevant description of the perceptual dimension. In fact, this can help improving baby diaper characteristics and adapting it to customers' needs. Baby diapers manufacturers with mediocre properties must make a great effort in designing their products to satisfy their consumers. Especially for hand touch and feeling dry properties considered as very weak and unacceptable. During this study, consumer have been imposed to ignore the price criterion and to judge baby diapers according to their functional properties. 90% of consumers distinguish comfort, absorption, feeling dry and softness as the most important criteria when purchasing a baby diaper. It has been also found that interest in product safety information is important for 100% of consumers. Other criteria such as structure and design are cited by 65% of consumers. The main results of the preference mapping study confirms that hedonic consumer maps are reliable with trained sensory profiles (descriptive side by tactile senses). Furthermore, regarding to the preference scores regression on the PAC axes, the hedonic test can be linked with the descriptive side and give a clear explanation of the customer's likings.

4 Conclusion

The main objective of this study was to consolidate the sensory specifications of baby diapers perceived by expert judges who best meet different consumer's expectations. For that purpose consumers have been asked to consider different brands of baby diapers in a comprehensive manner. As well as, expert judges (from a trained panel) were asked to judge the same baby diapers according to six descriptors. All collected dada permit the construction of the external preference mapping for the five brands of the baby diapers. It has been found that consumers have a maximum of preference terms for the B1 and B2 brands. These obtained results allow the manufacturers to adjust their products to the consumers' needs and therefore can position himself against competitors in a fast way with a subtle graphic representation.

References

Gentner, A., et al.: Mapping a multi-sensory identity territory at the early design phases. Int. J. Affect. Eng. **12**(2), 191–200 (2012). https://hal.archives-ouvertes.fr/hal-00965150. Accessed Aug 2019

Tilouche, R., et al.: Industrial tactile sensory panel analysis implementation: decision support tool for hygienic product company. In: Derjipark (ed.) Tekstil ve confection, pp. 294–303 (2018). Accessed 2018

Tilouche, R., et al.: Development of a new global quality index for evaluating baby diaper customer desirability: a useful tool for manufacturer. J. Text. Inst. **112**(12), 2021–2038 (2020)

Talebi, D., et al.: The role of quality benchmarking deployment to world-class manufacturing. Qual. Eng. **26**(2), 206–214 (2014). https://doi.org/10.1080/08982112.2013.803121

Environment Agency: Life-cycle assessment of disposable and reusable nappies in the UK (2010). www.environment-agency.gov.uk. Accessed 2018

A Modeling and Simulation Enhanced Lean Six Sigma Design Process in Clothing SME

Nedra Abbes(✉), Néjib Sejri, and Morched Cheikhrouhou

Textile Engineering Laboratory (LGTEX), Higher Institute of Technical Studies of Ksar Hellal (ISET), Monastir University, Monastir, Tunisia
nedra.abbes@enim.u-monastir.tn

Abstract. This study explores how Lean Six Sigma and Simulation can be integrated together based on Sim-Lean approach, using a process improvement effort in a Tunisian Clothing Small Medium Enterprise (SMEs). A structured framework integrating these research methodologies is developed, which might benefit a variety of future clothing process improvement efforts, and could inform quality improvement efforts in other industries. The aim is to allow a successful implementation of the approach in Tunisian Clothing Industry in order to improve the Lead Time, the daily output and the average staying times (min) of jobs waiting in queues.

Keywords: Lean Six Sigma · Discrete event simulation · Modeling · Sewing process · Process improvement

1 Introduction

Lean Six Sigma (LSS), as a business or management strategy, has delivered companies with a competitive advantage (Chaurasia et al. 2016; Sunder 2016) this method is widely used in manufacturing firms in the world and is applied in different industrial fields, that include manufacturing (Ben Ruben et al. 2017; Krueger et al. 2014; Timans et al. 2012; Anderson and Konvach 2014; Kumar 2006), services (Sunder and Antony 2015; Vijaya Sunder 2016), commercial (eJose Arturo Garza-Rayes et al. 2016), health care (Zhang et al. 2015; Vinodh et al. 2014) and logistics (Garza-Reyes et al. 2014). Few studies have been reported about the application of Lean Six Sigma in Clothing small- and medium-sized enterprises (SMEs) because of several problems.

Simulation techniques started their development in the late 1950's and have many usages in different practical areas. According to (Banks 1998), "simulation is the imitation of the operation of a real-word process or system over time". By using Simulation, several alternatives or "simulation scenarios" can be evaluated to choice the best solution, reduce the risk of wrong and costly decisions before putting them into practice (Pritsker and Alan 1989). (Robinson et al. 2012) define from a theoretical and empirical perspective the potential complementary roles of discrete event simulation and a Lean approach in healthcare systems during the kaizen event. They also present their model called Sim-Lean, to define three roles for discrete event simulation used with Lean:

S. Msahli and F. Debbabi (Eds.): CIRATM 2020, SPM 17, pp. 386–393, 2022.
https://doi.org/10.1007/978-3-031-08842-1_61

education, facilitation and evaluation. In 2014, they focused on Sim-Lean facilitate as an example of facilitated modeling using discrete-event simulation for better process understanding. Based on this model, (Baril et al. 2016) developed a simulation model to conduct a Lean project in an oncology clinic to validate improving ideas proposed by the Kaizen event participants in "back office". In their book, (El-Haik and Al-Aomar 2006) present a simulation-based Lean Six-Sigma approach. They define Lean tools integrated in DMAIC (Define, Measure, Analyze, Improve and Control) process, and where DES plays a key role in each phase. Other researchers discuss how DES can make a lean transformation more precisely (Miller et al. 2010; Richard et al. 2010). The DES with Six Sigma also is used by (Elhaouzi et al. 2009), to design and validate of a product-driven control system. Lean transformation processes are considerably different from one to another. (Feld 2001) discusses a streamlined road map to Lean manufacturing. This approach presented four phases for the implementation of a lean manufacturing project which are; the assessment of processes capabilities, the analysis of the current state gap in order to analyze and identify root cause, the design future state and the implementation state. This paper shows the mutual benefits that can be gained when combining Lean Six Sigma and simulation. A study in clothing SME for this integration as well as a simplified sewing process are presented. This paper is organized as follows. Section 2 presents the material and methods. Finally the results are discussed in Sect. 3 and Sect. 4 presents the conclusion of our research.

2 Material and Methods

In this work, the modeling and simulation of our process is achieved by integrating the Sim-Lean Model with an enhanced Lean Six Sigma process based on Feld's streamlined method using BPMN (Business Process Modeling Notation) language, in order to validate a certain number of indicators and avenues of the process improvement. Based on the Sim-Lean approach presented by (Robinson et al. 2012), we propose to support our Lean Six Sigma process by Sim-Lean approach with the following purposes (Fig. 1):

- Educational purpose: It can be the way to ensure the training of employees (Adams et al. 1999; Robinson et al. 2014).
- Facilitation purpose: In our model BPMN standard is used to offer better information and ensure that the different actors' participants can communicate clearly and effectively on the business processes by describing them by means of a notation simple graphics.
- Evaluation purpose: with discrete event simulation model, analytic and quantitative output can are offed to the current state.

Figure 1 presents a view of our proposed framework in which, Lean Six Sigma and Sim-Lean approach are combined. Each one of the Sim-Lean purposes described above is shown in their correspondent process steps. The process shows how Lean Six Sigma and simulation are supporting each process step. Moreover, we used several tools and the success factors for Lean Six Sigma implementation such as an input for the stages of process mapping and definition of the decision variables. 20 measurements were taken for each operator to calculate the real process time of every task. Then, data were analyzed and the distribution fit and goodness is used as input to drive the simulation model. Once we have determined the probability distributions for the processing times, the simulation model can be run. We added the future state validation before implementation using DES simulation. Once the future state is validated and the decision is taken, the implementation step is started and the Lean Six Sigma is applied.

3 Results and Discussions

Our approach was implemented in a Clothing Small Medium Enterprise in Tunisia, specialized in work wear.

Define

The define Stage allow understanding the process, and determining performance indicators. Interviews, meetings, observations, project charter and spreadsheet were realized as primary tools to achieve this stage. The project team should is composed by employees, who are responsible for performing the ship loading process (Furterer 2009). As management commitment is the most important key of LSS implementation success, a meeting was prepared by the top management in order to explain the employees about the importance of and the need for the Lean Six Sigma project and its deliverables.

Measure

In this step, data on measurable indicators of production processes are collected. The primary outputs from the measure Step were: detailed process maps, in our case we used BPMN standard to model our process, process activities, and the current systems simulation model. To achieve this, the used methods included interviews, observations, time study, and summarizing data in a spreadsheet. The trousers production cycle includes different steps of assembly operations to transform raw materials into finished trousers product. There are mainly four sequences of steps namely (i) prepreparation of pockets, (ii) production of back of trousers, (iii) production of front of trousers, and (iv) assembling of fabric parts. To collect enough data more than 100 h of observation were required, divided between the Production unit and Cut unit. An example of the real process time for the 'sew back pocket hem' is shown in Table 2. Same measurement is applied for all tasks. To calculate the real process system, some conjectures were considered:

- Only one worker is at each machine.
- Allowances are taken into consideration.
- Delay times (machine breakdowns, changing apparatus) are not taken into consideration.
- There is no energy problem in the system.
- Fabric loss is not taken into consideration.
- Raw material is unlimited.

Results of the reference layout model are shown in Table 1 according to the performance measures. As seen in Table 3, it can be observed that the average trousers Lead Time is 14.77 min, the average number of finished trousers in a day is 419.5, and the average staying time of jobs in queues is 657.8 min.

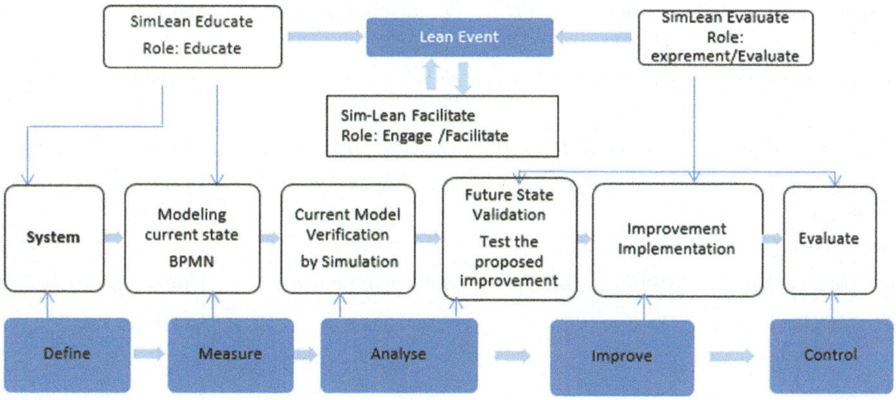

Fig. 1. Steps of modeling and simulation method.

Table 1. Results based on current layout model.

Performance measures	Average	St. Deviation	Min	Max
Lead time	14.77	0.2	14.2	14.95
Daily output	419.5	1.9	418.3	423.3
Waiting time	657.8	1.3	643.5	672.2

Analyze

In this stage, the causes of the non-value added activities were determined, the time study were analyzed, and data were analyzed and tested for distribution fit and goodness. This latter is used as input to drive, verify and validate the simulation model. The statistical distributions used in the simulation model are based on data collected from the production process. The mean-square error is used to select the best statistical distributions for our data. Our simulation model was developed to be as near as possible to the real process to ensure the calibration of our system. The Stat fit Student Version program is used to determine the statistical distributions. Table 2 summarizes the distributions estimated for all tasks.

Table 3 shows that the difference between the simulation model and the real process is at most 6% which is lower than the maximum allowable error (±10%) confirming the validity and calibration of model to be used to test scenarios.

Machines are busy with 98.9% 98.3% respectively. Therefore, in current model, post 2 and post 12 were identified as bottleneck. The usage of the resources indicates some

Table 2. Estimated distributions for tasks.

Task	Fit distribution	Task	Fit distribution
Prepare the launching	Lognormal (1.25; 0.12)	Close the sides and crotch	Lognormal (1.45; 0.3)
Sew back pocket hem	Lognormal (0.17; 0.014)	Put together waist and lining	Uniform (0.47; 0.21)
Sew pocket meter	Uniform (0.17; 0.020)	Attach waist	Weibull (0.80; 0.54)
Close sub-bridge	Normal (0.15; 0.01)	Execute waist	Gamma (1.22; 0.54)
Overcast left front/right front/crotch	Uniform (0.14; 0.2)	Sew down hem	Lognormal (0.17; 0.03
Sew back pocket	Uniform (0.7; 1)	Make boutonnieres	Uniform (1.16; 0.2)
Close pocket bag	Poisson (0.3)	Place button	Lognormal (0.25; 0.12)
Sew zipper	Beta (1.20)	Control	Uniform (2.29; 0.1)
Attach back pocket and pocket meter	Lognormal (1; 0.13)	Measure control	Lognormal (0.51; 0.43)
Attach trousers back	Gamma (0.45; 0.21)	Pack and ship	Exponential (1.50; 0.54)

Table 3. Validation of the discrete event simulation Model

	Lead time (waiting and service)			Waiting time		
	Real (min)	Simulated (min)	Gap (%)	Real (min)	Simulated (min)	Gap (%)
Total process time	14.77	15.86	5.73	657.8	698.34	6.1

sub and over-utilization. Therefore we confirm the hypothesis about a possible problem of resources capacity.

Improve

To improve the process efficiency, LSS techniques were applied. A kaizen event was planned and directed to ensure that each team member could participate in finding solutions and propose suggestions for the implementation. The improvements activities are begin with the presentation of the objectives by the team members. Lean approach and the results obtained after the Measure phase were presented to ensure that all the kaizen members receive the same information, then models were run, the results were analyzed and the developed system are validate. Several scenarios of the sewing process were tested and are tried by what-if analysis to check if the model still performs adequately. For brevity, only three are listed and tested for the sewing process as given in the following.

First improvement scenario: The first proposed scenario is to change the layout of sewing process; in fact the locations of sew pocket meter hem in production line are reformed. To this end, *Sew Pocket meter hem* post is transferred in parallel with post 2. The equipment of the sew pocket meter hem post is portable and it is not necessary to perform costly changes in the layout of the factory. The improved production line causes the capacity of '*Sew Back Pocket hem*' post to increase.

The second scenario: The second scenario is one extra sewing machine operator for the 12th post '*Put together waist and lining*' was added to the reference layout line, and to training the workers to improve their efficiency.

The third scenario is a total of two extra sewing machine operators were added to the reference layout for the '*Close the sides and crotch*' post and '*Attach waist*' post. The results of the three scenarios are provided in Table 4.

Table 4. The results based on scenarios.

Scenario	Performance measures	Average	St. Deviation	Min	Max
Scenario 1	Lead time	14.5	0.2	14.1	14.6
	Daily output	422.5	1.9	420	424.1
	Staying times (min) of jobs waiting in queues	653	1.3	652	654
Scenario 2	Lead time	14.2	0.1	14	14.5
	Daily output	450.3	1.8	419	458.3
	Staying times (min) of jobs waiting in queues	650.6	1.2	648	655
Scenario 3	Lead time	14.6	0.3	14.2	14.95
	Daily output	435.7	1.7	430.6	438.3
	Staying times (min) of jobs waiting in queues	653.7	1.1	655.7	659

Table 4 shows that all scenarios are better than the current process. Scenario 2 has the largest positive impact on the fabrication process. The daily output of the system is increased to 422.5 with scenario 1, and 435.7 with scenario 3. Furthermore; the average staying times with scenario 3 is decreased to 635.7 min from 657.8 min. To sum up, with these scenarios the efficiency of the line was increased. These scenarios can not only provide decision alternatives to manufacturers, but to suggest more comprehensive solution alternatives.

Results were presented to group members, who are interested in the projected flow time. The last step in the improve stage is the implementation. Scenario 2 is implemented because of its features such as: height daily output, low waiting and low Lead time. The group members built an action plan to quickly implement those changes with other Lean tools in 30 working days. Lean tools like 5S, FMEA were implemented. In addition a revision of the current organization of production workshops (physical flows, information flows) was carried out in order to modernize the management of the company and the work organization to ensure an increase of the industrial performances while controlling the costs and the delays of the production. Twenty weeks after implementing the modifications, new data was collected. Trousers production lead times were reduced from 14.77 to 14 min, (a 5% reduction). The daily production of trousers is increased to 450. (Up by according to the reference system), the average staying times (min) of jobs waiting in queues was found 560 min.

Control

The main purpose of the Lean Six Sigma methodology is not only improving the process performance but also having the improved results sustained in the long run (Kumar et al. 2006). The applied improvements have been fully integrated into the training regime and the process documentation, the information related to company performance was sharing with its employees. Visual management, Total Productive Maintenance and Process FMEA are to be implemented after completion of the project to provide a visual aid for controlling the key input and output variables pertaining, and to ensure the team could not revert back to old habits. According to (Pyzdek and Keller 2014), control charts are a powerful tool for achieving process control and stability. For this project, the implementation of control charts was important to check that the product meet the desired specifications.

4 Conclusion

The sewing process affects several improvement outcomes, but it has rarely been studied in the literature. In addition, Lean Six Sigma, and simulation have rarely been combined and used in a single sewing process improvement study. This study aimed to understand and improve the trousers sewing process as well as provide an integrated framework by using three approaches. Their use in the trousers sewing process not only demonstrated their complementary roles in process improvement, but also provided the sewing process studied with some good scenarios and insights as a result of using them together. A number of scenarios were proposed and the simulation was applied to demonstrate the impact of most of these scenarios. The simulation models and statistical results showed that all scenarios can result in a significant improvement in the sewing processes. The trousers process flow time is achieve an improvement from 14.77 to 14 min, a 5% reduction which considers as important in clothing SME. Therefore, the framework was created to guide future sewing process improvement efforts in textile industries. To achieve a success improvement implementation, several success factors have been identified such as: Necessity of involvement of managers and employees, necessity of acceptance of change culture and a culture of continuous improvement. There are a few limitations in this study. Other major stochastic variables (machine breakdown, repair, absenteeism, the work of the supervisor, maintenance etc.) not be used to detail the model in this study. Future work could investigate a wider range of performance measures and developed to be implemented in another SME clothing industries.

References

Adams, M., Componation, P., Czarnecki, H., Schroer, B.J.: Simulation as a tool for continuous process improvement. In: Farrington, P.A., Nembhard, H.B., Sturrock, D.T., Evans, G.W. (eds.) Proceedings of the 1999 Winter Simulation Conference, pp. 766–773 (1999)

Baril, C., Gascon, V., Miller, J., Côté, N.: Use of a discrete-event simulation in a Kaizen event: a case study in healthcare. Eur. J. Oper. Res. **249**(1), 327–339 (2016)

El-Hail, B., Al-Aomar, R.: Simulation-Based Lean Six-Sigma and Design for Six-Sigma. Wiley, Hoboken (2006)

Feld, W.M.: Lean Manufacturing: Tools, Techniques, and How to Use Them. St. Lucie Press: APICS, Boca Raton (2001)

El Haouzi, H., Pétin, J.-F., Thomas, A.: Design and validation of a product-driven control system based on a six sigma methodology and discrete event simulation. Prod. Plann. Control Manag. Oper. **20**(6), 510–524 (2009)

Kapur, K.C., Feng, Q.: Integrated optimization models and strategies for the improvement of the Six Sigma process. Int. J. Six Sigma Compet. Advant. **1**(2), 210–228 (2005)

Krueger, D.C., Parast, M.M., Adam, S.: Six Sigma implementation: a qualitative case study using grounded theory. Prod. Plann. Control **25**(10), 873–889 (2014)

Robinson, S., Radnor, Z.J., Burgess, N., Worthington, C.: SimLean: Utilising simulation in the implementation of lean in healthcare. Eur. J. Oper. Res. **219**(1), 188–197 (2012)

Garza-Reyes, J.A., Flint, A., Kumar, V., Antony, J., Soriano-Meier, H.: A DMAIRC approach to lead time reduction in an aerospace engine assembly process. J. Manuf. Technol. Manag. **25**(1), 27–48 (2014)

Kumar, M., Antony, J., Singh, R.K., Tiwari, M.K., Perry, D.: Implementing the Lean Sigma framework in an Indian SME: a case study. Prod. Plann. Control **17**(4), 407–423 (2006)

Anderson, N.C., Kovach, J.V.: Reducing welding defects in turnaround projects: a Lean Six Sigma case study. Qual. Eng. **26**(2), 168–181 (2014)

Pritsker, A., Alan, B.: Why simulation works. In: Proceeding of the 21st Conference on Winter Simulation (WSC 1989), pp. 1–9 (1989)

Ben Ruben, R., Vinodh, S., Asokan, P.: Implementation of Lean Six Sigma framework with environmental considerations in an Indian automotive component manufacturing firm: a case study. Prod. Plann. Control **28**(15), 1–19 (2017)

Robinson, S.: Simulation: The Practice of Model Development and Use, 2nd edn. Palgrave Macmillan, Basingstoke (2014)

Sunder, V., Antony, J.: Six-sigma for improving top-box customer satisfaction score for a banking call centre. Prod. Plann. Control **26**(16), 1291–1305 (2015)

Sunder, M.V.: Rejects reduction in a retail bank using Lean Six Sigma. Prod. Plann. Control **27**(14), 1–12 (2016)

Vinodh, S., Kumar, S.V., Vimal, K.E.K.: Implementing Lean Sigma in an Indian rotary switches manufacturing organisation. Prod. Plann. Control. **25**(4), 288–302 (2014)

Timans, W., Antony, J., Ahaus, K., van Solingen, R.: Implementation of Lean Six Sigma in small- and medium-sized manufacturing enterprises in The Netherlands. J. Oper. Res. Soc. **63**(3), 339–353 (2012)

Huang, Y., Klassen, K.J.: Using Six Sigma, Lean, and simulation to improve the phlebotomy process. Qual. Manag. J. **23**(2), 6–21 (2016)

Zhang, M., Wang, W., Goh, T.N., He, T.: Comprehensive Six Sigma application: a case study. Prod. Plann. Control **26**(3), 219–234 (2015)

Ergonomics Study and Analysis of Workstations of a Garment Manufacturing Factory in Tunisia

Nahed Jaffel Memmi[1](✉), Najeh Maâtoug[1], Aoutef Mahfouth Kraiem[2], Taoufik Khalfallah[2], and Faouzi Sakli[1]

[1] Laboratory of Textile Engineering, Research Unit 11ES42, University of Monastir, Monastir, Tunisia
Jaffel.nahed@hotmail.com

[2] Department of Occupational Medicine and Ergonomics, Faculty of Medicine Monastir, Research Unit 12SP39, University of Monastir, Monastir, Tunisia

Abstract. The operators' on sewing workstations performance in the garment-manufacturing sector is the key factor for the enhancement of productivity and quality. At present, it seems that the available information on the ergonomics and the working environment of the garment-manufacturing operators in our country is scarce. This research study the relationship between the working posture and method and the workstation design in garment manufacturing. An approach based on RULA methods using CATIA software is proposed in order to evaluate the ergonomic performance of workstations. The results showed that ergonomics has a determinant role in improving garment-manufacturing postures. Using the Rapid Entire Body Assessment (RULA) investigation tool, the examination of the workstations revealed uncomfortable and awkward work postures. In addition, data from the analysis conducted revealed that neither the workstations were ergonomic nor the operators complied with the ergonomic rules. In order to support requests for change in workstation layout, the virtual ergonomic study via CATIA software was convincing proof. She led to the drafting of recommendations and support during the implementation of the new development. The ergonomic design of work situations is an opportunity to improve productivity for the company and the comfort and safety of employees.

Keywords: Ergonomics assessment · RULA · Digital human modeling · CATIA · Industrial sewing workstation

1 Introduction

Musculoskeletal disorders (MSDs) are a major occupational health problem and a major cause of disability in various industrialized and developing countries (Widanarko et al. 2014). Their social and economic outcomes influence companies' business success (Zare et al. 2015). Tunisia is marked both by industrial development and an increase in pathology (Ghram et al. 2010). The link between MSDs and the exposure to risk factors related to physical constraints at work has been proven (Winnemuller et al. 2004). Working on

S. Msahli and F. Debbabi (Eds.): CIRATM 2020, SPM 17, pp. 394–397, 2022.
https://doi.org/10.1007/978-3-031-08842-1_62

a sewing workstation involves repetitive, high-speed gestures of the shoulder/arm system involving the coordination of both hands, usually performed in a seated position maintained throughout the work period.

In the Tunisian industry, few companies have ideal working conditions. Most of them suffer from poor environmental conditions where 80% of the employees are subject to harm (Boulila et al. 2018).

Hence, it is vital to consider the environment or the workspace in which the operators perform their tasks to increase productivity and improve quality. Accordingly, based on the previously discussed background, this study aims to correct the working method of the operators and to revise workstations.

2 Material and Methods

2.1 Study Protocol

To begin with, the purpose of the study was explained to all subjects and the interviewer answered all of their questions. Participation was strictly voluntary; operators were not required to complete the study and were not paid for their participation.

The sample consisted of assembly tasks taken from the production line of a garment manufacturer. The selected tasks represented a broad range of typical work activities in garment assembly. We had chosen assembly tasks that contains physical strain, uncomfortable workers' position, and high-assumed overall workload. Every task was composed of several subtasks that had to be carried out in a defined sequence.

We proceed to make video recordings of the different workstations followed by identification of the serious postures. Catia, a computer aided design/computer aided manufacturing (CAD/CAM) software was used to make a three-dimensional model of manikin and manufacturing machines. The designed digital models will be used to study RULA score to reduce its level. Technical solutions will be proposed in order to obtain an acceptable posture for the manikins (Fig. 1).

Fig. 1. Ergonomics methodology

3 Results and Discussions

The study seeks to point out the ergonomics problems in workstations of the sewing manufacturing in Tunisian industry. The first step is to collect all necessary workstation information, such as dimensions and workers data in order to identify the ergonomics problems.

In order to create the most accurate model of the real workplace on which analyses will be performed, the operators have been followed, by video recordings, in different stages manufacturing.

The gestures and body positions of the operators were observed as well as the scores recorded during work in accordance with RULA. The RULA scores for the different regions of the body, A and B scores, and High score. The means of the A, B and High scores are respectively; 4.82 (SD: 0.24, 4–5), 5.86 (SD: 0.66, 4–7) and 6.91 (SD: 0.16, 6–7). The upper part of the arms of the operators had on average a score of 3, which is explained by the fact that this part of the arms was in abduction, the arms resting and flexing between 45° and 90°. The Forearm score for the majority of the operators was also 3, and this was due to the need for the operators to extend their arms through the median line of the body with the flexion of the elbow more than 100°. The wrist score was between 3 and 4, with the cuffs in extension (sagittal plane) 15° or more (Table 1).

Table 1. RULA scores

	A				B				
	Arm	Forearm	Wrist	A	Neck	Trunk	Legs	B	High Score
Average	2.95	2.86	3.14	4.82	3.72	3.19	1.05	5.86	6.91
Sd	0.10	0.24	0.24	0.29	0.45	0.54	0.10	0.66	0.16
Min	2	2	3	4	2	2	1	4	6
Max	3	3	4	5	4	4	2	7	7

Thus, the improvements mainly affect the level of the machines tables and the methods and work habits. Based on these observations, it is necessary to plan improvement actions.

We performed the design of ergonomic sewing table. To minimize the effort, we have improved the layout of the work area. This eliminates handling tasks that require leaning posture or torsion (Fig. 2).

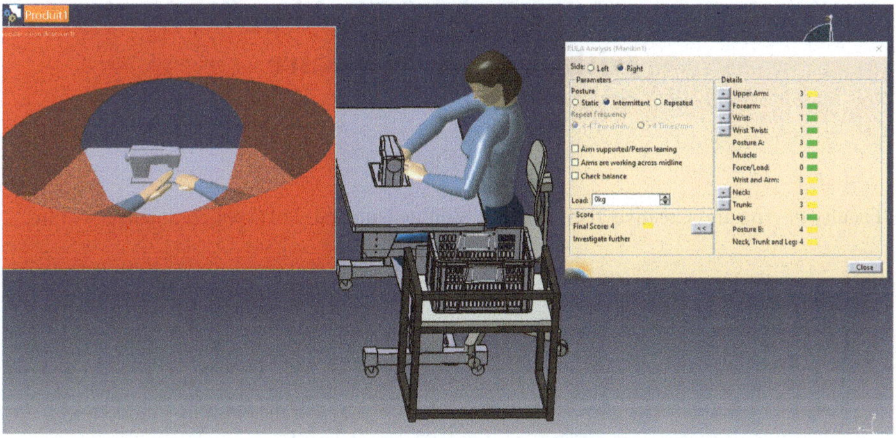

Fig. 2. Improved sewing workstation

4 Conclusion

This article focuses on one of the most important factors of the industry of garment manufacture, which is the human resources. For the study of the workstations ergonomics, a three-dimensional workstation model was designed with CATIA software to improve work situation productivity and quality.

References

Boulila, A., Ayadi, M., Mrabet, K.: Ergonomics study and analysis of workstations in Tunisian mechanical manufacturing. Human Factors Ergon. Manuf. Serv. Indust. **28**(4), 166–185 (2018)

Ghram, R., Fournier, C., Khalfallah, T., Six, F.: Analyse des facteurs socioculturels et survenue des troubles musculosquelettiques: Le cas des couturières en Tunisie. Perspectives interdisciplinaires sur le travail et la santé, 12–22 (2010)

Widanarko, B., Legg, S., Devereux, J., Stevenson, M.: The combined effect of physical, psychosocial/organisational and/or environmental risk factors on the presence of work-related musculoskeletal symptoms and its consequences. Appl. Ergon. **45**(6), 1610–1621 (2014)

Winnemuller, L.L., Spielholz, P.O., Daniell, W.E., Kaufman, J.D.: Comparison of ergonomist, supervisor, and worker assessments of work-related musculoskeletal risk factors. J. Occup. Environ. Hyg. **1**(6), 414–422 (2004)

Zare, M., Bodin, J., Cercier, E., Brunet, R., Roquelaure, Y.: Evaluation of ergonomic approach and musculoskeletal disorders in two different organizations in a truck assembly plant. Int. J. Ind. Ergon. **50**(1), 34–42 (2015)

Simulation and Analysis of Clothing Production with FlexSim Software

Strazdiene Eugenija(✉)

Faculty of Arts and Creative Technologies, Vilnius University of Applied Science, Vilnius, Lithuania
e.strazdiene@mtf.viko.lt

Abstract. FlexSim program provides virtual environment with the ability to simulate any process. The aim was to analyse the efficiency of the clothing production process on the basis of a FlexSim simulation model and animated visualization by determining the bottlenecks of the flow. Also, to take measures to increase the efficiency of the production by eliminating them.

Keywords: Production flow · Real-time simulation · FlexSim · Bottleneck

1 Introduction

Nowadays simulation is becoming a common tool for various processes analysis, whether it's a video game, local weather forecast, or any other area of daily life. FlexSim program provides virtual environment with the ability to simulate any process with as much information as the user needs. FlexSim simulations are commonly applied in areas such as manufacturing (Shivraj 2015), warehousing (Szczepański et al. 2019) distribution of goods and products, materials processing (Shuangping et al. 2020) etc. The aim of this work is to analyse the efficiency of the clothing production process on the basis of a FlexSim simulation model and animated visualization by determining the bottlenecks of the flow. Also, to take measures to increase the efficiency of the production by eliminating them.

2 Experimental

The object of investigation is the flow of garment (corporative trousers) production operations and its optimization by computational method and by applying real-time simulation modelling with FlexSim software. At the beginning, technological sequence of garment production is formed from micro-operations the duration of which predominates within seconds, e.g. such micro-operation as - to sew the waist with a tape and topstitch of 2 mm - will be accomplished with a universal sewing machine in 138 s. Work plan is set up taking into account process capacity, i.e. the number of garments ($PC = 800$ trousers) produced in one shift ($T_{\text{shift}} = 8$ h) and production time of one garment ($T_{\text{trousers}} = 1142$ s), which is defined from technological sequence. Average time of synchronised

S. Msahli and F. Debbabi (Eds.): CIRATM 2020, SPM 17, pp. 398–402, 2022.
https://doi.org/10.1007/978-3-031-08842-1_63

operation (process tact) is calculated as: $\tau = T_{shift}/PC = 28800/800 = 36$ s. The number of workplaces is calculated as: $D = T_{trousers}/\tau = 1142/36 = 31{,}72 - 32$. Production organization system of corporative trousers is straight-line, i.e. the work flows through workplaces, each of which is synchronized to the next by ensuring that the time spent at each workplace is exactly the same ($\tau = 36$ s). Taking this into account, work plan must be developed by grouping micro-operations in order to obtain their synchronization (Fig. 1).

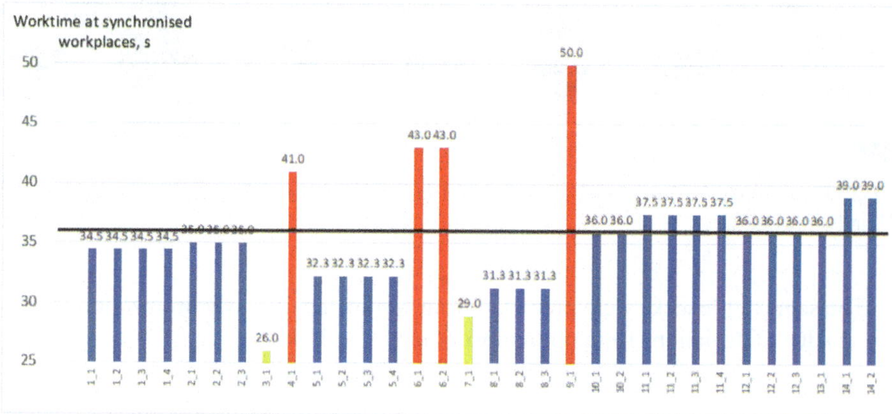

Fig. 1. Production flow of corporate trousers after micro-operations grouping and their distribution for 14 synchronised workplaces

Based on trouser production sequence, a flow algorithm is developed that reflects the entire production process. In order to create a simulation model, operations are coded, e.g. the 1st organizational operation is performed with a universal sewing machine, 4 workstations are assigned to the operation, so the coding is 1_1; 1_2; 1_3 and 1_4 (Fig. 2).

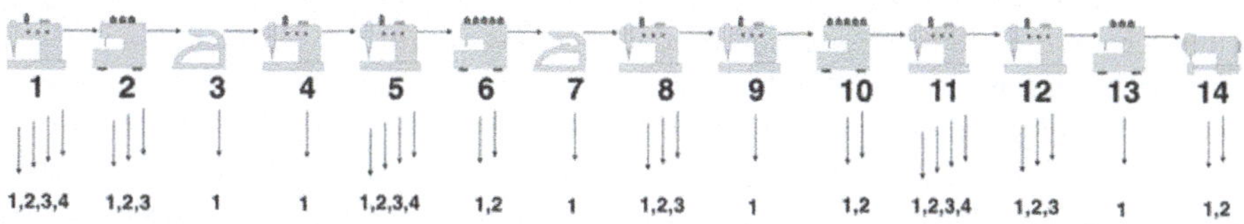

Fig. 2. The algorithm of corporative trousers production flow and the sample of operation coding

In simulation modelling the direction of semi-finished products flow from one fixed resource to the other is indicated by output-input port connections. FlexSim simulation model of corporate trousers production flow is presented in Fig. 3. Real-time simulation have shown that instead of 800 items only 554 will be produced during the shift of 8 h due to bottlenecks that appear in 4th, 6th and 9th workplaces (Fig. 3).

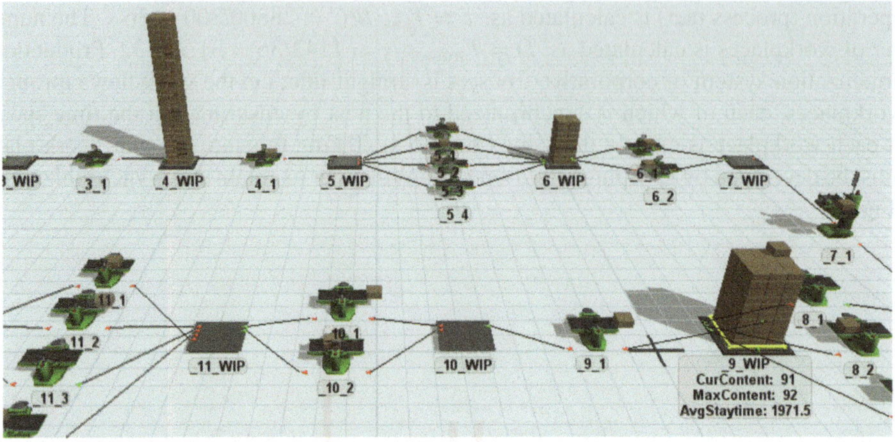

Fig. 3. FlexSim simulation model of corporate trousers production flow and bottlenecks at 4th, 6th and 9th workplaces

The same bottlenecks were predicted from workplace time synchronization graphic (Fig. 1) due to deviations from optimal work tack $\tau = 36$ s.

3 Results

In depth analysis of FlexSim simulation process shows that all planned 800 items have reached 4th workplace, but only 696 were processed there within the shift of 8 h. Accordingly, 692 were input into 6th workplace with 660 processed, and input at 9th workplace was 655 while output – 564 items. The rest of 20 items were still at different stages of production flow (Fig. 4). Also it must be stated that the efficiency of workplaces following directly after the bottlenecks were effected and did not reach their maximum capacities. Thus, processing efficiency of 5th workplace was 77,90%, of 7th workplace – 66,25% and of 10th workplace – 70,26%.

Different measurements can be taken in order to eliminate or to minimize the effect of bottlenecks and to reach planned production capacity for one shift. In this analysis universal sewing machines of higher speed were used in 4th and 9th workplaces. Meantime in 6th additional workplace was added so that the work was divided not for two, but for three work places. Taking into account that the number of workplaces must be kept constant, i.e. $D = 32$, the decision was made to join 3rd and 8th workplaces by combining their operations and investing into more advanced and efficient equipment - freely programmable automatic sewing unit for sewing pockets PFAFF 3588 (Fig. 4).

Real-time FlexSim simulations have shown which changes allow to improve the efficiency of corporate trousers production flow by reaching close to maximum capacities from 89,58% up to 97,94% at almost all workplaces as it is shown in Fig. 5. Still in two workplaces - 6th and 7th maximal processing efficiency will not be reached, but these workplaces will make a workforce reserve that could be used to address unforeseen process flow disruptions.

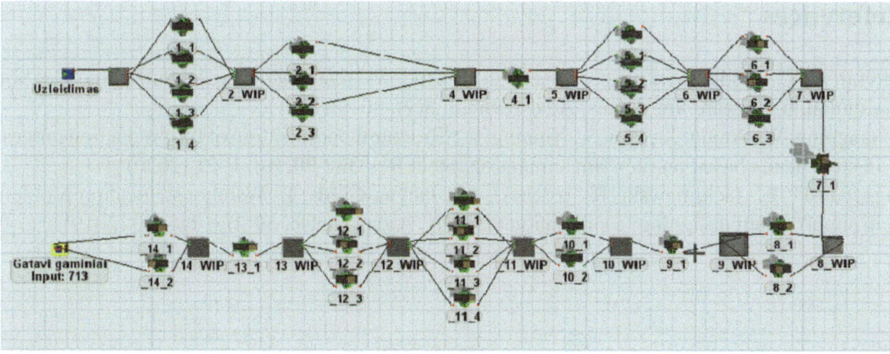

Fig. 4. Organisational changes in trousers production flow and their effect after FlexSim re-simulation of one shift work

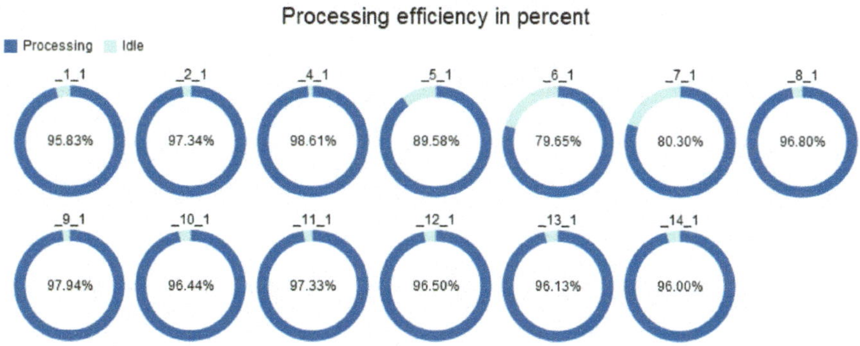

Fig. 5. The result of real-time FlexSim simulation after the changes in corporate trousers production flow

4 Conclusions

FlexSim software provides a powerful real-time simulations tool which makes it possible not only to check different solutions for the elimination of bottlenecks, which may appear in production flows due to the deviations in the synchronization of production operations, but also to check them through the whole virtual production period. In this research computational method together with real-time FlexSim simulations allowed in a very short time to find and to check the solution to improve the efficiency of corporate trousers production flow by reaching close to maximum capacities from 89,58% up to 97,94% at almost all workplaces.

References

Shivraj, N.Y.: productivity improvement in a windows manufacturing layout using FlexSim simulation software. Int. J. Res. Adv. Technol. **3**(9), 86–90 (2015)

Shuangping, W., Anjun, X., Wei, S., Xiuping, L.: Structural optimization of the production process in steel plants based on FlexSim simulation. Steel Res. Int. **90**(10), 1–21 (2020)

Szczepański, E., Jachimowski, R., Izdebski, M., Jacyna-Gołda, I.: Warehouse location problem in supply chain designing: a simulation analysis. Arch. Transp. **50**(2), 101–110 (2019)

Author Index

S. Msahli and F. Debbabi (Eds.): CIRATM 2020, SPM 17, pp. 403–405, 2022.
https://doi.org/10.1007/978-3-031-08842-1

The manufacturer's authorised representative in the EU is Springer Nature Customer Service Centre GmbH, Europaplatz 3, 69115 Heidelberg, Germany. If you have any concerns regarding our products, please contact ProductSafety@springernature.com

Printed and bound by CPI Group (UK) Ltd, Croydon, CR0 4YY
30/09/2026
02225301-0010